Advances in Diabetes Research and Management

Rana Noor

Editor

Advances in Diabetes Research and Management

Springer

Editor
Rana Noor
Faculty of Dentistry
Department of Biochemistry
Jamia Millia Islamia
New Delhi, India

ISBN 978-981-19-0026-6   ISBN 978-981-19-0027-3   (eBook)
https://doi.org/10.1007/978-981-19-0027-3

This Springer imprint is published by the registered company Springer Nature Singapore Pte Ltd.
The registered company address is: 152 Beach Road, #21-01/04 Gateway East, Singapore 189721,
Singapore

Preface

Diabetes mellitus is a major health problem affecting approximately 425 million people worldwide. Physiologically, the body is unable to control blood glucose levels in the blood which results in high blood glucose (hyperglycemia). The implications of hyperglycemia are frequent urination, weight loss, cardiovascular diseases, and damage to nerves, kidneys, and eyes. Diabetes mellitus is classified into Type I and Type II, while the former present in 5–10% of the population occurs when the immune system of the body damages insulin-producing B-cells of the pancreas leading to a deficiency of insulin in the body. Type 2 diabetes is the more common form of the disease characterized by the loss of ability to respond to insulin. This book discuss the latest research in the field of diabetes and management and also presents the new technologies that have been introduced to facilitate early diagnosis and the new potential therapies for these complications. It highlight the molecular mechanisms of the microvascular and macrovascular complications of diabetes 2 mellitus and review the most prevalent microvascular complications. The book discusses the molecular mechanisms involved in the development of diabetic complications. It also explore the molecular mechanisms of metabolic and immunological abnormalities affecting several organs due to diabetes and discuss the applications of nanotechnology for diagnostics, monitoring, and treatment for diabetes. It reviews the development of novel glucose measurement and insulin delivery strategies for diabetes management. The chapter also examine the point-of-care (POC) diagnostics for enhancing glucose measuring sensitivity, temporal response, and discusses a closed-loop system that continuously monitors blood glucose and releases insulin in a controlled and self-regulated fashion. Importantly, it reviews the immense potential of stem cell therapy in diabetes mellitus and molecular mechanisms that relate gut microbiota to the onset of insulin resistance and diabetes. It is a valuable source for students, researchers, and practitioners working in the area of glucose metabolism, diabetes, and endocrinologists.

New Delhi, India   Rana Noor

Contents

Editor and Contributors

About the Editor

Rana Noor is currently working at the Faculty of Dentistry, Jamia Millia Islamia, New Delhi, India. She pursued her Ph.D. in Biochemistry from Aligarh Muslim University, India, in the year 2003. Her research work is highly interdisciplinary, spanning a wide range in protein biochemistry, enzymology, and application of biochemistry in dentistry. She is an editorial board member of various reputed journals.

Contributors

Chandan Kumar Acharya Department of Botany, Dr CV Raman University, Bilsapur, Chhattisgarh, India
Department of Botany, Bajkul Milani Mahavidyalaya, Kolkata, West Bengal, India

Rajesh Nanda Amarnath Department of Obstetrics and Gynecology, Apollo Womens Hospital, Chennai, India

Sheerin Bashar School of Forensic Sciences, Centurion University of Technology and Management, Bhubaneswar, Odisha, India

Biplab Kumar Behera Department of Zoology, Siliguri College, Siliguri, West Bengal, India

Paramita Biswas Regional Research Sub Station (OAZ), Uttar Banga Krishi Viswavidyalaya, Malda, West Bengal, India

Titas Biswas Department of Chemistry, Gurudas College, Kolkata, West Bengal, India

Aditi Chakraborty SHRM Biotechnologies Pvt. Ltd., Kolkata, West Bengal, India

Dibyashree Chhetri Cancer Science Laboratory, Department of Biotechnology, School of Bioengineering, SRM Institute of Science and Technology, Chengalpattu, India

Aishwariya Das School of Biotechnology and Bioinformatics, D Y Patil University, Navi Mumbai, Maharashtra, India

Balaram Das Department of Physiology, Belda College, Medinipur, West Bengal, India

Manna De Department of Botany, Dr CV Raman University, Bilsapur, Chhattisgarh, India

Geevaprabhakaran Ganesan Chettinad Hospital & Research Institute, Chettinad Academy of Research and Education (Deemed to be University), Kelambakkam, Chennai, Tamil Nadu, India

Dhanavathy Gnanasampanthapandian Cancer Science Laboratory, Department of Biotechnology, School of Bioengineering, SRM Institute of Science and Technology, Chengalpattu, India

Kartik Jana SHRM Biotechnologies Pvt. Ltd., Kolkata, West Bengal, India

Selvaraj Jayaraman Centre of Molecular Medicine and Diagnostics (COMManD), Department of Biochemistry, Saveetha Dental College & Hospital, Saveetha Institute of Medical & Technical Sciences, Saveetha University, Chennai, India

Iyshwarya Bhaskar Kalarani Human Cytogenetics and Genomics Laboratory, Faculty of Allied Health Sciences, Chettinad Hospital and Research Institute, Chettinad Academy of Research and Education, Kelambakkam, Tamil Nadu, India

Chandrashekar Kirubhanand Department of Anatomy, All India Institute of Medical Sciences, Nagpur, Maharashtra, India

Langeswaran Kulanthaivel Molecular Cancer Biology Lab, Department of Biotechnology, Science Campus, Alagappa University, Karaikudi, Tamil Nadu, India

S. Gowtham Kumar Faculty of Allied Health Sciences, Chettinad Hospital & Research Institute, Chettinad Academy of Research and Education, Kelambakkam, Chennai, Tamil Nadu, India

K. Kumar Ebenezar Natural Medicine and Molecular Physiology Lab, Faculty of Allied Health Sciences, Chettinad Hospital and Research Institute, Chettinad Academy of Research & Education, Kelambakkam, Chennai, Tamil Nadu, India

K. Langeswaran Molecular Cancer Biology Laboratory, Department of Biotechnology, Science Campus, Alagappa University, Karaikudi, Tamil Nadu, India

Nithar Ranjan Madhu Department of Zoology, Acharya Prafulla Chandra College, New Barrackpore, West Bengal, India

Raktim Mukherjee Shree PM Patel Institute of PG Studies and Research in Science, Sardar Patel University, Anand, India

K. T. Nachammai Molecular Cancer Biology Laboratory, Department of Biotechnology, Science Campus, Alagappa University, Karaikudi, Tamil Nadu, India

V. Nithya Department of Animal Health and Management, Science Campus, Alagappa University, Karaikudi, India

Kanagaraj Palaniyandi Cancer Science Laboratory, Department of Biotechnology, School of Bioengineering, SRM Institute of Science and Technology, Chengalpattu, India

Swetha Panneerselvam Centre of Molecular Medicine and Diagnostics (COMManD), Department of Biochemistry, Saveetha Dental College & Hospital, Saveetha Institute of Medical & Technical Sciences, Saveetha University, Chennai, India

Bhuvaneswari Ponnusamy Centre of Molecular Medicine and Diagnostics (COMManD), Department of Biochemistry, Saveetha Dental College & Hospital, Saveetha Institute of Medical & Technical Sciences, Saveetha University, Chennai, India

D. Prabu Department of Microbiology, Dr. ALM PG IBMS, University of Madras, Taramani Campus, Chennai, Tamil Nadu, India

Natesan Sella Raja Membrane-Protein Interaction Lab, Department of Genetic Engineering, School of Bio-engineering, SRM Institute of Science and Technology, Kattankulathur, Chennai, Tamil Nadu, India

Ponnulakshmi Rajagopal Department of Central Research Laboratory, Meenakhsi Ammal Dental College and Hospitals, Meenakhsi Academy of Higher Education and Research, Chennai, India

Rupak Roy SHRM Biotechnologies Pvt. Ltd., Kolkata, West Bengal, India

Sunita Samal Department of Medical Gastroenterology, SRM Medical College Hospital and Research Centre, SRM Institute of Science and Technology, Chengalpattu, India

P. Sangavi Department of Bioinformatics, Science campus, Alagappa University, Karaikudi, India

S. Santhi Priya Natural Medicine and Molecular Physiology Lab, Faculty of Allied Health Sciences, Chettinad Hospital and Research Institute, Chettinad Academy of Research & Education, Kelambakkam, Chennai, Tamil Nadu, India

Bhanumati Sarkar Department of Botany, Acharya Prafulla Chandra College, New Barrackpore, West Bengal, India

Somnath Sau Department of Nutrition, Egra S.S.B. College, Medinipur, West Bengal, India
Department of Physiology, Belda College, Medinipur, West Bengal, India

Varsha Singh Membrane-Protein Interaction Lab, Department of Genetic Engineering, School of Bio-engineering, SRM Institute of Science and Technology, Kattankulathur, Chennai, Tamil Nadu, India

Subhashree Sivakumar Membrane-Protein Interaction Lab, Department of Genetic Engineering, School of Bio-engineering, SRM Institute of Science and Technology, Kattankulathur, Chennai, Tamil Nadu, India

R. Srinithi Department of Bioinformatics, Science campus, Alagappa University, Karaikudi, India

Gowtham Kumar Subbaraj Faculty of Allied Health Sciences, Chettinad Hospital and Research Institute, Chettinad Academy of Research and Education (Deemed to be University), Kelambakkam, Chennai, Tamil Nadu, India

Taniya Sur Department of Biotechnology, KIIT School of Biotechnology, KIIT University, Bhubaneswar, India

Sambit Tarafdar Amity Institute of Virology and Immunology, Amity University, Noida, India

Ramakrishnan Veerabathiran Human Cytogenetics and Genomics Laboratory, Faculty of Allied Health Sciences, Chettinad Hospital and Research Institute, Chettinad Academy of Research and Education, Kelambakkam, Tamil Nadu, India

Biochemical Assay for Measuring Diabetes Mellitus

Taniya Sur, Aishwariya Das, Sheerin Bashar, Sambit Tarafdar, Bhanumati Sarkar ⓘ, and Nithar Ranjan Madhu ⓘ

Abstract

Increased blood sugar brought on by inadequate insulin synthesis or activity is known as diabetes mellitus (DM), resulting in disrupted carbohydrate, lipid, and protein metabolism. Diabetes prevalence was reportedly expected to be 171 million (2.8%) in 2000 and might climb to 366 million (4.4%) by 2030. It is crucial to understand that DM is a wide term that refers to disorders that cause chronic hyperglycemia. The differences in the processes that cause the various kinds of diabetes serve as the foundation for their classification. Type 1 diabetes and type 2 diabetes are the most common types of diabetes. Diabetes type 1 is a chronic illness in which the pancreas produces very little or no insulin. Insulin is a hormone that must be present for glucose, or sugar, to enter cells and turn into energy. Whereas, fat liver and muscle cells become insulin-resistant in type

T. Sur
Department of Biotechnology, KIIT School of Biotechnology, KIIT University, Bhubaneswar, India

A. Das
School of Biotechnology and Bioinformatics, D Y Patil University, Navi Mumbai, Maharashtra, India

S. Bashar
School of Forensic Sciences, Centurion University of Technology and Management, Bhubaneswar, Odisha, India

S. Tarafdar
Amity Institute of Virology and Immunology, Amity University, Noida, India

B. Sarkar
Department of Botany, Acharya Prafulla Chandra College, New Barrackpore, West Bengal, India

N. R. Madhu (✉)
Department of Zoology, Acharya Prafulla Chandra College, New Barrackpore, West Bengal, India

© The Author(s), under exclusive license to Springer Nature Singapore Pte Ltd. 2023
R. Noor (ed.), *Advances in Diabetes Research and Management*,
https://doi.org/10.1007/978-981-19-0027-3_1

1

2 diabetes. Numerous biochemical assays are used to identify the type of diabetes and its severity to some extent. There has been some recent methodology that has also proven to be successful in determining the stages of diabetes. Some instruments are also available for the patient to measure their blood glucose/sugar (BS) level by sitting at home. These new trends and advancements have helped and are helping in the assaying of diabetes much faster and as well as less painful.

Keywords

Diabetes mellitus · Insulin · OGTT · HbA1C · FPG · Glucose urine · Serum creatinine · Urine microalbumin · Glucometer · Salivary glucose

1   Introduction

Diabetes mellitus (DM) characterized by high blood glucose levels, brought on by insufficient insulin synthesis or action and resulting in disturbed carbohydrate, lipid, and protein metabolism, is the second most common chronic illness affecting young people (if obesity is excluded, asthma is the most common) (Styne, 2016). According to estimates, 171 million people (2.8%) will have diabetes in 2000, and 366 million (4.4%) will have it by 2030 (Williamson et al., 2002). More people die from diabetes yearly than from HIV/AIDS, with one death every 10 s, making it the leading cause of chronic illness and early mortality (Ahmad, 2013). Type 1 and type 2 are the customary types of DM. Type 1 diabetes, also known as insulin-dependent diabetes, was once known as juvenile-onset diabetes because it frequently began in infancy. Type 1 diabetes is an autoimmune disease. In this case, insulin production has been halted. Previously, type 2 diabetes was known as non-insulin-dependent diabetes and adult-onset diabetes. However, it has grown increasingly frequent among children and teenagers over the last 15 to 20 years, partly due to obesity. The majority of them have type 2 diabetes. In this scenario, the pancreas generates insulin but either does not create enough or does not utilize it properly. Insulin resistance is most commonly present in the liver, fat, and muscle cells. Type 2 diabetes is frequently less severe than type 1 diabetes. Improvements in glucose control are often followed by a "honeymoon" phase in which the quantity of insulin required reduces for variable durations of time before increasing again as cell death persists (Joslin Diabetes Center and Joslin Clinic, 2006).

The nations most afflicted by this disease in 2025 will be India followed by China and the United States (WHO, 2011). Changes in life expectancy and a lack of innovation in healthcare are two variables that may be to blame for the startling rise in diabetes prevalence. As a result, diabetes is becoming increasingly prevalent, particularly in metropolitan areas. Diabetes patients are more likely to have both short- and long-term difficulties, as well as early mortality; thus governments throughout the world will face huge increases in healthcare expenses. Diabetes places a greater burden on the patient or family than most other chronic illnesses, and it can result in severe psychosocial stress for everyone involved. The treating physician must address each of these issues and support the kid and family as they navigate the complexities of this illness.

2   Types of DM

It is critical to recognize that diabetes is a broad word that refers to a group of illnesses that produce persistent hyperglycemia. The differences in the processes that cause the various kinds of diabetes serve as the foundation for their classification.

2.1   Type 1 Diabetes Mellitus

Activated CD4+ and CD8+ T lymphocytes, as well as macrophages, infiltrate and destroy beta cells in persons with T1DM, also known as insulin-dependent diabetes or juvenile diabetes (Phillips et al., 2009). It is generally understood that suscepti- bility is influenced by both hereditary and environmental variables. T1DM has been linked to the HLA (human leukocyte antigen) gene on chromosome 6. HLA proteins on cell surfaces are used by the immune system to distinguish between healthy body cells and external infectious and noninfectious substances. An abnormality in the HLA proteins causes an autoimmune reaction against the cells in T1DM (Choo, 2007). T1DM is significantly impacted by the DR gene, another HLA-related gene (Gorodezky et al., 2006). There is proof that certain viruses may be at fault for the development of T1DM (Jun & Yoon, 2002). Idiopathic diabetes, which excludes autoimmune, is a separate type of T1DM that some people develop. It is more widespread in African and Asian cultures than autoimmune T1DM. Although the etiology and pathogenesis are unknown, without antibodies against beta cells, humans cannot make insulin and risk getting ketoacidosis (Harris et al., 1998).

2.2   Diabetes Mellitus Type 2

T2DM is characterized by inadequate insulin secretion and synthesis as a result of insulin resistance. Nearly 90% of all diabetes cases worldwide are diagnosed during the fourth decade of life. There is evidence that T2DM incidence and prevalence increase with age (Masharani UaK, 2001). Diabetes with obesity and diabetes without obesity are the two types of T2DM. Obese T2DM patients commonly acquire resistance to endogenous insulin due to changes in cell receptors connected to the distribution of abdominal fat. Insulin synthesis and release are impaired in nonobese T2DM, and some insulin resistance occurs at the post-receptor level.

Because being overweight and having metabolic abnormalities are two of the major risk factors for developing type 2 diabetes, emerging nations with significant food and lifestyle changes have seen the greatest increase in diabetes prevalence.

3   Biosynthesis and Action of Insulin

Nobel Prize winner Fred Sanger discovered the primary structure of insulin, in the middle of the 1950s (Sanger, 1959). The discovery of the structure of a single-chain precursor molecule in 1967 spurred a fresh round of discussion over the production of two-chain hormones. This debate was rekindled because of the molecule's structure (Steiner & Oyer, 1967). This discovery and subsequent study unveiled a novel model for biosynthesizing physiologically active peptides and giving rise to a sizable portion of the processing of protein precursors we see today. Today, it is understood that posttranslational cleavage is a typical method for regulating the production and functionality of a variety of proteins. The Discovery of proinsulin also reactivated the insulin research field, opening up new research directions in beta-cell biology that are now being explored in the current setting of genome-focused solid research.

3.1   Insulin: Structure and Function

Beta cells of the pancreas are responsible for producing the peptide hormone insulin. Its primary function is to regulate the process by which glucose is transported from the blood into the body's cells. It does so by entering the circulation through the exocrine system. Insulin is made from a sequence of single-chain molecules, including preproinsulin, proinsulin, and other intermediate cleavage products (Steiner, 2001; Halban, 1991). Preproinsulin is the initial translation product of insulin mRNA, and it comprises a hydrophobic N-terminal signal peptide of 24 residues. This signal peptide is found in almost all secreted proteins derived from animals, plants, or microorganisms. The signal peptide interacts with the signal recognition particle (SRP) (Egea et al., 2005), a cytosolic ribonucleoprotein particle that aids in the segregation of the nascent proinsulin polypeptide chain from the cytosolic compartment into the secretory channel (Chan et al., 1976; Lomedico et al., 1977). The nascent pre-prohormone is delivered into the rough endoplasmic reticulum (RER) lumen via a peptide-conducting channel on the other side of the membrane. A signal peptidase found on the inner surface of the RER membrane cuts the signal sequence. Following that, the signal sequence degrades dramatically (Patzelt et al., 1978). For proinsulin to obtain its normal structure, it must first fold, and once it does, it immediately establishes its three disulfide bonds within the RER (Huang & Arvan, 1995). This activity is catalyzed by several chaperone proteins, the most noteworthy of which is the enzyme protein-thiol reductase, which is an endoplasmic reticulum (ER) resident protein with a C-terminal KDEL (Lys-Glu-Asp-Leu) ER-retention sequence (Munro & Pelham, 1987). Exit from the endoplasmic reticulum (ER) is usually the slowest stage in the intracellular transit of many secretory proteins, although proinsulin is one of the quickest (Lodish, 1988). After 10–20 min, it is transferred to the Golgi apparatus for additional processing before being packaged. When proinsulin is broken down into components by the Golgi apparatus and then within immature secretory vesicles, insulin and the 31-residue C-peptide are generated. The C-peptide lacks the basic residue pairs that attach it to the B and A chains at the molecule's ends. Insulin and the C-peptide are preserved in the secretion granules with modest amounts of intact proinsulin and its intermediate cleavage variants (Steiner et al., 1972), islet amyloid polypeptide (IAPP or amylin), and other less abundant beta-cell secretory products (Nishi et al., 1990). In this aspect, proproteins differ from the inert zymogen forms of many enzymes, which become active after being released. Intracellular proprotein processing is required by nearly all cells in the brain and endocrine system that create peptides. This procedure, however, occurs in several different organs. Regardless of whether or not extra protein synthesis happens, proinsulin conversion to insulin typically begins approximately 20 min after the preproinsulin chain is created and continues for an additional 1–2 h (Halban, 1991; Steiner et al., 1967, 1969; Sando et al., 1972). Intermediate cleavage products are produced during intracellular processing. Des-31,32 proinsulin and des-64,65 proinsulin are two examples of these compounds. Des-31,32 proinsulin is the most common intermediate form in humans (Given et al., 1985; Sobey et al., 1989; Sizonenko et al., 1993). The principal storage facility comprises a huge population of mature secretory granules that are found dispersed throughout the cytoplasm of beta cells. These evolve gradually over hours to days and include only tiny levels (1–2 percent) of proinsulin and other intermediary components. Consequently, precursor-related peptides may typically only be discovered in released insulin at extremely low concentrations. Although stored insulin accounts for the majority of the molecule emitted, newly created insulin is selectively released at a lower amount (Sando et al., 1972; Gold et al., 1982).

3.2   Insulin Biosynthesis: The Cell Biology

The beta cells of the islets of Langerhans resemble other neurosecretory cells in many ways. The Golgi apparatus is required for the synthesis of dense-core secretory granules in these cells (Farquhar & Palade, 1981). Although proinsulin is not known to be modified in this compartment, the medial and trans-Golgi compartments undergo posttranslational modifications such as glycosylation, sulfation, and, in certain circumstances, phosphorylation (Mains et al., 1990). Due to its dynamic nature, it is still unclear how the secretory product is transported and sorted inside the Golgi apparatus (Schekman & Orci, 1996). Individual cisternae migrate from trans to cis with their secretory contents, but various Golgi resident proteins and enzymes retrotranslocation from trans to cis in small, coated vesicles, according to recent research (Losev et al., 2006). According to this theory, secretory products develop in the same compartment before dispersing as immature secretory granules in the trans-Golgi network (TGN). COPI is found on a large number of small Golgi-associated vesicles. Both anterograde and retrograde protein flow through the Golgi are thought to be significantly regulated by this protein (Orci et al., 1997). Proinsulin is abundant in newly formed clathrin-clad granules in the trans-Golgi cisternal network (TGN), indicating that insulin conversion occurs mostly during the maturation of these secretory "progranules," according to immunocytochemical studies (Orci et al.,

1985). Energy poisons such as brefeldin A prevent newly generated proinsulin from being transported into secretory vesicles for proteolytic processing (Huang & Arvan, 1994; Steiner et al., 1970; Howell, 1972; Lee et al., 2004). However, it no longer needs the energy to transform into insulin once freshly produced proinsulin reaches the trans-Golgi and/or progranules (Steiner et al., 1970). This energy need appears to be connected to the budding and/or fusing of the small vesicles that transport secretory products from the ER to the Golgi cisternae (Steiner et al., 1970; Orci et al., 1971).

4   Biochemical Assay of Diabetes Mellitus

Accurate diabetes prevalence estimates are required for the right funding distribution for patient care and coverage monitoring. Researchers require cost-effective, accurate, and reliable methods for population-wide assessments of diabetes. Because up to half of all patients may go undiagnosed, one of three biochemical tests is employed to determine diabetes prevalence (International Diabetes Federation, 2014). Fasting plasma glucose (FPG) is a test that analyses blood glucose levels following a 12-h fast in which a person consumes nothing except water. The oral glucose tolerance test (OGTT) measures changes in blood sugar levels after a certain amount of glucose is administered. HbA1c does provide information on the typical blood glucose levels over the previous 2 to 3 months, despite not being a direct predictor of blood sugar (American Diabetes Association, 2014). Most people have no idea how these measures relate to various ethnic and geographic populations.

When comparing data from resource-constrained nations, it is crucial to consider whether these measurements yield equivalent results. Diabetes affects 5.1 percent of people in sub-Saharan Africa but 11.4 percent of those in North America and the Caribbean. The condition is thought to affect 8.3% of adults (aged 20 to 79) worldwide (International Diabetes Federation, 2014). Only 57% of the world's 221 countries and territories have trustworthy diabetes surveys, and only 19% of those have data based on OGTT (International Diabetes Federation, 2014). Instead, numerous low-income countries employ the WHO STEPwise approach to surveillance (STEPS) tool, launched in 2005. Demographic information, healthy lifestyle choices, BMI, waist circumference, and blood pressure are some crucial risk variables identified by STEPS. WHO advises wealthy nations to conduct an FPG analysis with an optional OGTT module (World Health Organization, 2005).

HbA1c was not standardized when coupled to FPG for the first time in the late 1970s (Koenig et al., 1976). However, by 2009, a worldwide panel organized by the American Diabetes Association, the European Association for the Study of Diabetes, and the International Diabetes Federation advised utilizing a lab-based HbA1c test to diagnose type 2 diabetes (The International Expert Committee, 2009). Diabetes cannot be ruled out with an HbA1c level below 65% (48 mmol/mol), but it cannot be diagnosed with a level below this.

5   Fasting Plasma Glucose (FPG) Test

A fasting plasma glucose test, often known as a fasting glucose test (FGT), is a test used to assist diagnose diabetes or prediabetes. After a several-hour fast, the test is a straightforward blood test. The FPG blood test determines how much glucose is in your blood at a certain time. For the most reliable results, do this test in the morning after a minimum of an 8-h fast. Only small amounts of water are consumed during a fast. The test calls for a blood sample from the patient's arm.

The World Health Organization reports the following fasting glucose test results:

Normal: less than 5.5 mmol/l (100 mg/dl).
Impaired fasting glucose: between 5.5 and 6.9 mmol/l (100–125 mg/dl).
Diabetic: 7.0 mmol/l or higher (126 mg/dl or higher).
Prediabetes is characterized by impaired fasting glycemia.

6   Glucose Urine Test

A urine sample is analyzed for glucose to determine the amount of sugar (glucose) present. Glycosuria or glucosuria is the presence of glucose in the urine. A blood test or a cerebrospinal fluid test can also be used to detect glucose levels.

Urine samples are scrutinized as soon as they are supplied. The medical practitioner uses a dipstick with a color-sensitive pad. The color of the dipstick can be used by the doctor to assess the quantity of glucose in the patient's urine. The physician may give the user instructions to collect urine at home for 24 h if necessary.

This test was once regularly used to detect and track diabetes. The glucose urine test has been replaced with quick and easy blood tests that evaluate blood glucose levels. Doctors who suspect renal glycosuria could run a glucose urine test. Even when the blood glucose level is normal, glucose is released from the kidneys into the urine in this unusual circumstance.

Urine glucose range is normal between 0 and 0.8 mmol/l (0 to 15 mg/dL).
These conditions may lead to higher than usual glucose levels:

- Diabetes: Small rises in urine glucose levels after a heavy meal are typically not a reason for concern.
- Pregnancy: Up to 50% of pregnant women have glucose in their urine at some time throughout their pregnancy. A woman may have gestational diabetes if there is glucose in her urine.
- Renal glycosuria: An uncommon condition in which blood glucose levels are normal but the kidneys excrete glucose through the urine.

7   Oral Glucose Tolerance Test (OGTT)

A glucose tolerance test assesses a person's capacity to handle a glucose load. The test can disclose a person's ability to digest a specific amount of glucose. There is a possibility of normal, impaired, and abnormal outcomes. A glucose tolerance test helps distinguish between type 1 diabetes, type 2 diabetes, and gestational diabetes mellitus. It is a blood test that collects many blood samples over a while, usually 2 h (Wei et al., 2019; Li et al., 2019; Maldonado-Hernández et al., 2019).

Before the glucose tolerance test, the patient should be instructed to consume at least 150 g of carbohydrates every day for at least 3 days. A patient must normally attend fasting on the day of the test. To establish a baseline glucose level, a fasting sample is acquired by phlebotomy or intravenous access. The patient will then consume glucose (which comes in 2 formulas, 75 or 100 g). The drug is given to children at a weight-based dose of 1.75 g/kg of body weight, with a maximum dose of 75 g for all nonpregnant patients. Except for ingesting the glucose, patients are recommended to fast throughout the test. Following that, samples are obtained at various intervals, and the study is finished 60 or 120 min after glucose administration. Patients must remain still throughout the examination, and excessive hydration with beverages is suggested because it could skew the results. A glucose tolerance test can be given or ordered in a variety of ways. The typical one-step glucose tolerance test involves obtaining two samples: one at the start and one after 60 min. This non-fasting test detects gestational diabetes mellitus in pregnant women between the ages of 24 and 28 weeks. If the test results show that the patient's glucose tolerance is impaired or atypical, a 2- or 3-h glucose tolerance test is required.

Another method for completing a glucose tolerance test is to collect a baseline sample from a fasting subject and a sample 120 min following glucola administration. If the blood glucose level is abnormal at the baseline or after 120 min, this test can be used to confirm the diagnosis of diabetes.

Multiple samples obtained at baseline, 30, 60, 90, and 120 min can also be used to complete a glucose tolerance test. This allows medical experts to assess impaired tolerance and establish whether there is a delayed reaction in insulin excretion or glucose absorption in the liver.

7.1   Specimen Requirement

In most cases, insulin samples are collected; however, glucose and c-peptide samples can also be requested if necessary. The materials must be centrifuged for the serum and platelets to be separated. After aliquoting the serum into a transport tube, the platelets are eliminated at the end of the procedure. To guarantee the viability and authenticity of insulin samples, they should be refrigerated as soon as feasible following separation. It is critical to label each sample with the time point of the drawing and the precise time when the draw was conducted.

7.2   Procedure for Testing

Normally, the patient should consume 150 g of carbs every day for 3 days before the test. The patient must fast for at least 8 h before the test. If the patient does not appear to be fasting, the test should be postponed. To do the test, a BC-shielded IV catheter or repeated phlebotomies may be used. Collecting the fasting sample and recording the date and time is necessary. The patient must next ingest the required amount of glucose (depending on weight, up to 75 g) within 5 min.

More samples must be obtained at 30-, 60-, 90-, and 120-min intervals, or as indicated by the provider. If a BC-shielded IV catheter is utilized, the line should be flushed with saline or heparin solution after each sample to ensure line patency. The test is finished after 120 min, and the blood samples must be processed and transported to the lab for examination.

7.3   Results

Normal Results for Type 1 Diabetes or Type 2 Diabetes

- Fasting glucose levels of 60 to 100 mg/dL
- A 1-h glucose level of less than 200 mg/dL
- Two-hour glucose level less than 140 mg/dL

  Impaired Results for Type 1 Diabetes or Type 2 Diabetes

- Fasting glucose level: 100 to 125 mg/dL
- Two-hour glucose level 140 to 200 mg/dL

  Abnormal (Diagnostic) Results for Type 1 Diabetes or Type 2 Diabetes

- Fasting glucose level greater than 126 mg/dL
- Two-hour glucose level greater than 200 mg/dL

7.4   Clinical Significance

The glucose tolerance test is designed to assess how quickly glucose is eliminated from the body. This test screens for illnesses of glucose metabolism such as acromegaly, reactive hypoglycemia, decreased beta-cell activity, diabetes, and insulin resistance. Other diseases of glucose metabolism that may be detected include insulin resistance and diabetes.

A glucose tolerance test is normally ordered by a medical doctor or an advanced-trained nurse practitioner. To execute the test successfully, multiple sorts of specialists must work together. It is the attending physician's or registered nurse's obligation to ensure that the patient is given correct instructions on how to prepare for the test and what will happen during the examination (Huhn et al., 2018; Benhalima et al., 2018).

8   HbA₁C Testing

When evaluating the clinical implications of diabetes and its consequences, it is critical to identify people who either have diabetes but are not diagnosed or are at risk of developing diabetes shortly. The American Diabetes Association (ADA) advises monitoring asymptomatic adults with an FPG test or a 2-h OGTT every 3 years. Both tests measure a person's ability to tolerate glucose in their mouth (American Diabetes Association, 2009). However, conducting the OGTT in primary care settings can be difficult, and whether FPG concentration alone can reliably identify diabetes is controversial. This is proven that 40% of the population has undiagnosed diabetes (Cowie et al., 2009).

A standardized test measures the hemoglobin A1c (A1C) level. This test's findings are compared to the global A1C-derived average glucose and the Diabetes Control and Complications Trial (UK Prospective Diabetes Study Group, 1998; Rohlfing et al., 2002). The A1C level is a reliable indicator of chronic glycemic control and closely correlates with long-term diabetes risk of complications and mortality. The patient is not obligated to fast before the test or to take the sample at a specific time (Khaw et al., 2004; Tapp et al., 2008). Several population-based studies have been undertaken to evaluate the A1C level's usefulness in detecting previously undiscovered diabetes and its potential as a type 2 diabetes screening tool (Buell et al., 2007; Nakagami et al., 2007). However, the American Diabetes Association's (ADA) new definition of diabetes diagnosis using an A1C threshold of 6.5 percent, which takes into consideration several elements of diagnostic tests as well as economic cost, raises worries about potential delays in diabetes detection (International Expert Committee, 2009; Kramer et al., 2010). As a direct consequence of this, there is a great deal of debate on the appropriate A1C cutoff value for the diagnosis of diabetes.

When glucose accumulates in the circulation, it binds to hemoglobin in red blood cells. This causes your blood sugar level to rise. The A1c test calculates the proportion of glucose in the blood that is bound. Since red blood cells may live for 3 months, the test evaluates the average amount of glucose in your blood during the last 3 months. Your hemoglobin A1c level will be greater if the glucose levels have been persistently high during the last few weeks (Table 1).

8.1   A1c and Blood Sugar

Adults who do not have diabetes typically have hemoglobin A1c readings that fall below the range of 4 to 5.6%. Readings of the hemoglobin A1c that fall between 5.7% and 6.4% suggest prediabetes and a higher chance of acquiring diabetes. At the level of 6.5% or more, diabetes is considered to be present.

Table 1 The A1C test results match the blood sugar levels shown below

A1c (%)	Average blood sugar (mg/dL)
4	68
5	97
6	126
7	152
8	183
9	212
10	240
11	269
12	298
13	326
14	355

When referring to those with diabetes, the ideal A1c score is lower than 7%. If your hemoglobin A1c is high, you have a larger chance of developing diabetes complications. Someone who has had diabetes for an extended period without receiving treatment may have a higher than 8% level.

8.2   Limitations

A few different circumstances could impact the precision of the A1C test findings. Among these are:

- Pregnancy
- Heavy blood loss
- Blood transfusion
- Anemia
- Variants of hemoglobin

Hemoglobin A is the type of oxygen-carrying hemoglobin protein that is most abundant. Incorporating extra protein variants can lead to inaccurate results in the A1C test. People with ancestry in Africa, the Mediterranean region, or Southeast Asia are more likely to carry variants of the hemoglobin gene.

9   Estimation of Serum Creatinine

Some different things can bring on albuminuria, one of which is an abnormality in the glomerular endothelial barrier (Stehouwer et al., 2004), which leads to an increase in the amount of albumin that is filtered out of the kidney and a decrease in the amount of albumin that is broken down and reabsorbs by renal tubular cells.

Several factors contribute to albuminuria, including glomerular hypertension, inflammation, and oxidative stress, with angiotensin II (Coresh et al., 2003) and mechanical stress factors having a role.

Diabetes-related renal impairment is associated with abnormalities in vasodilation and the formation of reactive oxygen species mediated by endothelial-generated nitric oxide (NO). This finding points to a connection between vascular and metabolic issues. Angiotensin II and aldosterone activate NADP oxidase, an oxidative stress mediator when they interact with pulse pressure and elevate systolic blood pressure. This activation of NADP oxidase takes place. By increasing the rate at which NO is metabolized into peroxynitrite, angiotensin II can suppress endothelial-derived vasodilation (Stehouwer et al., 2004).

The ability to create endothelial progenitor cells (EPCs), as measured by the biological marker CD 34, declines with age and contributes to a high risk of cardiovascular disease (CVD). These bone marrow-derived cells have a role to play in the process of replacing damaged endothelium, but the number of these cells is reduced in patients who have problems with endothelium-dependent vasodilation (Hill et al., 2003). Although angiotensin II and aldosterone can both limit EPC production, angiotensin-converting enzyme inhibitors (ACEIs) can boost it.

Charles Heilig and colleagues (Heilig et al., 2006) investigated the role of renal glucose transporters in the evolution of diabetic nephropathy. The primary glucose transporter in mesangial cells, GLUT1, is responsible for controlling the development of the extracellular matrix. When GLUT 1 is overexpressed by the mesangial cells, the result is a diabetes-like phenotype. This is due to an increase in the synthesis of type I and IV collagen, fibronectin, and laminin. When GLUT1 is overexpressed in the glomeruli, it causes a nephropathy phenotype similar to diabetic renal disease in mice. This phenotype of nephropathy is distinguished by the increased mean glomerular volume, mesangial hypertrophy, and sclerosis.

9.1   Creatinine

Most serum creatinine, a creatine metabolite, is found in skeletal muscle. Levels of blood creatinine range from 0.8 to 1.4 mg/dL. Because they have less muscle mass than men, women have lower creatinine levels (0.6 to 1.2 mg/dL) than men (Molitoris, 2007).

The rate of creatine breakdown and the quantity of creatine per unit of skeletal muscle mass are both constant. Thus, plasma creatinine concentration is a valid and typically consistent measure of skeletal muscle mass (Martin, 2003). Intriguingly, Nobuko Harita et al. (2009) proposed that lower serum creatinine levels are related to a greater risk of type 2 diabetes, which might be explained by a loss in skeletal muscle mass. Skeletal muscle loss leads to fewer insulin target sites, which increases insulin resistance since skeletal muscle is an important insulin target tissue. As a result, type 2 diabetes develops (DeFronzo et al., 1985). The pathogenesis of type 2 diabetes, which is linked to lower blood creatinine levels, may be better understood due to this.

10 Estimation of Urine Microalbumin

You may detect minute levels of the blood protein albumin in your urine with a urine microalbumin test. In those who are at risk of developing kidney disease, a microalbumin test is used to detect the first indications of kidney damage. Your kidneys are necessary because they eliminate waste from your blood while retaining beneficial substances such as albumin. Proteins may leak through your kidneys and exit your body through your urine if you have a renal illness. When the kidneys are injured, albumin is one of the first proteins to leave. Patients at high risk for renal disorders, such as diabetes or high blood pressure, are encouraged to have microalbumin tests.

A doctor may recommend a urine microalbumin test to detect early indications of renal damage. More severe renal disease advancement may be stopped or delayed by treatment. The likelihood of kidney damage and any underlying conditions dictate how frequently microalbumin tests are administered.

As an example:

Type 1 diabetes: Starting 5 years following diagnosis, the doctor may recommend a microalbumin test once a year.

Type 2 diabetes: Your doctor may recommend a microalbumin test once a year beginning soon after your diagnosis.

High blood pressure: The doctor could advise you to undergo more frequent microalbumin testing.

The doctor may advise therapy and more regular testing if the urine microalbumin level is high.

The microalbumin test is a straightforward urine examination. You may eat and drink normally before the exam. The volume of pee required by the doctor to test may vary; they may require a random sample or require you to collect 24 h worth.

There are several ways to carry out this test:

Your doctor may instruct you to collect the urine for 24 h in a specified container and submit it to a laboratory for testing.

Timed urine test: Your doctor may request a urine sample either immediately in the morning or 4 h after the last urination.

Anytime, a random urine test can be conducted. However, to increase the accuracy of the results, it is frequently paired with a urine test for creatinine, a waste product generally filtered by the kidneys.

The urine sample is forwarded to a laboratory for testing. You can immediately resume your normal activities after obtaining the urine sample.

10.1   Results

- Milligrams (mg) of protein leakage over 24 h is the outcome of the microalbumin test. Generally:
- <30 mg is considered normal.
- 30 to 300 mg might suggest the presence of early renal disease (microalbuminuria).
- >300 mg suggests that the renal condition has progressed (macroalbuminuria).

Higher-than-expected urine microalbumin levels can result from many factors, such as:

- Hematuria
- Several drugs
- Fever
- Recent strenuous exercise
- Infection of the urinary tract
- Other kidney conditions

---

11   New Assays for Diabetes

11.1   Glucometer at Home Measuring

Regardless of the schema employed, careful BS monitoring is necessary at least before breakfast, lunch, and dinner as well as before the nightly snack. The ability of the family and the patient to precisely regulate BS has changed, thanks to at-home glucometer BS measurement. Several glucometers can measure BS from blood samples taken from the fingertip, each with advantages and disadvantages. Although most people have recollections, they cannot completely substitute meticulous record-keeping that supports the patient's and doctor's investigation of BS control. Most glucometer memory can be downloaded to a computer or uploaded to web-based programs to create BS vs time-of-day graphs and other pertinent analytic data. Changing insulin dosages and preventing a dead battery from tricking or emptying the glucometer's memory require patients to observe patterns of BS variations, even though glucometers may recall data from the previous month or longer. The fact that some patients use multiple glucometers or neglect to care for the device properly is another reason to keep a written record. Suppose the glucometer does not have the date or time set (for instance, as a result of changing the battery or dropping the device). In that case, the memory will be useless because it will be impossible to tell when any glucose value occurred. Blood ketone levels can be determined by some glucose meters. This is a more accurate sign of impending ketosis that can help control (Styne, 2016).

11.2 Salivary Glucose as a Nonintrusive Type 2 Diabetes Biomarker

Newcastle University in Australia has developed a noninvasive blood sugar monitoring strip that monitors glucose levels in diabetic saliva samples. In persons with diabetes, a drop of blood is frequently placed on the testing strip following a daily lancet prick to assess blood sugar levels. Some diabetics avoid this uncomfortable process by limiting the number of tests. The new test employs an enzyme capable of detecting glucose in a transistor and communicating its presence.

Early diabetes identification can avoid or postpone long-term health impacts; as a result, a person who is not diagnosed or treated for diabetes is more likely to have poor health outcomes. However, there may be no symptoms or indicators of diabetes in its early stages. By promptly detecting elevated blood glucose levels, diabetes complications can be minimized. Blood glucose levels are currently used to diagnose diabetes; readings exceeding 126 mg/dL are considered diabetic (Pandey et al., 2015). A far easier and noninvasive technique for diagnosing and monitoring diabetes would be ideal because measuring blood glucose at regular intervals causes patients significant discomfort and mental injury. Because saliva is easy to obtain using noninvasive ways, researchers have attempted to employ it as a diabetes diagnostic tool to alleviate these issues (Javaid et al., 2016; Ravindran et al., 2015; Abikshyeet et al., 2012; Kadashetti et al., 2015; Gupta et al., 2015; Aitken et al., 2015; Belce et al., 2000; Lima-Aragão et al., 2016). Because saliva, like the serum, includes antibodies, enzymes, growth factors, hormones, and microorganisms and their products, researchers have observed various correlation rates between serological and saliva characteristics. As a result, it can often be seen as a reflection of the body's physiological state (Javaid et al., 2016; Lima-Aragão et al., 2016; Negrato & Tarzia, 2010; Soares et al., 2009). This study will outline recent advancements in creating patient saliva-based diabetes mellitus diagnostic tools.

Saliva contains a comparable set of substances to blood that can be used to identify diabetes mellitus. Furthermore, saliva is far easier to handle and requires less pre-analysis processing than blood samples, which clot easily. As a result, this salivary diagnostic approach has significant potential for the early identification of diabetes without the need for difficult and costly procedures (Dhanya & Hegde, 2016; Kadashetti et al., 2015; Arakawa et al., 2016; Soni & Jha, 2015; Numako et al., 2016; Fujii et al., 2014; Chee et al., 2016; Zhang et al., 2015; Aitken et al., 2015; Abdolsamadi et al., 2014; Barnes et al., 2014; Satish et al., 2014). Because virtually all studies show a highly substantial positive connection between salivary glucose and blood glucose in diabetes patients and controls, salivary glucose levels can be utilized as a noninvasive diagnostic technique. In addition to biochemical and metabolomics studies, a noninvasive glucose biosensor based on paper strips has been successfully developed for salivary analysis to identify diabetes (Soni & Jha, 2015). A mouth guard biosensor with a telemetry system was also created to detect diabetes using noninvasive, real-time saliva glucose monitoring (Arakawa et al., 2016). According to another study, dried saliva spots are a suitable and reliable sample technique for bioanalysis in diabetes diagnosis (Numako et al., 2016). A disposable on-chip nanobiosensor with the appropriate level of sensitivity was produced in another investigation (Zhang et al., 2015). As a result, salivary diagnostics has evolved into a sophisticated area of molecular diagnostics that is now acknowledged as an essential component of biomedical, basic, and clinical research.

12   Conclusion

Diabetes is a slowly fatal condition for which there is likely no known cure. With enough understanding and prompt care, such issues can be avoided. Three significant effects are heart attack, kidney damage, and blindness. To prevent issues, it's essential to control patients' blood glucose levels. One of the problems with stringent blood glucose regulation is that it can lead to hypoglycemia, which can have much more severe side effects than high blood sugar levels. As a result, alternative diabetic treatment modalities are currently being sought after by researchers.

References

Abdolsamadi, H., Goodarzi, M. T., Ahmadi Motemayel, F., Jazaeri, M., Feradmal, J., Zarabadi, M., Hoseyni, M., & Torkzaban, P. (2014). Reduction of melatonin level in patients with type 2 diabetes and periodontal diseases. *Journal of Dental Research Dental Clinics Dental Prospects, 8*, 160–165.

Abikshyeet, P., Ramesh, V., & Oza, N. (2012). Glucose estimation in the salivary secretion of diabetes mellitus patients. *Diabetes, Metabolic Syndrome and Obesity, 5*, 149–154.

Ahmad, S. I. (2013). Introduction to Diabetes mellitus. *Advances in Experimental Medicine and Biology, 771*, 1–11. https://doi.org/10.1007/978-1-4614-5441-0_1

Aitken, J. P., Ortiz, C., Morales-Bozo, I., Rojas-Alcayaga, G., Baeza, M., Beltran, C., & Escobar, A. (2015). α-2-macroglobulin in saliva is associated with glycemic control in patients with type 2 diabetes mellitus. *Disease Markers, 2015*, 1. https://doi.org/10.1155/2015/128653

American Diabetes Association. (2009). Standards of medical care in diabetes: 2009. *Diabetes Care, 32*(Suppl 1), S13–S61.

American Diabetes Association. (2014). Diagnosis and classification of diabetes mellitus. *Diabetes Care, 37*, S81–S90.

Arakawa, T., Kuroki, Y., Nitta, H., Chouhan, P., Toma, K., Sawada, S., Takeuchi, S., Sekita, T., Akiyoshi, K., Minakuchi, S., & Mitsubayashi, K. (2016). Mouthguard biosensor with telemetry system for monitoring of saliva glucose: A novel cavitas sensor. *Biosensors & Bioelectronics, 84*, 106–111.

Barnes, V. M., Kennedy, A. D., Panagakos, F., Devizio, W., Trivedi, H. M., Jönsson, T., Guo, L., Cervi, S., & Scannapieco, F. A. (2014). Global metabolomic analysis of human saliva and plasma from healthy and diabetic subjects, with and without periodontal disease. *PLoS One, 9*, e105181.

Belce, E., Uslu, M., Kucur, M., Umut, A., Ipbüker, H. O., & Seymen, H. O. (2000). Evaluation of salivary sialic acid level and Cu-Zn superoxide dismutase activity in type 1 diabetes mellitus. *The Tohoku Journal of Experimental Medicine, 192*, 219–225.

Benhalima, K., Minschart, C., Ceulemans, D., Bogaerts, A., Van Der Schueren, B., Mathieu, C., & Devlieger, R. (2018). Screening and management of gestational diabetes mellitus after bariatric surgery. *Nutrients, 10*(10).

Buell, C., Kermah, D., & Davidson, M. B. (2007). Utility of A1C for diabetes screening in the 1999–2004 NHANES population. *Diabetes Care, 30*, 2233–2235.

Chan, S., Keim, P., & Steiner, D. (1976). Cell-free synthesis of rat preproinsulins: Characterization and partial amino acid sequence determination. *Proceedings of the National Academy of Sciences of the United States of America, 73*, 1964–1968.

Chee, C. S., Chang, K. M., Loke, M. F., Angela Loo, V. P., & Subrayan, V. (2016). Association of potential salivary biomarkers with diabetic retinopathy and its severity in type-2 diabetes mellitus: A proteomic analysis by mass spectrometry. *PeerJ, 4*, e2022. https://doi.org/10.7717/peerj.2022

Choo, S. Y. (2007). The HLA system: Genetics, immunology, clinical testing, and clinical implications. *Yonsei Medical Journal, 48*, 11–23.

Coresh, J., Astor, B. C., Greene, T., Eknoyan, G., & Levey, A. S. (2003). Prevalence of chronic kidney disease and decreased kidney function in the adult US population: Third National Health and Nutrition Examination Survey. *American Journal of Kidney Diseases, 41*, 1–12.

Cowie, C. C., Rust, K. F., Ford, E. S., et al. (2009). Full accounting of diabetes and pre-diabetes in the U.S. population in 1988-1994 and 2005-2006. *Diabetes Care, 32*, 287–294.

DeFronzo, R. A., Gunnarsson, R., Björkman, O., Olsson, M., & Wahren, J. (1985). Effects of insulin on peripheral and splanchnic glucose metabolism in noninsulin-dependent (type II) diabetes mellitus. *The Journal of Clinical Investigation, 76*, 149–155.

Dhanya, M., & Hegde, S. (2016). Salivary glucose as a diagnostic tool in Type II diabetes mellitus: A case-control study. *Nigerian Journal of Clinical Practice, 19*, 486–490.

Egea, P. F., Stroud, R. M., & Walter, P. (2005). Targeting proteins to membranes: Structure of the signal recognition particle. *Current Opinion in Structural Biology, 15*, 213–220.

Farquhar, M. G., & Palade, G. E. (1981). The Golgi apparatus complex (1954–1981) from artifact to center stage. *J Cell Biol., 91*, 77s–103s.

Fujii, S., Maeda, T., Noge, I., Kitagawa, Y., Todoroki, K., Inoue, K., Min, J. Z., & Toyo'oka, T. (2014). Determination of acetone in saliva by reversed-phase liquid chromatography with fluorescence detection and the monitoring of diabetes mellitus patients with ketoacidosis. *Clinica Chimica Acta, 430*, 140–144.

Given, B. D., Cohen, R. M., Shoelson, S. E., Frank, B. H., Rubenstein, A. H., & Tager, H. S. (1985). Biochemical and clinical implications of proinsulin conversion intermediates. *The Journal of Clinical Investigation, 76*, 1398–1405.

Gold, G., Gishizky, M. L., & Grodsky, G. M. (1982). Evidence that glucose marks cells resulting in the preferential release of newly synthesized insulin. *Science, 218*, 56–58.

Gorodezky, C., Alaez, C., Murguia, A., et al. (2006). HLA and autoimmune diseases: Type I diabetes (TID) as an example. *Autoimmunity Reviews, 5*(3), 87–94.

Gupta, S., Sandhu, S. V., Bansal, H., & Sharma, D. (2015). Comparison of salivary and serum glucose levels in diabetic patients. *Journal of Diabetes Science and Technology, 9*, 91–96.

Halban, P. A. (1991). Structural domains and molecular lifestyles of insulin and its precursors in the pancreatic beta cell. *Diabetologia, 34*, 767–778.

Harita, N., Hayashi, T., Sato, K. K., Nakamura, Y., Yoneda, T., Endo, G., & Kambe, H. (2009). Lower serum creatinine is a new risk factor of type 2 diabetes: The Kansai healthcare study. *Diabetes Care, 32*, 424–426.

Harris, M. I., Flegal, K. M., Cowie, C. C., et al. (1998). Prevalence of diabetes, impaired fasting glucose, and impaired glucose tolerance in U.S. adults. The Third National Health and Nutrition Examination Survey, 1988-1994. *Diabetes Care, 21*(4), 518–524.

Heilig, C. W., Brosius, F. C., 3rd, & Cunningham, C. (2006). Role for GLUT1 in diabetic glomerulosclerosis. *Expert Reviews in Molecular Medicine, 8*, 1–18.

Hill, J. M., Zalos, G., Halcox, J. P., Schenke, W. H., Waclawiw, M. A., Quyyumi, A. A., & Finkel, T. (2003). Circulating endothelial progenitor cells, vascular function, and cardiovascular risk. *The New England Journal of Medicine, 348*, 593–600.

Howell, S. L. (1972). Role of ATP in the intracellular translocation of proinsulin and insulin in the rat pancreatic beta cell. *Nature: New Biology, 235*, 85–86.

Huang, X. F., & Arvan, P. (1994). Formation of the insulin-containing secretory granule core occurs within immature β-granules. *The Journal of Biological Chemistry, 269*, 20838–20844.

Huang, X. F., & Arvan, P. (1995). Intracellular transport of proinsulin in pancreatic beta cells. *The Journal of Biological Chemistry, 270*, 20417–20423.

Huhn, E. A., Rossi, S. W., Hoesli, I., & Göbl, C. S. (2018). Controversies in screening and diagnostic criteria for gestational diabetes in early and late pregnancy. *Front Endocrinol (Lausanne), 9*, 696.

International Diabetes Federation. (2014). *IDF Diabetes Atlas 6th edition update for 2014*. International Diabetes Federation.

International Expert Committee. (2009). International expert committee report on the role of the A1C assay in the diagnosis of diabetes. *Diabetes Care, 32*, 1327–1334.

Javaid, M. A., Ahmed, A. S., Durand, R., & Tran, S. D. (2016). Saliva as a diagnostic tool for oral and systemic diseases. *Journal of Oral Biology and Craniofacial Research, 6*, 66–75.

Joslin Diabetes Center & Joslin Clinic. (2006, 20 Oct). *Clinical guideline for adults with diabetes*.

Jun, S.-K., & Yoon, Y.-W. (2002). A new look at viruses in Type I diabetes. *Diabetes/Metabolism Research and Reviews, 19*, 8–31.

Kadashetti, V., Baad, R., Malik, N., Shivakumar, K. M., Vibhute, N., Belgaumi, U., Gugawad, S., & Pramod, R. C. (2015). Glucose level estimation in diabetes mellitus by saliva: A bloodless revolution. *Romanian Journal of Internal Medicine, 53*, 248–252.

Khaw, K. T., Wareham, N., Bingham, S., Luben, R., Welch, A., & Day, N. (2004). Association of hemoglobin A1c with cardiovascular disease and mortality in adults: The European prospective investigation into cancer in Norfolk. *Annals of Internal Medicine, 141*, 413–420.

Koenig, R. J., Peterson, C. M., Jones, R. L., Saudek, C., Lehrman, M., & Cerami, A. (1976). Correlation of glucose regulation and hemoglobin AIc in diabetes mellitus. *The New England Journal of Medicine, 295*, 417–420.

Kramer, C. K., Araneta, M. R., & Barrett-Connor, E. (2010). A1C and diabetes diagnosis: The rancho Bernardo study. *Diabetes Care, 33*, 101–103.

Lee, M. C., Miller, E. A., Goldberg, J., Orci, L., & Schekman, R. (2004). Bi-directional protein transport between the ER and Golgi. *Annual Review of Cell and Developmental Biology, 20*, 87–123.

Li, L. J., Yu, Q., Tan, K. H., & IPRAMHO-INTERNATIONAL Study Group. (2019). Clinical practice of diabetic pregnancy screening in Asia-Pacific countries: A survey review. *Acta Diabetologica, 56*(7), 815–817.

Lima-Aragão, M. V., de Oliveira-Junior, J., Maciel, M. C., Silva, L. A., do Nascimento, F. R., & Guerra, R. N. (2016). Salivary profile in diabetic patients: Biochemical and immunological evaluation. *BMC Research Notes, 9*, 103.

Lodish, H. F. (1988). Transport of secretory and membrane glycoproteins from the rough endoplasmic reticulum to the Golgi. *The Journal of Biological Chemistry, 263*, 2107–2110.

Lomedico, P. T., Chan, S. J., Steiner, D. F., & Saunders, G. F. (1977). Immunological and chemical characterization of bovine preproinsulin. *The Journal of Biological Chemistry, 252*, 7971–7978.

Losev, E., Reinke, C. A., Jellen, J., Strongin, D. E., Bevis, B. J., & Glick, B. S. (2006). Golgi maturation visualized in living yeast. *Nature, 441*, 939–940.

Mains, R. E., Dickerson, I. M., May, V., Stoffers, D. A., Perkins, S. N., Ouafi, K. L., Huster, E. J., & Eipper, B. A. (1990). Cellular and molecular aspects of peptide hormone biosynthesis. *Frontiers in Neuroendocrinology, 11*, 52–89.

Maldonado-Hernández, J., Martínez-Basila, A., Rendón-Macías, M. E., & López-Alarcón, M. (2019). Accuracy of the ^{13}C-glucose breath test to identify insulin resistance in non-diabetic adults. *Acta Diabetologica, 56*(8), 923–929.

Martin, R. F. (2003). Renal function. In L. A. Kaplan, A. J. Pesce, & S. C. Kazmierczak (Eds.), *Clinical chemistry: Theory, analysis, correlation* (4th ed., pp. 483–484). Saunders Elsevier.

Masharani UaK, J. H. (2001). Diabetes mellitus and hypoglycemia. In L. M. Tierney Jr., S. J. McPhee, & M. Papadakis (Eds.), *Current medical diagnosis and treatment*. McGraw Hill.

Molitoris, B. A. (2007). Acute kidney injury. In L. Goldman & D. Ausiello (Eds.), *Cecil medicine* (23rd ed., p. 121). Saunders Elsevier.

Munro, S., & Pelham, H. R. B. (1987). A C-terminal signal prevents secretion of luminal ER proteins. *Cell, 48*, 899–907.

Nakagami, T., Tominaga, M., Nishimura, R., et al. (2007). Is the measurement of glycated hemoglobin A1c alone an efficient screening test for undiagnosed diabetes? Japan National Diabetes Survey. *Diabetes Research and Clinical Practice, 76*, 251–256.

Negrato, C. A., & Tarzia, O. (2010). Buccal alterations in diabetes mellitus. *Diabetology and Metabolic Syndrome, 2*, 3.

Nishi, M., Sanke, T., Nagamatsu, S., Bell, G. I., & Steiner, D. F. (1990). Islet amyloid polypeptide. A new beta cell secretory product related to islet amyloid deposits. *The Journal of Biological Chemistry, 265*, 4173–4176.

Numako, M., Takayama, T., Noge, I., Kitagawa, Y., Todoroki, K., Mizuno, H., Min, J. Z., & Toyo'oka, T. (2016). Dried saliva spot (DSS) as a convenient and reliable sampling for bioanalysis: An application for the diagnosis of diabetes mellitus. *Analytical Chemistry, 88*, 635–639.

Orci, L., Lambert, A. E., Kanazawa, Y., Amherdt, M., Rouiller, C., & Renold, A. E. (1971). Morphological and biochemical studies of B cells in fetal rat endocrine pancreas in organ culture. Evidence for proinsulin biosynthesis. *The Journal of Cell Biology, 50*, 565–582.

Orci, L., Ravazzola, M., Amherdt, M., Madsen, O., Vassalli, J. D., & Perrelet, A. (1985). Direct identification of prohormone conversion site in insulin-secreting cells. *Cell, 42*, 671–681.

Orci, L., Stamnes, M., Ravazzola, M., Amherdt, M., Perrelet, A., Sollner, T. H., & Rothman, J. E. (1997). Bidirectional transport by distinct populations of COPI-coated vesicles. *Cell, 90*, 335–349.

Pandey, R., Dingari, N. C., Spegazzini, N., Dasari, R. R., Horowitz, G. L., & Barman, I. (2015). Emerging trends in optical sensing of glycemic markers for diabetes monitoring. *Trends in Analytical Chemistry, 64*, 100–108.

Patzelt, C., Labrecque, A., Duguid, J., Carroll, R., Keim, P., Heinrikson, R., & Steiner, D. (1978). Detection and kinetic behavior of preproinsulin in pancreatic islets. *Proceedings of the National Academy of Sciences of the United States of America, 75*, 1260–1264.

Phillips, J. M., Parish, N. M., Raine, T., et al. (2009). Type I diabetes development requires both C D4+ and CD8+ T-cell s and can be reversed by non-depleting antibodies targeting both T-cell populations. *The Review of Diabetic Studies, 6*(2), 97–103.

Ravindran, R., Gopinathan, D. M., & Sukumaran, S. (2015). Estimation of salivary glucose and glycogen content in exfoliated buccal mucosal cells of patients with type II diabetes mellitus. *Journal of Clinical and Diagnostic Research, 9*, ZC89-93.

Rohlfing, C. L., Wiedmeyer, H. M., Little, R. R., England, J. D., Tennill, A., & Goldstein, D. E. (2002). Defining the relationship between plasma glucose and HbA(1c): Analysis of glucose profiles and HbA(1c) in the diabetes control and complications trial. *Diabetes Care, 25*, 275–278.

Sando, H., Borg, J., & Steiner, D. F. (1972). Studies on the secretion of newly synthesized proinsulin and insulin from isolated rat islets of Langerhans. *The Journal of Clinical Investigation, 51*, 1476–1485.

Sanger, F. (1959). Chemistry of insulin. *Science, 129*, 1340–1344.

Satish, B. N., Srikala, P., Maharudrappa, B., Awanti, S. M. P., Kumar, P., & Hugar, D. (2014). Saliva: A tool in assessing glucose levels in diabetes mellitus. *Journal of International Oral Health, 6*, 114–117.

Schekman, R., & Orci, L. (1996). Coat proteins and vesicle budding. *Science, 271*, 1526–1533.

Sizonenko, S., Irminger, J. C., Buhler, L., Deng, S., Morel, P., & Halban, P. A. (1993). Kinetics of proinsulin conversion in human islets. *Diabetes, 42*, 933–936.

Soares, M. S., Batista-Filho, M. M., Pimentel, M. J., Passos, I. A., & Chimenos-Küstner, E. (2009). Determination of salivary glucose in healthy adults. *Medicina Oral, Patología Oral y Cirugía Bucal, 14*, e510–e513.

Sobey, W. J., Beer, S. F., Carrington, C. A., Clark, P. M. S., Frank, B. H., Gray, I. P., Luzio, S. D., Owens, D. R., Schneider, A. E., Siddle, K., Temple, R. C., & Hales, C. N. (1989). Sensitive and specific two-site immunoradiometric assays for human insulin, proinsulin, 65–66 split, and 32–33 split proinsulins. *The Biochemical Journal, 260,* 535–541.

Soni, A., & Jha, S. K. (2015). A paper strip based non-invasive glucose biosensor for salivary analysis. *Biosensors & Bioelectronics, 67,* 763–768.

Stehouwer, C. D., Henry, R. M., Dekker, J. M., Nijpels, G., Heine, R. J., & Bouter, L. M. (2004). Microalbuminuria is associated with impaired brachial artery, flow-mediated vasodilation in elderly individuals without and with diabetes: Further evidence for a link between microalbuminuria and endothelial dysfunction—The Hoorn Study. *Kidney International. Supplement, 66,* S42–S44.

Steiner, D. F. (2001). The prohormone convertases and precursor processing in protein biosynthesis. In R. E. Dalbey & D. S. Sigman (Eds.), *The enzymes* (Vol. XXII, pp. 163–198). Academic Press.

Steiner, D. F., & Oyer, P. E. (1967). The biosynthesis of insulin and a probable precursor of insulin by a human islet cell adenoma. *Proceedings of the National Academy of Sciences of the United States of America, 57,* 473–480.

Steiner, D. F., Cunningham, D. D., Spigelman, L., & Aten, B. (1967). Insulin biosynthesis: Evidence for a precursor. *Science, 157,* 697–700.

Steiner, D. F., Clark, J. L., Nolan, C., Rubenstein, A. H., Margoliash, E., Aten, B., & Oyer, P. E. (1969). Proinsulin and the biosynthesis of insulin. *Recent Progress in Hormone Research, 25,* 207–292.

Steiner, D. F., Clark, J. L., Nolan, C., Rubenstein, A. H., Margoliash, E., Melani, F., & Oyer, P. E. (1970). The biosynthesis of insulin and some speculation regarding the pathogenesis of human diabetes. In E. Cerasi & R. Luft (Eds.), *The pathogenesis of diabetes mellitus, Nobel symposium 13* (pp. 57–80). Almqvist & Wiksell.

Steiner, D. F., Kemmler, W., Clark, J. L., Oyer, P. E., & Rubenstein, A. (1972). The biosynthesis of insulin. In D. F. Steiner & N. Freinkel (Eds.), *Handbook of physiology—Section 7 Endocrinology I* (pp. 175–198). Williams & Wilkins.

Styne, D. M. (2016). Diabetes mellitus, Chapter 11. In *Pediatric endocrinology* (pp 263–304). Springer. https://doi.org/10.1007/978-3-319-18371-8_11

Tapp, R. J., Tikellis, G., Wong, T. Y., Harper, C. A., Zimmet, P. Z., Shaw, J. E., & Australian Diabetes Obesity and Lifestyle Study Group. (2008). Longitudinal association of glucose metabolism with retinopathy: Results from the Australian Diabetes Obesity and Lifestyle (AusDiab) study. *Diabetes Care, 31,* 1349–1354.

The International Expert Committee. (2009). International Expert Committee report on the role of the A1C assay in the diagnosis of diabetes. *Diabetes Care, 32,* 1327–1334.

UK Prospective Diabetes Study Group. (1998). Effect of intensive blood-glucose control with metformin on complications in overweight patients with type 2 diabetes (UKPDS 34). *Lancet, 352,* 854–865.

Wei, Y. M., Liu, X. Y., Shou, C., Liu, X. H., Meng, W. Y., Wang, Z. L., Wang, Y. F., Wang, Y. Q., Cai, Z. Y., Shang, L. X., Sun, Y., & Yang, H. X. (2019). Value of fasting plasma glucose to screen gestational diabetes mellitus before the 24th gestational week in women with different pre-pregnancy body mass index. *Chinese Medical Journal, 132*(8), 883–888.

WHO. (2011). [cited 20 II]. Accessed from http://www.who.int/diabetes/en/

Williamson, A. R., Hunt, A. E., Pope, J. F., & Tolman, N. M. (2002). Recommendations of dietitians for overcoming barriers to dietary adherence in individuals with diabetes. *The Diabetes Educator, 26*(2), 272–279.

World Health Organization. (2005). *WHO non communicable diseases and mental health WHO STEPS surveillance manual: The WHO STEPwise approach to chronic disease risk factor surveillance.* World Health Organization.

Zhang, W., Du, Y., & Wang, M. L. (2015). Noninvasive glucose monitoring using saliva nano biosensor. *Sensing and Biosensing Research, 4,* 23–29.

Diabetes and Other Comorbidities: Microvascular and Macrovascular Diseases Diabetes and Cancer

V. Nithya, P. Sangavi, R. Srinithi, K. T. Nachammai, S. Gowtham Kumar, D. Prabu, and K. Langeswaran

Abstract

Diabetes is characterized by unusually elevated blood glucose or levels of sugar. Dietary glucose is produced, and insulin is the hormone that aids glucose entry into cells to provide energy. The occurrence of type 1 diabetes occurs when insulin production does not take place in the body, and type 2 diabetes happens when the body does not appropriately utilize insulin. Persistent abnormal functions and continuous injury to vital organs such as eyes, blood vessels, nerves, kidneys, and heart are the foremost characteristic features of chronic diabetes. Genes related to diabetes risk include TCF7L2, which distress insulin secretion and glucose production. The development of diabetes is mainly associated with various pathogenic processes. In recent years, the prevalence of

P. Sangavi, S. Gowtham Kumar and D. Prabu contributed equally with all other contributors.

V. Nithya
Department of Animal Health and Management, Science Campus, Alagappa University, Karaikudi, India

P. Sangavi · R. Srinithi
Department of Bioinformatics, Science campus, Alagappa University, Karaikudi, India

K. T. Nachammai · K. Langeswaran (✉)
Molecular Cancer Biology Laboratory, Department of Biotechnology, Science Campus, Alagappa University, Karaikudi, Tamil Nadu, India

S. Gowtham Kumar
Faculty of Allied Health Sciences, Chettinad Hospital & Research Institute, Chettinad Academy of Research and Education, Kelambakkam, Chennai, Tamil Nadu, India

D. Prabu
Department of Microbiology, Dr. ALM PG IBMS, University of Madras, Taramani Campus, Chennai, Tamil Nadu, India

R. Noor (ed.), *Advances in Diabetes Research and Management*,
https://doi.org/10.1007/978-981-19-0027-3_2

21

diabetes and cancer has risen dramatically. Obesity, a sedentary lifestyle, smoking, and aging are frequent hazard issues for diabetes and cancer. According to a large body of epidemiological data, diabetes is a self-governing risk factor for amplified rates of several types of cancer incidence and death. People with diabetes slightly increase the incidence and mortality of several malignancies, namely, pancreatic, liver, colorectal, breast, endometrial, and bladder cancers.

Keywords

Diabetes · Cancer · Diabetes genes · Lifestyle disease · Pancreatic β cells · Metabolic disorder

1   Introduction

Diabetes mellitus is a metabolic condition characterized by hyperglycemia, which causes insulin secretion to be insufficient. It is a chronic illness that causes malfunction and damage to the eyes, nerves, kidneys, heart, foot, and other organs. Insulin deficiency and differentiation in insulin action have been produced via the autoimmune destruction of cells in the pancreas (Sen & Chakraborty, 2015). Abnormalities in glucose, lipid, and protein metabolism occur due to insulin insufficiency. Insulin secretion destruction and insulin action deficit sometimes occur in the same person, and it's not always evident which aberration is the primary cause of hyperglycemia (Alam et al., 2014). The mainly widespread categories of diabetes are type 1 and type 2. Type 1 diabetes mellitus (T1DM) is typified by the autoimmune death of insulin-producing beta cells, culminating in a complete lack of insulin. T2DM is associated with metabolic disorders in which cells become insulin-insensitive, resulting in a relative insulin deficit. Although both T1DM and T2DM have been connected with amplified jeopardy of cancer, some studies have shown that T2DM has a stronger epidemiological and biological relationship with cancer. Obesity, smoking, and growing older are potential risk factors for cancer and T2DM. Diabetes, especially type 2, has been linked to several cancers, including pancreatic, liver, colorectal, breast, endometrial, bladder, and prostate cancers. However, many lines of evidence point to the insulin/insulin-like growth factor (IGF) axis, hyperglycemia, inflammatory cytokines, and sex hormones as possible causes. Consequently, this chapter aims to show how diabetes and cancer are linked and the essential machinery.

2   Causes of Diabetes

The demolition of insulin-producing beta cells is creating type I diabetes. Still, it cannot recognize the reason for the demolition of the autoimmune system. Heredity comprises the history of total family and our prenatal background; it can be able to

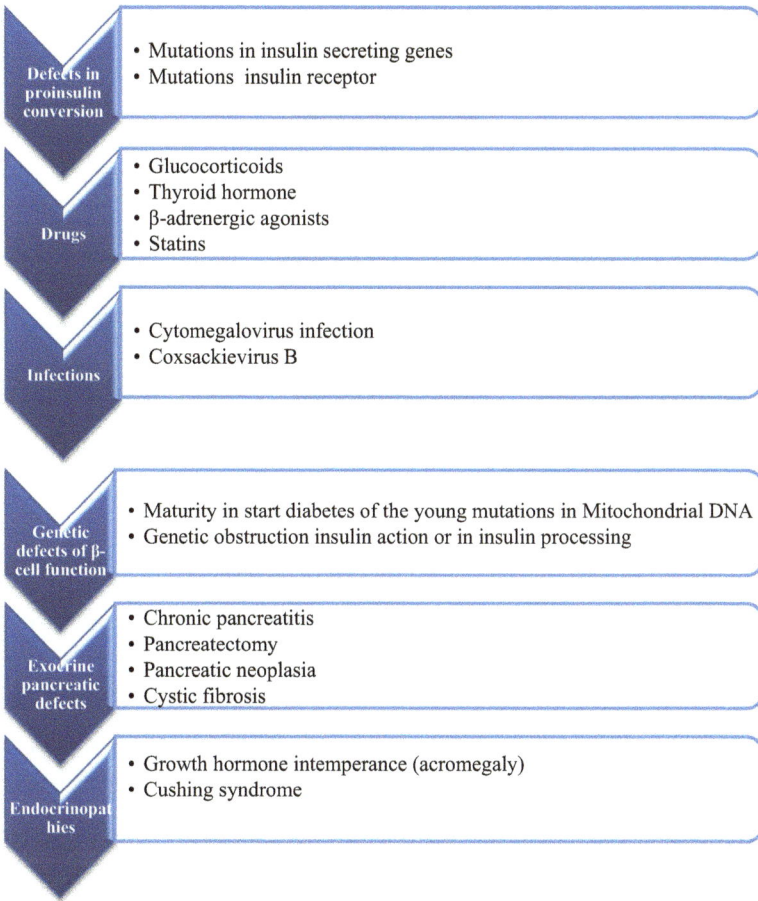

Fig. 1 Schematic representation of causes of diabetes

make type 1 diabetes. Endocrine upset is due to the reveal of chemicals present in plastics. Moreover, a viral contagion can elicit the autoimmune process. Recent studies show that the overture of certain foods is causing type I diabetes in infants (Fig. 1). Consume fruits before 5 months of age, or wait till afterward 7 months to bring in grains like rice and oats, which boost the possibility of getting diabetes. On the other hand, breastfeeding helps nourish and diminish these risks (Chen et al., 2011).

3 Diabetes Types

Autoimmune disorder diabetes is broadly classified into three types, namely, type 1 diabetes, type 2 diabetes, and gestational diabetes (Fig. 2).

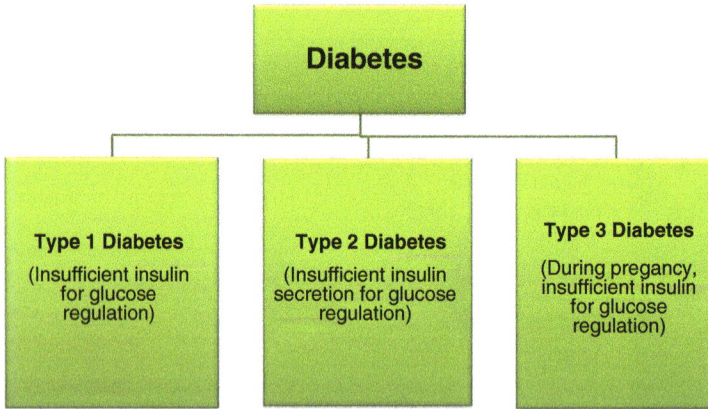

Fig. 2 Portrays the types of diabetes

1. **Type 1 Diabetes**
   Type 1 diabetes has been considered a chronic disorder caused by immunological reactions. Insulin-producing beta cells in the pancreas are assaulted and killed by the immune system. This type 1 diabetes has been marked as an irreversible hazard (Li et al., 2017).

2. **Type 2 Diabetes**
   Type 2 diabetes is mainly known due to insulin resistance. This implies that your body is powerless to use insulin adequately, prompting your pancreas to create more insulin in anticipation that it can no longer keep up with the stipulate. Insulin production declines, increasing blood sugar levels (Javeed & Matveyenko, 2018).

3. **Gestational Diabetes**
   Diabetes at the time of pregnancy because of insulin blocking substances is termed gestational diabetes. Pregnant women are the only ones who get this kind of diabetes. Preexisting prediabetes and a family history of diabetes increase the risk of developing it (Coustan, 2013).

3.1   Type 1 Diabetes

Juvenile-onset diabetes is another name for it. A total lack of insulin secretion brings it on. Serological facts and molecular markers may be used to identify it in pancreatic islets. Individuals with type I diabetes account for 10–20% of diabetic patients. Autoantibodies to the tyrosine phosphatases IA-2 and IA-2 autoantibodies, insulin autoantibodies, glutamic acid decarboxylase (GAD) autoantibodies, and autoantibodies to the glutamic acid decarboxylase (GAD) autoantibodies are all part of the cell devastation. When fasting hyperglycemia is identified, autoantibodies are found in 80–90 percent of people. It might be different in different people, and it can be fast in children and sluggish in adults. Ketoacidosis may be one of the first indications and symptoms of the illness in particular children and adolescents. Adults with residual cell function may avoid ketoacidosis for many years; nonetheless, they become insulin-dependent for survival. Plasma C-peptide stumpy or untraceable levels indicate that insulin production is low or absent in this stage of diabetes (Acharjee et al., 2013). Type I diabetes is caused by a genetic predisposition and is defined by the failure of the insulin-producing cell found in pancreatic islets, resulting in insulin shortage. It might be immune-mediated or idiopathic. In immune-mediated autoimmune illness, T cell-mediated autoimmune disease causes cell and insulin defects. Some diabetic individuals have persistent insulinopenia, which leads to ketoacidosis, but no evidence of autoimmune has been found. Hepatitis, Hashimoto's thyroiditis, myasthenia gravis, Graves' disease, and pernicious anemia are patients' autoimmune disorders (Trefz et al., 2019). Type I diabetic individuals have strong hereditary autoimmunity deficient in cells and do not connect with HLA. Insulin replacement therapy for afflicted patients may come and go.

3.2 Type 2 Diabetes

Type II diabetes can be caused by lack of insulin action and inadequate compensation of insulin secretion. This type of diabetes is the most commonly occurring. In this stage, hyperglycemia can be reversed by various measures and medicines that can be improved in the insulin-deficient. The numerous factors are the major complication of type II diabetes, such as obesity, physical unfitness, poor diet, and urbanization (Zaccardi et al., 2016). This type of diabetes is associated with obesity due to insulin deficiency. The patients who are obese frequently have serum insulin concentrations that is higher. However, these persons are unable to make sufficient insulin. Type 2 diabetes can manage the glucose level in the blood due to a balanced diet and exercise and, if required, through taking injections or taking medicines (Brunton, 2016).

3.3 Gestational Diabetes Mellitus (GDM)

Gestational diabetes mellitus depends on the conditions at the diagnosis, and many diabetic patients are not easily fitted to a single class. Due to insulin resistance, diabetes activated through pregnancy is known as gestational diabetes. It can be frequently diagnosed in middle or late pregnancy. During the pregnancy, when the blood glucose level is high in the mother, it circulates through the placenta to move to the baby. This diabetes has controlled the growth and development of the baby (Coustan, 2013).

4   Diabetes and Cancers

Diabetes and cancer are both widespread diseases having a number of behavioral risk factors in common, such as obesity, smoking, dietary variables, and physical inactivity, as well as disease pathways such as hyperglycemia and hyperinsulinemia. Diabetes raises the risk of developing solid tumors such as pancreatic cancer, liver cancer, colon cancer, breast cancer, and endometrial cancer (Fig. 3). Diabetes patients who also have cancer have unique challenges. Diabetes patients have a higher cancer-related death rate, and diabetes patients are at a higher risk of various illnesses and infection-related morbidity and mortality.

(a) **Pancreatic Cancer**

  With a 5-year survival rate of fewer than 10%, pancreatic cancer (PC) is one of the deadliest malignancies. For about 200 years, the link between diabetes and PC has been established (Chaudhry et al., 2013), and two theories have recently been offered to explain the link between these two conditions. On the one hand, epidemiological examinations have indicated that people with diabetes have a substantially higher frequency of PC than nondiabetics, implying that diabetes is a risk factor for PC. New-onset diabetes is linked to a 2.3-fold enlarged incidence of PC compared to long-term diabetes, indicating a precursor to PC (Setiawan et al., 2019). More study has shown a bidirectional connection between diabetes and PC, with the risk of diabetes and PC being inversely proportionate to time length. The risk of PC is highest in the first 2 years after a diabetes diagnosis, and the incidence steadily decreases with time. Those who have had diabetes for more than 5 years have a much lower probability of emergent PC. Consequently, long-term diabetes is now considered a risk factor for PC, whereas new-onset diabetes is considered a sign of the condition.

(b) **Liver Cancer**

Fig. 3 Symbols interlinks between diabetes and cancer

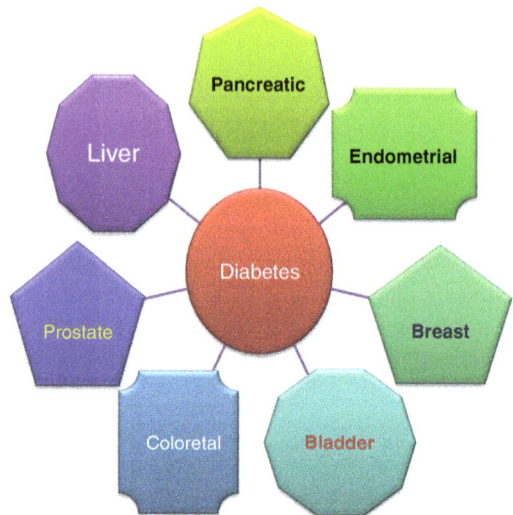

Globally, primary liver cancer hepatocellular carcinoma (HCC) is the fifth most frequent cancer in males and the seventh most widespread cancer in women (Bosetti et al., 2014), with a high prevalence in Asia's eastern part and Africa. This tumor is also known to be a disease with a high fatality rate. A recent study has connected diabetes to HCC, demonstrating a sovereign risk factor for the disease. Previous to delving into the link between diabetes and HCC, consider that HBV infection, aflatoxin exposure, and nonalcoholic fatty liver disease (NAFLD) are all critical risk factors for HCC. Because of their connections to hepatitis viruses and NAFLD, diabetes and HCC are closely related. NAFLD is a term used to describe a collection of progressive hepatic illnesses that range from pure steatosis to steatohepatitis. NAFLD is also caused by insulin resistance in more than 70% of diabetics, rendering diabetics more prone to major hepatic illnesses like HCC. Many systematic reviews and meta-analyses have revealed NAFLD as a focal point of the diabetes-HCC connection (Mantovani & Targher, 2017). Consequently, diabetes is a preventable risk feature, and the rapport between diabetes and an increased risk of HCC should not be neglected.

(c) **Colorectal Cancer**

Worldwide, colorectal cancer has been marked as the fourth most widespread cancer and second cancer with the high mortality rate. Furthermore, the fatality rate from CRC in developed countries is about 33% (Soltani et al., 2019). Much research has shown numerous shared hazard features between diabetes and CRC, namely, lifestyle, obesity, age, sedentary, and smoking confirmed in many investigations. Diabetes, on the other hand, is an autonomous menace factor for CRC. Furthermore, CRC patients with diabetes have a higher death rate (Gutiérrez-Salmerón et al., 2021). Surprisingly, sex dissimilarity has been firmly observed in several studies, with only a slightly augmented risk in women with diabetes compared to a considerably expanding risk of diabetics in males. Finally, nutrition plays a role in developing diabetes and CRC. However, studies have shown that only women, not men, can lessen CRC hazard, even if women and men eat an analogous healthy diet.

(d) **Breast Cancer**

Due to the speedily mounting western lifestyle in developing countries, breast cancer incidence and mortality have become magnified every year. Metabolic disorder diabetes is closely connected to an elevated risk of breast cancer. Diabetes is linked to a greater frequency and breast cancer mortality rate, according to a large body of epidemiological research. Furthermore, according to a meta-analysis, the link between diabetes and breast cancer seems limited to postmenopausal women (Boyle et al., 2012). However, this finding contradicts another research, which found that diabetes amplifies the occurrence of breast cancer in premenopausal women. This suggested that a history of diabetes is strongly linked to breast cancer (Wu et al., 2007). Furthermore, two studies came to similar results, stating that diabetes might make it difficult to concentrate on other health issues and lead to a low likelihood of breast cancer detection. Furthermore, hyperglycemia has been linked to tumor development.

As a result of the delayed diagnosis and limited treatment options caused by diabetes, breast cancer becomes more aggressive and has a higher fatality rate.

(e) **Endometrial Cancer**

Endometrial cancer (EC) is the most prevalent gynecological cancer and the fourth most common cancer in women worldwide. EC is frequently diagnosed early and has a better prognosis than other forms of cancer. However, during the last 20 years, the mortality rate of EC has increased dramatically. Longer life expectancy and lifestyle modifications might explain this phenomenon since diabetes is connected to aging and physical activity (Esposito et al., 2021). As a result, diabetes has been linked to EC, as shown by cohort studies, case-control studies, and meta-analyses. Diabetes, as an autonomous risk factor, has been shown to increase EC mortality in several investigations. More research is needed to confirm the link between diabetes and EC-specific mortality (Wartko et al., 2017).

(f) **Bladder Cancer**

Bladder cancer (BC) is one of the most common cancers worldwide, and its morbidity and mortality are thought to be linked to age, smoking, and occupational exposure (Gill et al., 2021). Researchers have recently focused their efforts on determining the impact of diabetes on BC. Consequently, the current findings do not adequately reflect the worldwide relationship between diabetes and BC. Furthermore, this meta-analysis discovered a negative relationship between BC and diabetes duration, with those with diabetes for fewer than 5 years having a greater menace of BC (Hu et al., 2018). Although the results of epidemiological research are mixed, most meta-analyses agree that diabetes is a risk factor for BC and that both the prevalence and mortality rates of BC rise in people with diabetes.

(g) **Prostate Cancer**

Prostate cancer is the second-largest cause of cancer mortality in American men (Kelkar et al., 2021). Even though diabetes seems to be a threat factor for various cancers, several examinations have shown an inverse relationship between diabetes and prostate cancer (Grossmann & Wittert, 2012). However, according to various studies, the frequency of the advanced stage is unrelated to a diabetes diagnosis. Furthermore, a meta-analysis found that the negative relationship between diabetes and prostate cancer is restricted to prevalence rather than mortality. Prostate cancer patients with diabetes had a poorer prognosis (Hua et al., 2016).

5   Diabetes and Cancer Biological Interlinks

Insulin is a peptide hormone that promotes glucose absorption and affects carbohydrate and fat metabolism. Insulin resistance develops in diabetics when Insulin's capacity to promote the uptake of cellular glucose and exploitation is impaired. Beta cells produce more insulin to compensate, resulting in hyperinsulinemia (Godsland, 2009). Hyperinsulinemia is defined as a high amount of insulin that induces liver cells to create IGF-1 when it attaches to the insulin receptor on the surface of target cells. By attaching to the IGF-1 receptor (IGF-1R), a receptor tyrosine kinase, IGF-1 affects cancer cell proliferation, differentiation, and death by activating numerous metabolic and mitogenic signaling pathways (Liao et al., 2019). The phosphoinositide-3-kinase-protein kinase B signaling pathway regulates cancer cell survival and migration. The rat sarcoma-mitogen-activated protein kinase/extracellular signal-regulated kinase signaling network, on the other hand, controls cancer cell metabolism and proliferation (Dong et al., 2020). As a result, diabetics have increased IGF-1 levels, increasing their risk of colorectal, breast, and prostate malignancies (Ferguson et al., 2013). Rising insulin levels on HER2-mediated primary tumor development and lung metastasis were studied in a mouse model of HER2-mediated breast cancer in a hyperinsulinemia environment. Another research investigated the link between high insulin levels and cancer mortality in obese and nonobese individuals. Finally, the insulin/IGF-1 axis promotes cancer cell development and dissemination.

6 Diabetes and Microvascular Complications

(a) **Diabetic Retinopathy**

Diabetic retinopathy (DR) is a microvascular condition that may impair the peripheral retina, the macula, or both, and is the most important reason for loss of vision and sightlessness in diabetics. The degree of DR varies from non-proliferative and pre-proliferative to high cruelly proliferative DR, in which new vasculature forms unusually (Duh et al., 2017). A vitreous hemorrhage or retinal detachment may cause whole or fractional loss of vision, while retinal vascular leakage and subsequent macular edema can lead to loss of vision in the central. Diabetes duration increases the risk of developing DR. Diabetic retinopathy was recently discovered in around 10% of patients with insulin resistance (prediabetes) and was linked to the occurrence of hypertension and a higher BMI. Other investigations of DR found links to onset at a younger age, smoking, insulin therapy, abnormal levels of blood lipid, pregnancy, renal illness, and a level of homocysteine are high.

(b) **Diabetic Neuropathy**

Around one-half of persons with diabetes develop peripheral neuropathy (PN), which may be monodiabetic or polydiabetic. Diabetes is also associated with autonomic neuropathy, including cardiovascular autonomic dysfunction characterized by aberrant heart rate (HR) and vascular control (Sugimoto et al., 2000). Poor glycemic control, age, diabetic duration, using a cigarette, dyslipidemia, and hypertension are all risk factors for PN, just as they are for DR. Increased height, the presence of cardiovascular disease (CVD), severe ketoacidosis (i.e., high-fat metabolism byproducts in the blood), and microalbuminuria (i.e., albumin in the urine, suggesting early renal failure) are all independent risk factors for PN.

(c) **Diabetic Nephropathy**

Diabetic nephropathy (DN) is a significant and adverse side effect in type 1 and type 2 diabetes. The major symptom of DN is microalbuminuria which develops into overt albuminuria, leads to renal failure, and is the major reason for end-stage renal disease (ESRD). One-fourth of persons with type 2 diabetes develop microalbuminuria, a more advanced form of diabetes that progresses at a rate of 2% to 3% every year (Lim, 2014). Glomerular basement membrane thickening and hyperfiltration are further features of DN, which contribute to mesangial extracellular matrix expansion and increased urine albumin excretion, finally leading to tubular and glomerular sclerosis and renal failure.

(d) **Diabetes Microvascular and Macrovascular Diseases: Common Mechanisms**

Chemical interactions between sugar and protein byproducts occur over days to weeks, eventually forming irreversible cross-linked protein derivatives known as AGE, which are one of the most common pathogenic mechanisms causing microvascular disease. These compounds have a broad variety of impacts on neighboring tissues, including collagen and endothelial alteration (e.g., thickening). In particular, in DR, AGE may impede retinal pericyte development and cause programmed cell death (i.e., apoptosis), increase vascular inflammation, and enhance pathologic angiogenesis; all of these effects raise the risk of formation of micro-thrombosis, capillary obstruction, and retinal ischemia. Vitreous hemorrhage, neovascularization, and elevated vascular endothelial growth factor levels may all contribute to retinal fibrosis, detachment, and vision loss (Cade, 2008). Furthermore, endothelial cell dysfunction may occur from AGE binding to immunoglobulin protein receptors for AGE. The pathophysiology of macrovascular disease in diabetes is multifaceted; nonetheless, the vascular endothelium is the most prevalent receiver of harm. Diabetes first inhibits the vascular endothelium's capacity to vasodilate by inhibiting the nitric oxide pathway. Hyperglycemia reduces the enzyme responsible for NO synthesis and increases ROS generation, resulting in additional inhibition of eNOS (Giacco & Brownlee, 2010).

7   Genetics of Diabetes

7.1   Diabetes Type 1

The risk of developing type 1 diabetes (T1D) is shown to be increased trends in relatives of the individuals compared with common populations. It is 6% vs < 1%, respectively. These findings indicate that in the development of T1D, genetic factors also have critically participated. Recently, a large body of evidence reported that over 20 regions of the genome had been identified as genetically accessible spots to the T1D. Nevertheless, genetic influence on T1D risk factors contributed by the gene placed in HLA region fragments of chromosome 6 is unexplored. This part of the genetic fragment possesses more than 100 genes believed to participate in the body's immunity. Among them, closely related diseases are HLA class II genes (HLA-DR, DQ, and DP); they are represented as IDDM1 genes (Dorman & Bunker, 2000).

7.2   IDDM1 Genes

The HLA class II genes exerted inheritable risk factors that range from 40% to 50% to the development of T1D. In assessing haplotypes, DQA1*-0501-DQB1* 0201 and DQA1*0301-DQB1*0302 are closely coupled with the disease observed in the Caucasian population. Such haplotypes remain and exhibit linkage disequilibrium with genes such as DRB1-03 and DRB1*04, correspondingly (Hirschhorn, 2003). Among abovementioned, particular DRB1*4 alleles also alter the risk-related haplotypes such as DQA1*0301-DQ31*0302. Similarly, one related study has also reported that risk haplotypes for the T1 diabetes represented by DRB1*07-DQA1*0301-DQB1*0201 prevail in African American races and DrB1*09-DQA1-0401-DQB1*0303 found among Japanese. Moreover, DRB1*04 DQA1-0401-DQB1*0302 haplotypes were found.

DRB1*-15-DQA1*-0602-DQB1*0102 haplotypes keep safe from harm and strong relation and reducing the risk of T1D in different populations. Current studies have underlined that other genes present in central class I and extrude class I may also express and enhance the T1D risk factors independently, apart from HLA class II genes. Possessing two high-risk DRB1-DBA1-DQB1 haplotypes caused a significant vulnerability to the high risk for T1D compared with no high-risk haplotype possessing individual. Among them, T1D risk is increased with one accessible haplotype, and its effectiveness seems to be vigorous. It has been estimated that the relative risk of T1D is found to be the range of age between 10 and 45 and three to seven individuals, respectively (Anjos & Polychronakos, 2004). Similar to IDDM1, other novel two genes are known to be targeted T1D risks, such as INS and CTLA-4.

8   Insulin (INS) Expression

Insulin gene (INS) appeared on chromosome 11p 15.5, otherwise known as IDDM2. A close positive interaction has been recorded with a non-transcribed, variable number of tandem repeats (VNTR) at the 5-flanking area (Pugliese et al., 1997). Two common variants exist, such as the shorter class I variant susceptible to T1D (relative elevation 1–2) and larger class III variants, exhibiting a highly dominant protector. The biological credibility of those interconnections may be related to the insulin expression present in mRNA in the thymus. Similarly, class III variants contribute to a higher insulin mRNA level than class I. The difference caused significant immune tolerability for class III positive individuals by enhancing the adverse selection of autoreactive T-cell clones. Insulin effects represent diverse based on ethnicity, with low expression (Billert et al., 2017).

9   Cytotoxic T Lymphocyte-Associated 4 Genes (CITLA-4)

CTLA-4 gene is seen on chromosome 2q31–35, whereas T1D genes are also present. CTLA-4 variants have been related to T1D and autoimmune disorders. CITLA-4 negatively controls T-cell performance. Although the impairing process was related to the Thr17Ala variant, which induces a higher risk of T1D, the entire inheritance in risk for CTLA-4Ala 17 has been assessed as ~1.5 (Gunavathy et al., 2019).

9.1   Type 2 Diabetes

At an early date, T2D was believed to be inherited partly. Family studies indicated that first-degree relative individuals affected with T2D showed three folds prone to develop T2D with no positive family history of T2D (Fendrick et al., 2019). The concordance rate for monozygotic twins was high, 60–90% when comparing dizygotic twins. This indicated the involvement of a vital genetic contribution in T2D development. A recent strategy is employed to recognize ailment vulnerable genes based on identifying promising/candidate genes (Morris et al., 2012). They have been selected due to their participation in pancreatic B cell role, in the action of insulin/metabolism of glucose, and other levels of metabolic actions that caused an increased risk of T2D were widely studied in several populations. Despite this, the outcome for all candidate genes has shown to be conflicting and uncertain. The possibilities for divergent findings may be due to the small sample size and changes in T2D sensitivity throughout ethnic groups. Changes in genetic attributes in the exposure to the environment, gene, and environment relation are unknown, and this chapter describes a few most critical candidate genes covering PPAR, ABCC8, KCNJ11, and CALPN10 genes (Sethi et al., 2016).

9.2   PPARy Gene

PPARy gene is referred to as peroxisome proliferators activated receptor-gamma – this gene is common and popularly studied due to its participation in adipocyte and lipid metabolism. Moreover, it has been a target for therapeutic agents for hyperglycemia, i.e., thiazolidinediones (Sabaratnam et al., 2019). One type of PPARy gene (Pro) reduces insulin susceptibility and enhances T2D risk to many folds and is predominantly found to occur in several populations. About 98% of Europeans possess one copy of progene.

9.3   ABBCCS Gene

This gene is otherwise known as ATP binding cassette, subfamily (members). It has been reportedly coded for sulfonylurea receptor with increased affinity (SUR 1) attached with Kir 6.2 subunit, KVCNJ11U, a potassium channel, coded that genes altogether participate in controlling the discharge of hormones insulin and glucagon in beta cells of Langerhans. Mutation in any one gene eventually causes impairing in potassium channels regulatory and secretion of Insulin that leads to the development of T2D. It is noteworthy that the genes such as ABCC8 and KCNJ11 are located at a 4.5 kb distance from the INS gene. Variant types of KCNJ11 (Lys) and ABCC8 (alanine) genes have been interconnected with T2D and other diabetes-related disease traits due to the proximity location (Proks et al., 2006). Many current studies have been focusing on its mechanism of action, whether it functions independently or synergistically with each other to the effect on T2D sensitivity. Popular drugs target PPARy, ABCC8, and KCNJ11 for the creatives of T2D, resulting in a positive glycemic reduction. Sensitivity to hypoglycemic therapy is associated with the genotypes of individuals. Therefore, genetic testing cannot determine individual risk for T2D but can be used for clinical treatment options (Babenko et al., 2006).

10   Calpain10 or CAPN10 Gene

CAPN10 gene targets and codes a calcium-dependent cysteine protease overexpressed. A haplotype linked with 720 contains A to G mutation at 43 positions that reportedly participated in CAPN10 gene transcription. In addition, two amino acid polymorphic (Thr 504 Ala and Phe 200 Thr) are also related to T2D risks, although coding and noncoding polymorphisms cannot act independently and confer T2D risks but caused at an earlier diagnosis. Several physiological investigations have revealed differences in calpine-10 activity on insulin secretion, resulting in increased vulnerability to T2D (Zhang et al., 2019).

11   Maturation: Onset of T2D in Young

MODY is a rare type of T2D (less than 5% in all cases) caused by young (below 25 years). This has been expressed as unique characteristics such as the slow onset of indications, loss of obesity, no ketosis, and no symptoms of beta-cell autoimmunity. Usually, it has been regulated without exogenous insulin administration; MODY represented autosomal dominant pattern inheritance. Due to the improved genetic technology, it has been found that six types of MODY caused by a mutation that occurs in various genes contribute directly to the beta-cell functions. It has been underlined that -15% of MODY gene-affected persons do not have mutations in one among three genes. MODY will be identified the genes that are causative agents for 6072D to be discovered shortly (Kim et al., 2004).

12   GCK (Glucokinase)

GCK is one of the MODY genes that cannot participate in controlling the expression of other genes. At the same time, the GCK gene plays a crucial role in glucose and insulin secretion metabolism. Hence, the prognosis associated with MODY2 is distinct from other MODY patients, prone to mild fasting hyperglycemia, indicated from birth to end in their lifestyle. There is a slight reduction of age-related normoglycemia, whereas patients with MODY2 mutations are commonly asymptomatic (Zhou et al., 2019). In medical screening, large bodies have been detected in regular tests. MODY2 mutation in women has been identified commonly during pregnancy. However, pregnancy outcomes depend upon whether the mother or child possesses mutation. While mother and child possess MODY2 positive, no impact on birth weight was observed. When a cheerful mother carries a negative MODY2 child, gestational age has been extended due to maternal hyperglycemia. On the contrary, the fetus possesses positive, and the mother carries negative, resulting reduction of weight by 500 gm due to minimum insulin secretion, which leads to the suppression of the growth of the fetus (Urakami, 2019).

13   HNF4A Gene

It is otherwise regarded as hepatocyte nuclear factor 4-a while sudden change occurs in the region of promoter and coding part of HNF4A gene that influence MODY1 effects. HNF4A is overexpression in several tissues covering liver and pancreas organs. HNF4A gene controls hepatic gene expression and stimulates the expression of their MODY genes, including HNF1A, resulting in MODY3. It was reported it directly involved in gene expression in insulin . The sudden change caused by HNF4A is closely associated with T2D (Ozsu et al., 2018).

14   HNF1A Gene

Hepatocyte nuclear factors 1-a is one of the genes associated with the risk of T2D. Indeed, MODY3 is often the prevalent cause of the disease; the mutation outcome occurred in the HNF14A expression, which links MODY1 and MODY3. It was demonstrated that transcription factors from MODY controlling network maintain glucose homeostasis. HNF1A mutations are linked with the development of T1D (Ozsu et al., 2018).

15   IPF1 Gene

It is otherwise referred to as insulin promoter factor MODY4, a scarce occurrence of the disease attributable to the specific gene 1PF1. Homozygosity of such sudden changes is interlinked with newborn pancreatic agencies and neonatal diabetes.

Hence, the fetus, which possesses MODY4 mutation, exhibited a small duration for gestational age. MODY4 may also influence T2D to develop; the IPF1 gene controls the expression of glucokinase, insulin, and other genes to participate in the mechanism of metabolisms (Cockburn et al., 2004).

16 HNF1B

Hepatocyte nuclear factor 1-B is otherwise MODY5; it is a rare MODY type, and close links with MODY1 since HNF1B controls HNF4a. Although dissimilar to MODY1 and MODY5, this type of gene links with renal cysts, proteinuria, and renal failure (Wang et al., 2019).

17 NEUROD1

This is also one of the genes associated with T1D. The mutations in NEUROD1 are answerable for MODY6, which is rare. Combined with MODY4, MODY5 and MODY6 constitute less than 3% of entire MODY cases. This gene expresses in the pancreatic B cells, the intestine, and the brain. In the pancreas, it regulates the expression of insulin secretion. In concise, MODY genes are expressed with pancreatic islets cells and participate in the glucose metabolism process, controlling insulin and another side, regulating glucose transport, and are involved in the development of the pancreas (Romer et al., 2019). Due to MODY phenotypes' diverse nature, various genes have been involved. Genetic examination helps to treat such genetic disorders.

18 Conclusion

Diabetes is one of the most common diseases in most developing countries. There is mounting confirmation that the original determinants of diabetes are an indication of the significant forces through social, economic, and cultural change as globalization, urbanization, inhabitant's aging, and a healthy environment. The disease's facts and vigorously contributing to the action are essential since the problem is far less severe in people with well-managed blood sugar levels. Cancer is a metabolic illness that both internal and external sources may cause. Many types of research have proven the link between diabetes and increased cancer incidence and death.

Furthermore, due to lifestyle changes and higher life expectancy, the prevalence of diabetes and cancer is rapidly increasing globally. To enhance diabetes and cancer outcomes, preventative measures such as physical activity and frequent cancer screening are required. Furthermore, since diabetes and cancer are global issues, worldwide health professionals or organizations should produce standards for diabetes and cancer prevention, diagnosis, and treatment to lessen the societal cost. Because research on diabetes and cancer is difficult to undertake due to their inherent variability, there are still many unsolved questions: Do T1DM and T2DM have the same effect on cancer? How do we characterize each person's overall and particular cancer risks? Also, how can we comprehend the biological systems at work entirely? Additional research is needed to resolve these problems so that diabetes and cancer patients have more preventative and treatment options.

Acknowledgments Author Dr. K. LANGESWARAN mercifully acknowledges Alagappa University for the infrastructure facility, and RUSA phase 2.0 grant sanctioned vide letter No.F.24–51/2014-U policy (TN Multi-Gen), Dept. of Govt. of India Dt, 09.10.2018.

References

Acharjee, S., Ghosh, B., Al-Dhubiab, B. E., & Nair, A. B. (2013). Understanding type 1 diabetes: Etiology and models. *Canadian Journal of Diabetes, 37*(4), 269–276.

Alam, U., Asghar, O., Azmi, S., & Malik, R. A. (2014). General aspects of diabetes mellitus. In *Handbook of clinical neurology* (Vol. 126, pp. 211–222). Springer.

Anjos, S., & Polychronakos, C. (2004). Mechanisms of genetic susceptibility to type 1 diabetes: Beyond HLA. *Molecular Genetics and Metabolism, 81*, 187–195.

Babenko, A. P., Polak, M., Cavé, H., et al. (2006). Activating mutations in the ABCC8 gene in neonatal diabetes mellitus. *The New England Journal of Medicine, 355*, 456–466.

Billert, M., Skrzypski, M., Sassek, M., Szczepankiewicz, D., Wojciechowicz, T., Mergler, S., Strowski, M. Z., & Nowak, K. W. (2017). TRPV4 regulates insulin mRNA expression and INS-1E cell death via ERK1/2 and NO-dependent mechanisms. *Cellular Signalling, 35*, 242–249.

Bosetti, C., Turati, F., & La Vecchia, C. (2014). Hepatocellular carcinoma epidemiology. *Best Practice & Research. Clinical Gastroenterology, 28*, 753–770.

Boyle, P., Boniol, M., Koechlin, A., Robertson, C., Valentini, F., Coppens, K., Fairley, L. L., Boniol, M., Zheng, T., Zhang, Y., Pasterk, M., Smans, M., Curado, M. P., Mullie, P., Gandini, S., Bota, M., Bolli, G. B., Rosenstock, J., & Autier, P. (2012). Diabetes and breast cancer risk: A meta-analysis. *British Journal of Cancer, 107*, 1608–1617.

Brunton, S. (2016). Pathophysiology of Type 2 Diabetes: The evolution of our understanding. *The Journal of Family Practice, 65*(4 Suppl), supp_az_0416.

Cade, W. T. (2008). Diabetes-related microvascular and macrovascular diseases in the physical therapy setting. *Physical Therapy, 88*(11), 1322–1335. https://doi.org/10.2522/ptj.20080008

Chaudhry, Z. W., Hall, E., Kalyani, R. R., Cosgrove, D. P., & Yeh, H. C. (2013). Diabetes and pancreatic cancer. *Current Problems in Cancer, 37*, 287–292.

Chen, L., Magliano, D. J., & Zimmet, P. Z. (2011). The worldwide epidemiology of type 2 diabetes mellitus–present and future perspectives. *Nat Rev Endocrinol, 8*(4), 228–236.

Cockburn, B. N., Bermano, G., Boodram, L. L., Teelucksingh, S., Tsuchiya, T., Mahabir, D., Allan, A. B., Stein, R., Docherty, K., & Bell, G. I. (2004). Insulin promoter factor-1 mutation and diabetes in Trinidad: Identification of a novel diabetes-associated mutation (E224K) in an indo-Trinidadian family. *The Journal of Clinical Endocrinology and Metabolism, 89*, 971–978.

Coustan, D. R. (2013). Gestational diabetes mellitus. *Clinical Chemistry, 59*(9), 1310–1321.

Dong, R., Tan, Y., Fan, A., Liao, Z., Liu, H., & Wei, P. (2020). Molecular dynamics of the recruitment of immunoreceptor signaling module DAP12 homodimer to lipid raft boundary regulated by PIP2. *The Journal of Physical Chemistry B, 124*, 504–510.

Dorman, J. S., & Bunker, C. H. (2000). HLA-DQ locus of the human leukocyte antigen complex and type 1 diabetes mellitus: A HuGE review. *Epidemiologic Reviews, 22*, 218–227.

Duh, E. J., Sun, J. K., & Stitt, A. W. (2017). Diabetic retinopathy: current understanding, mechanisms, and treatment strategies. *JCI Insight., 2*(14), e93751. https://doi.org/10.1172/jci. insight.93751

Esposito, G., Bravi, F., Serraino, D., Parazzini, F., Crispo, A., Augustin, L. S. A., Negri, E., La Vecchia, C., & Turati, F. (2021). Diabetes risk reduction diet and endometrial cancer risk. *Nutrients, 13*(8), 2630.

Fendrick, A. M., Buxbaum, J. D., Tang, Y., Vlahiotis, A., McMorrow, D., Rajpathak, S., & Chernew, M. E. (2019). Association between switching to a high-deductible health plan and discontinuation of type 2 diabetes treatment. *JAMA Network Open, 2*(11), e1914372.

Ferguson, R. D., Gallagher, E. J., Cohen, D., Tobin-Hess, A., Alikhani, N., Novosyadlyy, R., Haddad, N., Yakar, S., & LeRoith, D. (2013). Hyperinsulinemia promotes metastasis to the lung in a mouse model of Her2-mediated breast cancer. *Endocrine-Related Cancer, 20*, 391–401.

Giacco, F., & Brownlee, M. (2010). Oxidative stress and diabetic complications. *Circulation Research, 107*(9), 1058–1070. https://doi.org/10.1161/CIRCRESAHA.110.223545

Gill, E., Sandhu, G., Ward, D. G., Perks, C. M., & Bryan, R. T. (2021). The Sirenic links between diabetes, obesity, and bladder cancer. *International Journal of Molecular Sciences, 22*(20), 11150.

Godsland, I. F. (2009). Insulin resistance and hyperinsulinaemia in the development and progression of cancer. *Clinical Science (London, England), 118*, 315–332.

Grossmann, M., & Wittert, G. (2012). Androgens, diabetes and prostate cancer. *Endocrine-Related Cancer, 19*(5), F47–F62.

Gunavathy, N., Asirvatham, A., Chitra, A., & Jayalakshmi, M. (2019). Association of CTLA-4 and CD28 gene polymorphisms with type 1 diabetes in south Indian population. *Immunological Investigations, 48*(6), 659–671.

Gutiérrez-Salmerón, M., Lucena, S. R., Chocarro-Calvo, A., García-Martínez, J. M., Martín Orozco, R. M., & García-Jiménez, C. (2021). Metabolic and hormonal remodeling of colorectal cancer cell signalling by diabetes. *Endocrine-Related Cancer, 28*(6), R191–R206.

Hirschhorn, J. N. (2003). Genetic epidemiology of type 1 diabetes. *Pediatric Diabetes, 4*(2), 87–100.

Hu, J., Chen, J. B., Cui, Y., Zhu, Y. W., Ren, W. B., Zhou, X., Liu, L. F., Chen, H. Q., & Zu, X. B. (2018). Association of metformin intake with bladder cancer risk and oncologic outcomes in type 2 diabetes mellitus patients: A systematic review and meta-analysis. *Medicine (Baltimore), 97*(30), e11596.

Hua, Q., Zhu, Y., Liu, H., & Ye, X. (2016). Diabetes and the risk of biochemical recurrence in patients with treated localized prostate cancer: a meta-analysis. *International Urology and Nephrology, 48*(9), 1437–1443.

Javeed, N., & Matveyenko, A. V. (2018). Circadian etiology of Type 2 Diabetes Mellitus. *Physiology (Bethesda, MD), 33*(2), 138–150.

Kelkar, S., Oyekunle, T., Eisenberg, A., Howard, L., Aronson, W. J., Kane, C. J., Amling, C. L., Cooperberg, M. R., Klaassen, Z., Terris, M. K., Freedland, S. J., & Csizmadi, I. (2021). Diabetes and prostate cancer outcomes in obese and nonobese men after radical prostatectomy. *Journal of the National Cancer Institute Cancer Spectrum, 5*(3), pkab023.

Kim, S. H., Ma, X., Weremowicz, S., Ercolino, T., Powers, C., Mlynarski, W., Bashan, K. A., Warram, J. H., Mychaleckyj, J., Rich, S. S., Krolewski, A. S., & Doria, A. (2004). Identification of a locus for maturity-onset diabetes of the young on chromosome 8p23. *Diabetes, 53*, 1375–1384.

Li, W., Huang, E., & Gao, S. (2017). Type 1 diabetes mellitus and cognitive impairments: A systematic review. *Journal of Alzheimer's Disease, 57*(1), 29–36.

Liao, Z., Tan, Z. W., Zhu, P., & Tan, N. S. (2019). Cancer-associated fibroblasts in tumor microenvironment - Accomplices in tumor malignancy. *Cellular Immunology, 343*, 103729.

Lim, A. K. H. (2014). Diabetic nephropathy - complications and treatment. *International Journal of Nephrology and Renovascular Disease, 7*, 361–381. https://doi.org/10.2147/IJNRD.S40172/

Mantovani, A., & Targher, G. (2017). Type 2 diabetes mellitus and risk of hepatocellular carcinoma: Spotlight on nonalcoholic fatty liver disease. *Annals of Translational Medicine, 5*, 270.

Morris, A. P., Voight, B. F., Teslovich, T. M., Ferreira, T., Segrè, A. V., Steinthorsdottir, V., Strawbridge, R. J., Khan, H., Grallert, H., Mahajan, A., Prokopenko, I., Kang, H. M., Dina, C., Esko, T., Fraser, R. M., Kanoni, S., Kumar, A., Lagou, V., Langenberg, C., Luan, J., Lindgren, C. M., Müller-Nurasyid, M., Pechlivanis, S., Rayner, N. W., Scott, L. J., Wiltshire, S., Yengo, L., Kinnunen, L., Rossin, E. J., Raychaudhuri, S., Johnson, A. D., Dimas, A. S., Loos, R. J., Vedantam, S., Chen, H., Florez, J. C., Fox, C., Liu, C. T., Rybin, D., Couper, D. J., Kao, W. H., Li, M., Cornelis, M. C., Kraft, P., Sun, Q., van Dam, R. M., Stringham, H. M., Chines, P. S., Fischer, K., Fontanillas, P., Holmen, O. L., Hunt, S. E., Jackson, A. U., Kong, A., Lawrence, R., Meyer, J., Perry, J. R., Platou, C. G., Potter, S., Rehnberg, E., Robertson, N., Sivapalaratnam, S., Stančáková, A., Stirrups, K., Thorleifsson, G., Tikkanen, E., Wood, A. R., Almgren, P., Atalay, M., Benediktsson, R., Bonnycastle, L. L., Burtt, N., Carey, J., Charpentier, G., Crenshaw, A. T., Doney, A. S., Dorkhan, M., Edkins, S., Emilsson, V., Eury, E., Forsen, T., Gertow, K., Gigante, B., Grant, G. B., Groves, C. J., Guiducci, C., Herder, C., Hreidarsson, A. B., Hui, J., James, A., Jonsson, A., Rathmann, W., Klopp, N., Kravic, J., Krjutškov, K., Langford, C., Leander, K., Lindholm, E., Lobbens, S., Männistö, S., Mirza, G., Mühleisen, T. W., Musk, B., Parkin, M., Rallidis, L., Saramies, J., Sennblad, B., Shah, S., Sigurðsson, G., Silveira, A., Steinbach, G., Thorand, B., Trakalo, J., Veglia, F., Wennauer, R., Winckler, W., Zabaneh, D., Campbell, H., van Duijn, C., Uitterlinden, A. G., Hofman, A., Sijbrands, E., Abecasis, G. R., Owen, K. R., Zeggini, E., Trip, M. D., Forouhi, N. G., Syvänen, A. C., Eriksson, J. G., Peltonen, L., Nöthen, M. M., Balkau, B., Palmer, C. N., Lyssenko, V., Tuomi, T., Isomaa, B., Hunter, D. J., Qi, L., Wellcome Trust Case Control Consortium; Meta-Analyses of Glucose and Insulin-related traits Consortium (MAGIC) Investigators; Genetic Investigation of ANthropometric Traits (GIANT) Consortium; Asian Genetic Epidemiology Network–Type 2 Diabetes (AGEN-T2D) Consortium; South Asian Type 2 Diabetes (SAT2D) Consortium, Shuldiner, A. R., Roden, M., Barroso, I., Wilsgaard, T., Beilby, J., Hovingh, K., Price, J. F., Wilson, J. F., Rauramaa, R., Lakka, T. A., Lind, L., Dedoussis, G., Njølstad, I., Pedersen, N. L., Khaw, K. T., Wareham, N. J., Keinanen-Kiukaanniemi, S. M., Saaristo, T. E., Korpi-Hyövälti, E., Saltevo, J., Laakso, M., Kuusisto, J., Metspalu, A., Collins, F. S., Mohlke, K. L., Bergman, R. N., Tuomilehto, J., Boehm, B. O., Gieger, C., Hveem, K., Cauchi, S., Froguel, P., Baldassarre, D., Tremoli, E., Humphries, S. E., Saleheen, D., Danesh, J., Ingelsson, E., Ripatti, S., Salomaa, V., Erbel, R., Jöckel, K. H., Moebus, S., Peters, A., Illig, T., de Faire, U., Hamsten, A., Morris, A. D., Donnelly, P. J., Frayling, T. M., Hattersley, A. T., Boerwinkle, E., Melander, O., Kathiresan, S., Nilsson, P. M., Deloukas, P., Thorsteinsdottir, U., Groop, L. C., Stefansson, K., Hu, F., Pankow, J. S., Dupuis, J., Meigs, J. B., Altshuler, D., Boehnke, M., McCarthy, M. I., & DIAbetes Genetics Replication And Meta-analysis (DIAGRAM) Consortium. (2012). Large-scale association analysis provides insights into the genetic architecture and pathophysiology of type 2 diabetes. *Nature Genetics, 44*(9), 981–990.

Ozsu, E., Cizmecioglu, F. M., Yesiltepe Mutlu, G., Yuksel, A. B., Calıskan, M., Yesilyurt, A., & Hatun, S. (2018). Maturity onset diabetes of the young due to Glucokinase, HNF1-A, HNF1-B, and HNF4-A mutations in a cohort of Turkish children diagnosed as type 1 diabetes mellitus. *Hormone Research in Pædiatrics, 90*(4), 257–265.

Proks, P., Arnold, A. L., Bruining, J., Girard, C., Flanagan, S. E., Larkin, B., Colclough, K., Hattersley, A. T., Ashcroft, F. M., & Ellard, S. (2006). A heterozygous activating mutation in the sulphonylurea receptor SUR1 (ABCC8) causes neonatal diabetes. *Human Molecular Genetics, 15*, 1793–1800.

Pugliese, A., Zeller, M., & Ferndandez, J. A. (1997). The insulin gene is transcribed in human thymus and transcription levels correlated with allelic variation at the INS VNTR-IDDM 2 susceptibility locus for type 1 diabetes. *Nature Genetics, 15*, 293–297.

Romer, A. I., Singer, R. A., Sui, L., Egli, D., & Sussel, L. (2019). Murine perinatal Beta cell proliferation and the differentiation of human stem cell derived insulin expressing cells require NEUROD1. *Diabetes, 68*, 2259.

Sabaratnam, R., Pedersen, A. J., Eskildsen, T. V., Kristensen, J. M., Wojtaszewski, J. F. P., & Hojlund, K. (2019). Exercise induction of key transcriptional regulators of metabolic adaptation in muscle is preserved in type 2 diabetes. *The Journal of Clinical Endocrinology and Metabolism, 104*(10), 4909–4920.

Sen, S., & Chakraborty, R. (2015). Treatment and diagnosis of diabetes mellitus and its complication: Advanced approaches. *Mini Reviews in Medicinal Chemistry, 15*(14), 1132–1133.

Sethi, I., Bhat, G. R., Singh, V., Kumar, R., Bhanwer, A. J., Bamezai, R. N., Sharma, S., & Rai, E. (2016). Role of telomeres and associated maintenance genes in type 2 diabetes mellitus: A review. *Diabetes Research and Clinical Practice, 122*, 92–100.

Setiawan, V. W., Stram, D. O., Porcel, J., Chari, S. T., Maskarinec, G., Le Marchand, L., Wilkens, L. R., Haiman, C. A., Pandol, S. J., & Monroe, K. R. (2019). Pancreatic cancer following incident diabetes in African Americans and Latinos: The multiethnic Cohort. *Journal of the National Cancer Institute, 111*, 27–33.

Soltani, G., Poursheikhani, A., Yassi, M., Hayatbakhsh, A., Kerachian, M., & Kerachian, M. A. (2019). Obesity, diabetes and the risk of colorectal adenoma and cancer. *BMC Endocrine Disorders, 19*(1), 113.

Sugimoto, K., Murakawa, Y., & Sima, A. A. (2000). Diabetic neuropathy: A continuing enigma. *Diabetes/Metabolism Research and Reviews, 16*(6), 408–433. https://doi.org/10.1002/1520-7560(200011/12)16:6<408::aid-dmrr158>3.0.co;2-r

Trefz, P., Obermeier, J., Lehbrink, R., Schubert, J. K., Miekisch, W., & Fischer, D. C. (2019). Exhaled volatile substances in children suffering from type 1 diabetes mellitus: Results from a cross-sectional study. *Scientific Reports, 9*(1), 15707.

Urakami, T. (2019). Maturity-onset diabetes of the young (MODY): Current perspectives on diagnosis and treatment. *Diabetes, Metabolic Syndrome and Obesity, 12*, 1047–1056.

Wang, J., He, C., Gao, P., Wang, S., Lv, R., Zhou, H., Zhou, Q., Zhang, K., Sun, J., Fan, C., Ding, G., & Lan, F. (2019). HNF1B-mediated repression of SLUG is suppressed by EZH2 in aggressive prostate cancer. *Oncogene, 39*(6), 1335–1346.

Wartko, P. D., Beck, T. L., Reed, S. D., Mueller, B. A., & Hawes, S. E. (2017). Association of endometrial hyperplasia and cancer with a history of gestational diabetes. *Cancer Causes & Control, 28*(8), 819–828.

Wu, A. H., Yu, M. C., Tseng, C. C., Stanczyk, F. Z., & Pike, M. C. (2007). Diabetes and risk of breast cancer in Asian-American women. *Carcinogenesis, 28*, 1561–1566.

Zaccardi, F., Webb, D. R., Yates, T., & Davies, M. J. (2016). Pathophysiology of type 1 and type 2 diabetes mellitus: A 90-year perspective. *Postgraduate Medical Journal, 92*(1084), 63–69.

Zhang, X., Shi, C., Wei, L., Sun, F., & Ji, L. (2019). The association between the rs2975760 and rs3792267 single nucleotide polymorphisms of Calpain 10 (CAPN10) and gestational diabetes mellitus. *Medical Science Monitor, 25*, 5137–5142.

Zhou, W., Chen, M., Zhou, H., & Zhang, Z. (2019). Heterozygous lys169Glu mutation of glucokinase gene in a Chinese family having glucokinase-maturity-onset diabetes of the young (GCK-MODY). *Journal of Postgraduate Medicine, 65*(4), 241–243.

Diabetes and Cardiovascular Disorder

S. Santhi Priya and K. Kumar Ebenezar

Abstract

Diabetes mellitus especially type 2 diabetes mellitus (T2DM) is becoming more common at an alarming rate, resulting in increased disability, morbidity, and premature mortality. Chronic hyperglycemia, dyslipidemia, insulin resistance, oxidative stress, hypercoagulability, vascular calcification, inflammation, obesity, and genetic polymorphism in T2DM individuals can lead to macrovascular diseases development and progression like cardiovascular diseases (CVD). CVD is the primary cause of death among T2DM patients, and T2DM patients are at twofold more risk of developing CVD. This chapter focuses on the major mechanisms involved in cardiovascular complications among T2DM individuals and also discusses the various pharmacological interventions and agents developed to delay cardiovascular events and thereby the quality and duration of the patients.

Keywords

Type 2 diabetes mellitus · Cardiovascular diseases · Hyperglycemia · Insulin · Inflammation · Oxidative stress

S. Santhi Priya · K. Kumar Ebenezar (✉)
Natural Medicine and Molecular Physiology Lab, Faculty of Allied Health Sciences, Chettinad Hospital and Research Institute, Chettinad Academy of Research & Education, Kelambakkam, Chennai, Tamil Nadu, India

41

1   Introduction

The International Diabetes Federation (IDF) has identified diabetes mellitus as one of the fastest-growing health crisis worldwide with 537 million reported cases and 6.7 million deaths in 2021. An exponential increase in the number of diabetes cases has been reported over the years and is expected to rise to 643 million in the year 2030 and by 783 million in the year 2045. Among the different types, type 2 diabetes mellitus (T2DM) is the most widespread with 90% of all reported diabetes cases documented worldwide (IDF, 2021). Chronic diabetes can cause severe life-threatening complications like cardiovascular diseases (CVD), microvascular and macrovascular complications, chronic kidney diseases (CKD), retinopathy, etc. (IDF, 2019). CVD is reported to be the most frequent T2DM complication with higher prevalence and increased mortality and morbidity. A twofold increase in CVD development and deaths was reported among T2DM individuals, and the risk was found to increase progressively with increasing fasting plasma glucose levels (Yun & Ko, 2021). According to the findings of a systematic comprehensive review, the relative risk of CVD among diabetes ranges from 1.6 to 2.6, with the risk being higher among younger people and women (Sarwar et al., 2010; Rao Kondapally Seshasai et al., 2011). Coronary heart diseases (CHD), cerebrovascular diseases, peripheral artery disease, congestive heart failure, myocardial infarction (MI), acute coronary syndrome (ACS), etc. are the most prevalent CVD complications in T2DM individuals. Chronic diabetes leads to increased CVD risk through several common pathophysiologic mechanisms, and traditional cardiometabolic risk factors like hypertension, dyslipidemia, obesity, smoking, and physical inactivity also play an important link between diabetes and CVD (Paneni et al., 2013). Earlier studies like the Framingham and the Multiple Risk Factor Intervention Trial (MRFIT) reported that the incidence and the mortality by cardiovascular events were threefold higher among diabetes patients compared to normal healthy subjects (Kannel & McGee, 1979; Stamler et al., 1993). A cohort study with 5.7-year follow-up of 271,174 DM patients reported that increased glycated hemoglobin, low-density lipoprotein-cho-lesterol (LDL-c), systolic blood pressure, smoking, and low physical activity were a strong predictors of CVD risk among diabetes patients (Rawshani et al., 2018). These factors are majorly considered for developing novel treatment and prevention strategies for CVD treatment among DM patients. Renin-angiotensin system (RAS) blockade, statins, PCSK9 inhibitors, and antiplatelet treatment are now considered a treatment options to reduce the CVD risk (Patoulias et al., 2020). However large-scale clinical trials are required to better understand the mechanisms and potential of these treatment options. Therefore, it is vital to study the risk factors and mechanism for the better understanding of the pathophysiology underlying cardiovascular risk, development and progression among diabetes individuals. This chapter reviews the current development and understanding of risk factors and pathophysiology involved in the development of CVD conditions among diabetes individuals. A structured analysis of the published literature including clinical trials, meta-analysis, systematic reviews, research and review articles was considered. T2DM is a common metabolic syndrome and CVD is the major complication of diabetes. T2DM is highly prone to CVD, and several potential mechanisms are involved behind diabetic vascular complications.

2   Mechanism of CVD Risk in T2DM Patients

Patients with diabetes were reported to have an increased atherosclerotic burden compared to patients without diabetes (Nicholls et al., 2008). CVD and diabetes are linked through several cellular and molecular pathophysiological factors such as genetic predisposition, insulin resistance, inflammation, endothelial dysfunction, dyslipidemia, oxidative stress, etc. (Paneni et al., 2013; Poznyak et al., 2020). The roles of these factors in CVD development among T2DM patients have been described below.

2.1   Role of Hyperglycemia

Hyperglycemia is the hallmark of diabetes and is also considered a major contributing factor in the development of CVD among diabetes patients. About 1 mmol/L rise in both fasting plasma blood glucose and fasting serum blood glucose level was associated with 12% and 13% increased CVD risk and vascular death, respectively (Wang et al., 2016). A reduction in glucose level by inhibitors like glucagon-like peptide 1 receptor agonists (GLP-1RA) and sodium-glucose cotransporter-2 (SGLT2) showed reduced CVD endpoints. When lifestyle interventions were implemented, patients with hyperglycemia showed a slower progression of DM and a reduced risk of CVD (Schwarz et al., 2018). There is evidence supporting the role of increased glucose availability with increased oxidative stress due to the reactive oxygen species (ROS) generated from mitochondrial or non-mitochondrial sources and reduced antioxidant availability. Apart from oxidative stress, hyperglycemia has also been associated with increased flux by polyol pathway manifested by increased aldose reductase, NADPH depletion, glutathione reductase attenuation, PKC activation, etc. which causes ischemia reperfusion. The amplified sorbitol synthesis by the polyol pathway leads to decreased NADPH availability required for glutathione (GSH) regeneration which is an important antioxidant thereby increasing the oxidative stress (Mapanga & Essop, 2016). Likewise, it was estimated that for every 1% rise in the glycated hemoglobin 1c (HbA1c) levels; there was an 11–16% increases in cardiovascular risk. The HbA1c was found to be a stronger predictor of stroke and acute myocardial infarction (AMI) among DM patients (Rawshani et al., 2018). The SCORE study conducted on diabetes patients also confirmed that an HbA1c level of >6% increased all-cause mortality risk and hospitalization by CHD. A linear association was reported between the increasing HbA1c levels and risk of CHD hospitalization with 11.41 events/10,000 person to 70.26 events/10,000 for a rise from 6–6.5% to 7%, respectively (Navarro-Pérez et al., 2018). The effect of hyperglycemia on CVD development is shown in Fig. 1.

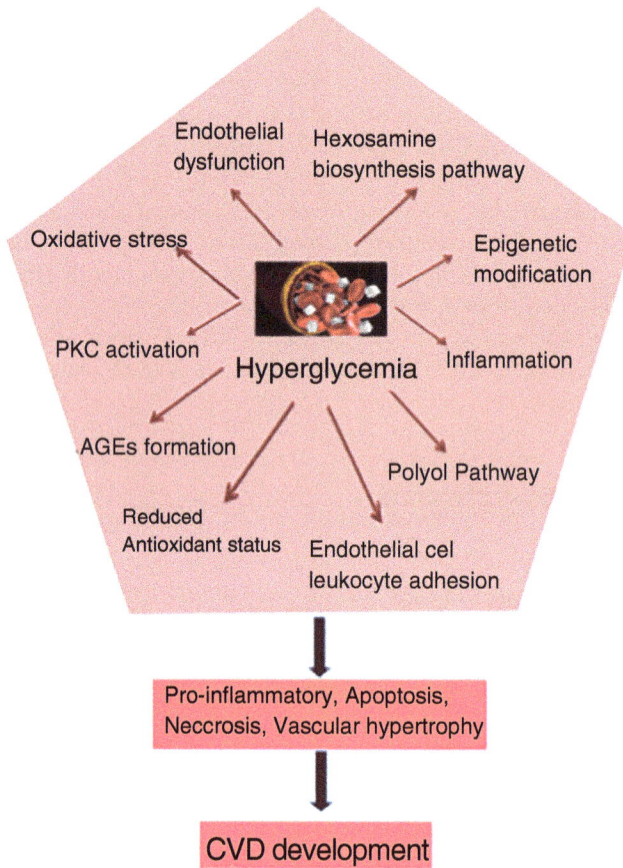

Fig. 1 Chronic hyperglycemia can induce cardiovascular events in diabetes individuals through the activation of several pathways. [PKC, protein kinase C; AGEs, advanced glycation end products; CVD, cardiovascular diseases]

Hyperglycemia can trigger advanced glycation end products (AGEs) generation causing inflammation and increased risk of atherosclerosis (Yen et al., 2022). Chronic hyperglycemia may cause the monosaccharides to react non-enzymatically with major biomolecules in the body to form AGEs which affects its structural integrity and function by increasing the oxidative stress, inflammatory, thrombotic, and fibrotic reactions. AGEs induce the expression of the receptor for AGEs (RAGE) which forms positive feedback for increased AGEs expression. AGEs were associated with an increased level of inflammatory markers, aortic plus wave velocity (arterial stiffness), decreased antioxidant capacity, reverse cholesterol transport, etc. suggesting the role of AGE in atherosclerotic development in DM. An increased association between cardiovascular mortality and circulating AGEs was reported in both T1DM and T2DM patients (Yamagishi & Matsui, 2018). Apart from RAGE, soluble receptors for AGEs (sRAGE) may act as a useful

Fig. 2 Advanced glycation end products (AGEs) interact with receptors for AGEs (RAGE) to induce several pathophysiological mechanisms. [NO, nitric oxide; ADMA, asymmetric dimethylarginine]

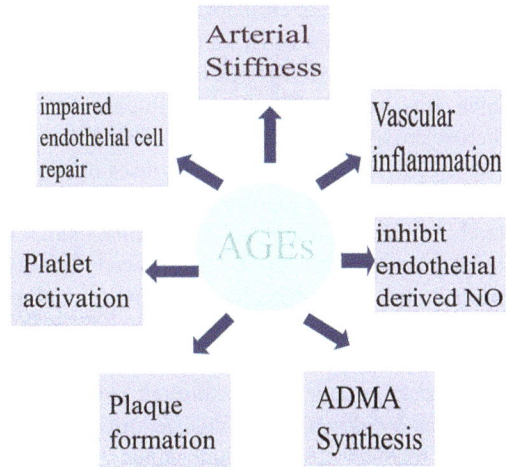

enzyme for risk stratification of heart failure and also as a predictor of aortic valve calcification (Mapanga & Essop, 2016). The role of AGE in the development of CVD has been depicted in Fig. 2.

Hyperglycemia was found to be associated with atherosclerotic lesion formation as the vascular smooth muscle cells (VSMC) undergo activation, proliferation, and differentiation (Wang et al., 2016). An accelerated macrophage adhesion to endothelial cells which promotes atherosclerotic lesion formation has been associated with enhanced blood glucose levels (Samanta, 2021). Hyperglycemia is one of the major inducers of epigenetic modifications in genes involved in vascular inflammation and may also activate several pathways leading to endothelial dysfunction (De Rosa et al., 2018). Thus hyperglycemia can promote and accelerate cardiovascular events through a cascade of several mechanisms and pathways.

2.2   Role of Insulin Resistance/Hyperinsulinaemia

Cells in the heart, skeletal muscles, liver, and adipose tissues use insulin to take up metabolic substrate under normal physiological conditions, but during insulin resistance (IR), these cells do not respond well to insulin, and as a consequence, the pancreas compensates by increasing the insulin secretion leading to hyperinsulinemia. A permanently elevated level of insulin is detrimental to the normal physiological process as it may affect normal cellular physiology and function (Kolb et al., 2020). IR increased the risk of coronary artery diseases (CAD) by three times compared to those who did not develop IR and normalizing IR reduced the risk of CVD by 55%. IR accounted for 42% of MI among young adults indicating the role of IR in CVD development among DM independently of other risk factors. Non-diabetic individuals with CVD showed altered IR suggesting that altering glucose metabolism can lead to vascular diseases or injury (Adeva-

Andany et al., 2019). Homeostasis Model Assessment Insulin Resistance (HOMA-IR) is a valid marker of IR and a meta-analysis of HOMA-IR with CHD risk found that an increase of HOMA-IR by one standard deviation increased CHD risk by 46% which was higher than fasting glucose and insulin concentration. IR promotes atherosclerosis by elevated glucose/insulin, dyslipidemia, inflammation, etc. (Gast et al., 2012). In cardiac cells, IR causes dyslipidemia which promotes a non-oxidative pathway leading to mitochondrial dysfunction and endoplasmic reticulum stress thereby enhancing apoptosis (Ormazabal et al., 2018). The non-atherosclerosis mice model crossed with Apoe−/− showed decreased anti-atherogenic properties of insulin such as nitric oxide (NO) synthase (NOS) activation, decreased acetylcholine-induced vasodilation, and VCAM suppression suggesting the role of insulin in CVD development (Bornfeldt & Tabas, 2011). Hyperinsulinemia was suggested to be atherogenic through several mechanisms such as loss of intact IR signalling by down-regulating IR, formation of pro-atherogenic Insulin growth factor-1 receptor (IGF1R), imbalance of IGF1R signalling and enhanced VSMC apoptosis which needs to be confirmed by future research (Bornfeldt & Tabas, 2011). The hyperinsulinemia was also associated with IR through altered gene expression of estrogen receptors causing the formation of atherosclerotic plaque in the animal model (Ormazabal et al., 2018). Apart from cardiomyocytes, the IR present in macrophages with defective IR was found to promote atherosclerosis by the formation of foam cells which plays an important role in the development of atherosclerosis (Bornfeldt & Tabas, 2011). Hyperinsulinaemia and IR enhanced endothelial dysfunction, and suppression of antioxidant and calcium influx in VSMC which led to CVD development (Kolb et al., 2020). IR in association with hyperglycemia also increases CVD risk by activating the renin-angiotensin-aldosterone system (Yen et al., 2022). These mechanisms explain the atherogenesis mechanism of IR and can be concluded that IR is a prominent marker of CVD development.

2.3   Role of Dyslipidemia

Another common manifestation of diabetes is dyslipidemia which includes fasting and postprandial hypertriglyceridemia, low-high-density lipoprotein-cholesterol (HDL-c), increased LDL, and VLDL that is highly atherogenic and determinants of CVD, and the serum cholesterol levels are a strong predictor of CVD mortality (Chen & Tseng, 2013). The incidences of dyslipidemia among T2DM individuals are high with a prevalence of 72–85% significantly increasing the risk of CVD among T2DM (Verges, 2015; Jialal & Singh, 2019). Atherogenic dyslipidemia complex (ADC) is a collection of dyslipidemia condition such as hypertriglyceridemia, smaller and denser LDL, decreased HDL levels, increased remnant lipoprotein and postprandial hyperlipidemia (Stahel et al., 2018). The role of traditional lipid profile in dyslipidemia leading to CVD in DM patients is shown in Fig. 3.

Fig. 3 The role of dyslipidemia in cardiovascular development in individuals with type 2 diabetes mellitus [VLDL, very low-density lipoprotein; FFA, free fatty acids; TRL, triglyceride-rich lipoprotein; LPL, lipoprotein lipase; HDL, high-density lipoprotein; CETP, cholesteryl ester transfer protein; LDL, low-density lipoprotein; Apo, Apo protein]

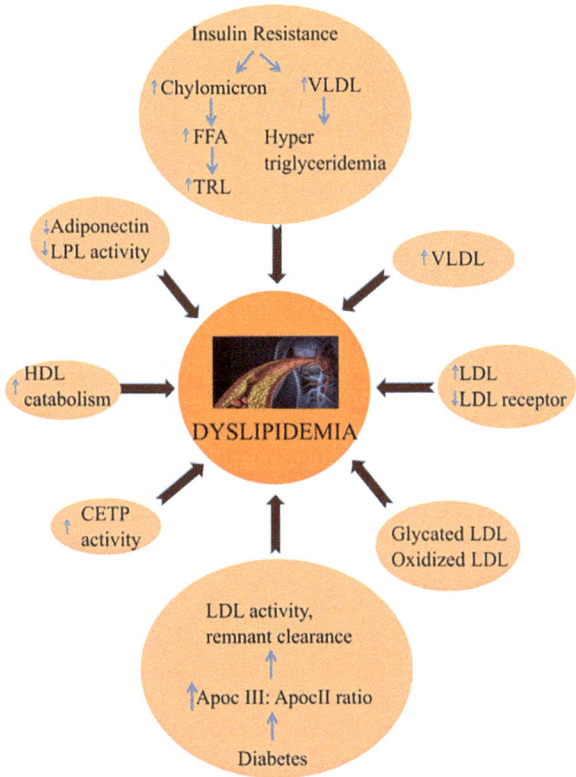

Cholesterol signalling dysregulation contributes to vascular inflammation and atherogenesis development through activation of toll-like receptor-4 (TLR4), NLR family pyrin domain 3 (NLRP3) inflammasome and NF-Κβ. The hypertriglyceridemia activates cholesteryl ester transfer protein (CETP) resulting in high triglycerides (TG) content in HDL and LDL. TG-rich HDL is subjected to catabolism leading to decreased HDL concentration, whereas TG-rich LDL undergoes hydrolysis leading to decreased particle size (Hasheminasabgorji & Jha, 2021). Among these lipids, LDL is considered the dominant form of atherogenic cholesterol, and the 2019 guidelines for dyslipidemia management by the ESC/EAS (European Society of Cardiology (ESC) and the European Atherosclerosis Society (EAS) suggested a linear relation between LDL and atherosclerosis risk (Mach et al., 2020). Similarly, UKPDS study indicated LDL as the number 1 predictor of CVD in T2DM even after adjusting for age and gender (Jialal & Singh, 2019). Small dense LDL (sdLDL) is also a known risk factor for atherosclerosis, and a cross-sectional study reported that an increased concentration of sdLDL was found among individuals with T2DM (Poznyak et al., 2020).

Apart from the traditional lipid components, postprandial hyperlipidemia promotes atherosclerosis by increasing postprandial triacylglycerol (TAG) and pro-inflammatory cytokines like TNFα, IL6, and VCAM-1 and reduced flow-mediated dilation which causes inflammation and endothelial dysfunction. The T2DM patient shows qualitative, quantitative, and kinetic changes in the lipid abnormalities. The quantitative changes include an increased plasma concentration of lipoproteins, while the qualitative changes include changes in the structure, proportion, content, and modifications such as glycations (Verges, 2015; Jialal & Singh, 2019).

2.4 Role of Oxidative Stress

Oxidative stress occurs due to the imbalance in the pro-oxidants and antioxidant levels causing a detrimental effect on the cells. Significant research over the last decade has proved the involvement of oxidative stress in vascular disease development among diabetes individuals. T2DM increases the intravascular ROS through activation of several pathways like NF-kB, Keap1-Nrf2, PKC (protein kinase C), macrophage activation, arachidonic acid mobilization, endothelial dysfunction and AGEs and thrombosis, inflammation, vascular homeostasis and cellular proliferation increasing the CVD risk (Robson et al., 2018). The pathogenesis of atherosclerosis in diabetes has been linked to the activation of xanthine oxidase (XO), mitochondrial respiratory chain enzymes, nicotinamide adenine dinucleotide phosphate (NADPH) oxidases (NOX), cyclooxygenase and uncoupled endothelial nitric oxide synthase (eNOS), as well as protein kinase C (PKC) and thioredoxin-interacting protein (TXNIP) pathway. Another major pathway is the AGE-RAGE pathway which causes increased cytokine secretion and endothelial dysfunction showing the link of the AGE-RAGE axis with atherosclerosis in T2DM (Hasheminasabgorji & Jha, 2021). The major pathway involved in vascular disease development among diabetes patients has been depicted in Fig. 4.

DM patients showed increased levels of 8-hydroxy-deoxyguanosine (8-ohdG), 8-iso-prostaglandin F2a, peroxidized fat, and oxidized LDL which are biomarkers of oxidative stress. The ROS involved in the cardiovascular development of diabetes evolves from several sources like mitochondrial electron transport chain (ETC) and non-mitochondrial sources like Nox (NADPH oxidases), xanthine oxidase, Lox (12/15 lipoxygenase or Arachidonic acid) and enzymatic and non-enzymatic antioxidants (Kayama et al., 2015). Oxidative stress leads to an increase in the cell adhesion molecules such as P-selectin, E-selectin, VCAM, and PCAM where P-selectin is released from activated platelets, endothelial cells, and E-selectin from adhesion molecules thereby inducing endothelial dysfunction (Robson et al., 2018). Oxidative stress promotes inflammation (majorly through TNFα, NF-Kβ, IL), fibrosis, enhanced RAAS, apoptosis, calcium accumulation, diminished contractility, arrhythmias vascular vessel wall and endothelium disruption (Kayama et al., 2015). ROS stimulate angiogenesis through endothelial cell regeneration, proliferation, and migration through the vascular endothelial growth factor (VEGF) pathway, and major ROS like hydrogen peroxide (H_2O_2) is involved in coronary endothelial-dependent and independent vasodilatation. Hyperglycemia which is a hallmark of

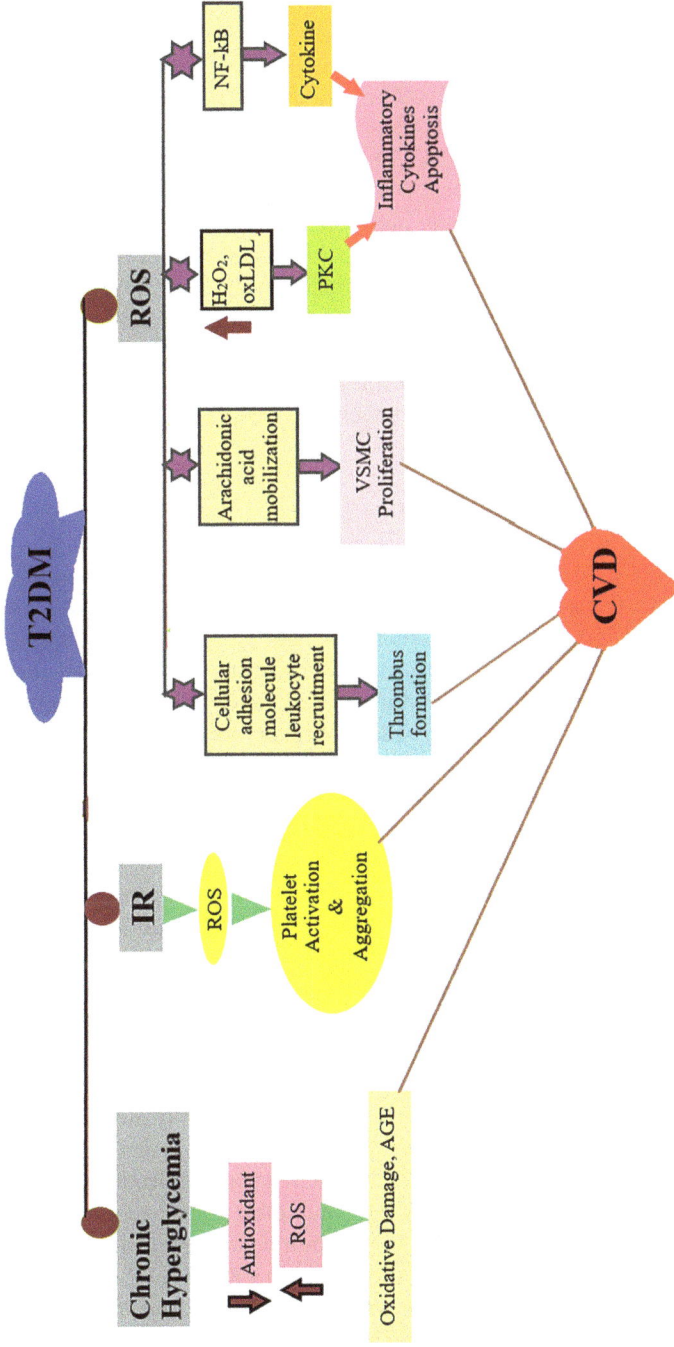

Fig. 4 Oxidative stress is a major risk factor involved in cardiovascular disease development in diabetes individuals through several pathways and mechanisms [ROS, reactive oxygen species; AGE, advanced glycation end products; VSMC, vascular smooth muscle cells; oxLDL, oxidized low-density lipoprotein; PKC, protein kinase C]

diabetes causes increased pro-oxidants and decreased antioxidant levels leading to oxidative damage, and nitrosylation of biomolecules causing atherosclerosis and vascular inflammation (Kayama et al., 2015). Prevalent oxidative stress has been associated with increased ischemia-modified albumin (IMA) and low HDL; however, the prolonged inflammation, glycation, etc. could reduce the effect of HDL leading to atherogenesis (Hasheminasabgorji & Jha, 2021). Even pre-diabetes individuals are at an elevated risk of CVD due to insulin resistance and impaired glucose tolerance (Petrie et al., 2018). Diabetes Control and Complications Trial/ Epidemiology of Diabetes Interventions and Complications (DCCT/EDIC) reported that an enhanced antioxidant activity reduced cardiovascular event risk among type 1 DM in a cohort study conducted on 349 participants (Tang et al., 2018). Apart from these factors, environmental factors like diet and smoking can also contribute to ROS production and therefore causes oxidative stress. Thus among T2DM individuals, oxidative stress plays a vital role in the initiation, and progression of vascular damage, atherosclerosis, and endothelial dysfunctions.

2.5   Role of Hypercoagulability

Several factors in T2DM contribute to the progression of inflammation as indicated by biomarkers like C-reactive protein (CRP), interleukins, tumour necrosis factors (TNF), and tissue factors which are closely linked with dyslipidemia and atherosclerosis. Dysregulated inflammation results in the formation of clots in the form of dense matted deposits, whereas in T2DM it was discovered to be amyloid in nature due to structural changes in the fibrin. T2DM patients were found to express an increased level of inflammatory markers like IL-1β, IL-6, IL-8, sP-selectin, platelet activation, and fibrinogen interaction which indicates a denser clot formation (Pretorius et al., 2018). Increased expression of these markers can amplify the inflammatory response exacerbating diabetes-related vascular complications by enhancing the expression of pro-coagulants and supresses the expression of anticoagulant molecules (Domingueti et al., 2016). Platelet activation significantly increased the spreading of platelet-derived micro-particle formation in individuals with T2DM which is involved in various diabetes complications like MI (Pretorius et al., 2018). Apart from these molecules, the expression of growth factors like VEGF and fibroblast growth factor (FGF) is enhanced which stimulates the blood vessel remodelling and thickening of the basement membrane favouring sclerosis and impaired vasodilation (Domingueti et al., 2016). Inflammation also causes monocytosis and neutrophilia, which results in an increase in circulating monocytes and activated neutrophils, which release endothelial damage molecules and neutrophil extracellular traps (Eckel et al., 2021). Thus these factors ultimately cause endothelial dysfunction which promotes both pro-inflammatory and pro-coagulant states. Besides initiating CVD, inflammatory molecules also contribute to the progression of atherosclerosis through platelet adhesion aggregation and rapid thrombus aggravation (Domingueti et al., 2016). The role of different inflammatory and hypercoagulability in vascular disease development among DM individuals is shown in Fig. 5.

Patients with T2DM showed an increased level of VWF which is a biomarker of endothelial dysfunction and the plasma levels of VWF has been well-established with the development of CAD and cerebrovascular events (Domingueti et al., 2016). According to animal studies, inflammatory mediators interact with RAGE, causing myelopoiesis and thrombocytosis, which aggravate atherosclerosis in diabetic mice (Eckel et al., 2021). These suggest that pro-inflammatory and pro-coagulants cause endothelial damage which is a vital mechanism in the development of cardiac and vascular events. However, further research to clarify the downstream activation of inflammatory molecules is required.

2.6   Role of Vascular Calcification

Vascular calcification (VC) is a hallmark of T2DM that manifests as CVD and is known to cause blood vessel hardening and dysfunction. Several factors like oxidative stress, endothelial dysfunction, irregular mineral metabolism, and inflammatory cytokines were found to be associated with VC development among T2DM individuals (Harper et al., 2016). VC is classified morphologically as atherosclerotic intimal calcification and arterial medial calcification where the former develops in atheromatic plaques and the latter in the media artery layer. Intimal calcifications are normally found in large arteries, while arterial medial calcifications occur in small arteries (Roumeliotis et al., 2019). Hyperglycemia stimulates the trans-differentiation of VSMC into cells capable of expressing bone morphogenesis molecules and transcription of VC-related molecules (Harper et al., 2016). Soluble RAGE (sRAGE) was also found to be a predictor of aortic valve calcification (Mapanga & Essop, 2016). The OPG/RANKL/RANK (osteoprotegerin/receptor activator of nuclear factor kappa-B ligand/receptor activator of nuclear factor kappa-B) signalling axis is vital for VC development (Harper et al., 2016). The OPG and RANKL were found to be contradictory in function, and their pathophysiological role in VC needs to be fully understood. An increased serum RANKL level causes increased mineralization of median arterial wall by accelerating the OPG/RANKL/RANK signalling pathway (Harper et al., 2016). Apart from OPG/RANKL/RANK signalling pathway, matrix Gla protein (MGP) secreted by chondrocytes and VSMCs is recognized as an inhibitor of VC both in vivo and in vitro (Roumeliotis et al., 2019). Coronary artery calcifications (CAC) help in the long-term prognosis of CVD risk among T2DM, while Coronary Artery Calcification in Type I Diabetes (CACTI) study found that T1DM showed a faster CAC progression compared to non-diabetic individuals. CAD risks are inversely associated with CAC density (CAC volume ≥ 100) and baseline CAC or CAC progression improved risk prediction (Guo et al., 2019). The degree of CAC is based on the hypoglycemia levels and was based on the degree of glycaemic control. In patients with controlled diabetes/glycaemic index, and atherosclerotic lesions were reduced, while patients with uncontrolled diabetes showed progressive

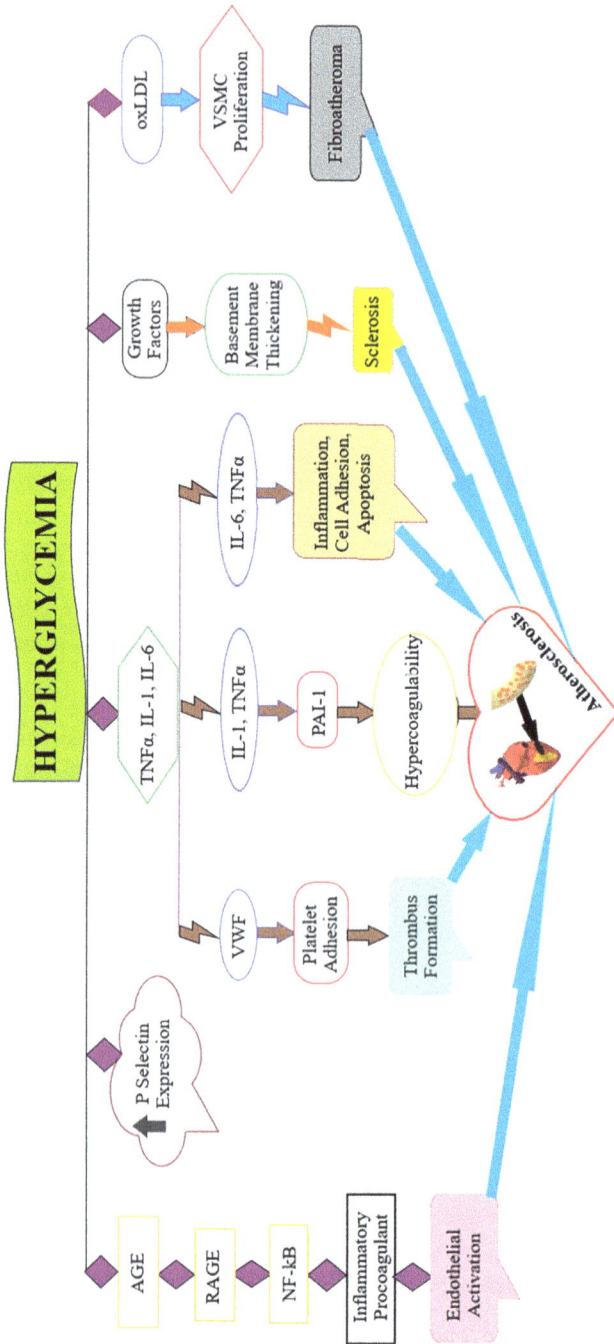

Fig. 5 Hypercoagulability in vascular disease development among diabetes individuals [AGE, advanced glycation end products; RAGE, receptor for AGEs; VWF, von Willebrand factor; PAI, plasminogen activator inhibitor-1; IL, interleukin; TNF, tumour necrosis factor; oxLDL, oxidized LDL; VSMC, vascular smooth muscle cells]

atherosclerotic lesions (Gajos, 2018). The coronary calcium score is an independent risk factor for CVD events when analysed using electron beam computed tomography (Wang et al., 2016). Thus VC plays a vital role in cardiovascular development among T2DM patients.

2.7 Role of Blood Pressure

High blood pressure (BP) is one of the major modifiable risk factors of CVD, and systemic arterial hypertension (HTN) is defined by a resting systolic BP (SBP) of 140 mmHg and diastolic BP (DBP) of 90 mmHg. DM and HTN are inter-relatable risk factors with overlapping etiology and disease pathology (Sunkara & Ahsan, 2017). When T2DM patients were compared to healthy people, the prevalence of HTN was found to be twofold higher. The risk for cardiovascular events like MI, stroke and HTN was increased among T2DM individuals with HTN (Arnold et al., 2020). The meta-analysis of randomized control trials (RCT) on intensive BP-lowering treatment showed that intensive BP treatment significantly reduced cardiovascular events and all-cause mortality (Wang et al., 2019). HTN by several mechanisms like arterial stiffness, carotid intimal media thickness, insulin resistance, AGE, inflammation and oxidative stress contribute to cardiovascular events among DM individuals (Strain & Paldánius, 2018; Haas & McDonnell, 2018). Environmental, behavioural and genetic factors are also contributing factors the cardiometabolic syndrome (Sunkara & Ahsan, 2017). Control of BP has been known to suppress the development and progression of cardiovascular complications among diabetes individuals (Wang et al., 2019).

2.8 Role of Obesity

Obesity is defined by excess body mass index (BMI), and it has been reported that a J-shaped relationship exists between BMI and mortality/morbidity risk. Obesity increases the risk of cardiovascular risk among diabetes and is often linked with hypertension and dyslipidemia. Obesity causes excessive adipose tissue formation increasing systemic resistance and also causes the adipose tissue to be insulin-resistant due to the increased flux of glucose. Obesity has also been associated with increased inflammatory markers worsening vascular injury (Glovaci et al., 2019). Increased abdominal fat known as the visceral adipose tissue (VAT) evaluated using the waist to hip ratio (WHR) was found to be a better predictor of glucose intolerance, insulin resistance, hypertension, and hypertriglyceridemia compared to BMI. T2DM patients have a larger waistline and VAT than non-T2DM patients, and the presence of VAT, combined with a larger waistline, increases the risk of CVD. VAT increased the vulnerability of atherosclerotic lesions within the arteries and aorta, indicating that VAT plays a major role in the pathogenesis of atherosclerosis (Piché et al., 2020). In contrast to the traditional view that lowering body fat decreases CVD risk, it was found that obese class I individuals had a lower risk of adverse cardiovascular events which was attributed to the obesity paradox (Pagidipati et al., 2020). Among acute heart failure and MI individuals, an inverse association was reported between BMI and mortality rate indicating the obesity paradox. This association was also reported in individuals with T2DM with CVD as comorbidity as weight loss was associated with increased mortality (Doehner et al., 2012). The obesity paradox may be attributed to epidemiological factors rather than its true association, and hence further studies are required to confirm its association (Pagidipati et al., 2020). Obesity is one of the modifiable risk factors, but the role of obesity in CVD development among T2DM is unclear because obesity influences both CVD and DM factors. Similarly, more research is needed to confirm the obesity paradox's role in CVD mortality.

2.9   Role of Genetic Polymorphism

Aside from these classical risk factors, genetic factors also predispose individuals with T2DM to CVD. The close relationship between DM and CVD indicates common genetic and epigenetic factors, and many single nucleotide polymorphisms (SNPs) are associated with CVD development among T2DM patients (Sharma et al., 2020). Many common SNPs have been found to be associated with both T2DM and CVD, and some important SNPs have been shown in Table 1. The genes predisposing to the association of both T2DM and CVD are identified through previously conducted candidate gene studies, linkage analyses, and genome-wide association studies (GWAS). GWAS combined with integrative pathway and network analysis identified eight major pathways involved in both T2DM and CVD development among all major ethnic groups (De Rosa et al., 2018). Certain candidate genes like glutathione S-transferases (GST) have reported the association of its polymorphism with CVD development among T2DM individuals. Polymorphism in major GST types – GSTM1, GSTT1, and GSTP1 – was identified to increase cardiovascular events among T2DM patients. However, for candidate genes, several interethnic population studies need to be conducted to confirm the association of genetic polymorphism with disease development (Sobha & Ebenezar, 2022). The study of SNPs helps to identify the susceptible genes associated with vascular complications in T2DM and helps to develop a possible treatment strategies for early detection of the complication. The SNPs may act as a biomarker of CVD prediction and can be of great significance regarding the increased survival period of the patients and thereby improving the quality of life. These factors also help in developing alternative management of disease progression (Yang et al., 2021).

3   Pharmacological Therapy

The control of hyperglycemia within its target levels may have some effect in delaying the major cardiovascular event among T2DM patients. The role of several other risk factors in CVD development among T2DM individuals demands

Table 1 List of genetic variants of genes, function and risk involved in cardiovascular events and its complications among type 2 diabetes mellitus individuals (Wang et al., 2018; Kaur et al., 2018; Rattanatham et al., 2021; Pérez-Hernández et al., 2016; Król & Kepinska, 2020; Sobha & Ebenezar, 2022)

Gene	Chromosome location	SNP	Function	Risk
Glutathione S-transferase (GSTM1/T1/P1)	1/22/11	Deletion/ deletion/ rs1695	Multigene superfamily of antioxidant enzymes involved in cellular detoxification	Increased CVD risk among DM
Nitric oxide synthase (NOS3)	17	27-bp VNTR	Nitric oxide synthesis	Atherosclerosis, coronary artery disease
Superoxide dismutase 2 (SOD2)	17	rs4880	Antioxidant enzyme which reduces oxidative stress	Increased CAD risk among T2DM
Apo B	2	5453 SNP rs693 rs6725189	LDL receptor-impaired intracellular transport of cholesterol. Increased CAS	Increased CAD risk
Paraoxonase	7q21.3	rs662 rs854560	Synthesizes an enzyme bound to HDL	Increased CAD risk
Transcription factor 7-like 2 gene (TCF7L2)	10q25.2	rs7903146	Regulate pancreatic β-cell development, vascular development and insulin signalling	Increased CAD risk among T2DM
Potassium voltage-gated channel subfamily Q member 1 (*KCNQ1*)	11p15.5	rs2237892	Encodes a pore-forming subunit of voltage-gated potassium (K+) channel vital for cardiovascular system	Increased CAD risk among T2DM
High-mobility group A1. (HMGA1)	17	rs146052672	Involved in glucose metabolism and cell growth differentiation	Increased CVD risk among T2DM
Protein phosphatase and actin regulator 1 (PHACTR1)	6	rs2026458 rs9349379	A molecule that inhibits protein phosphatase-1 (PP1) and binds actin via the C-terminal domain	CAC, coronary artery stenosis, vascular calcification

development of a therapeutic approach in order to suppress the onset of cardiovascular events thereby the improve the quality of life and also the life span of the patients. The rising number of diabetes further increases the concern about developing a therapeutic approach in the management of CVD in diabetic individuals. Table 2 highlights some of the common and recent therapeutic advances for DM and CVD (Schmidt, 2019; Wang et al., 2016).

Apart from these therapeutic interventions, lifestyle modifications by controlling BP, obesity, smoking, and alcoholism were found to be effective as demonstrated in earlier studies (Sharma et al., 2020; Yang et al., 2022; Regassa et al., 2021). The SCOUT trial found that a modest weight loss among diabetes individuals also reduced the risk of CVD-related deaths. Obesity treatment and BMI maintenance could significantly improve the condition and reduce cardiovascular risk factors in diabetic patients. Weight loss may reduce multiple risk factors like BP and HbA1c which require further confirmation. Lifestyle interventions (physical activity and diet) result in an excess loss of VAT and ectopic fat, which reduces the CV risk profile (Sharma et al., 2020). A meta-analysis of aerobic exercise as a lifestyle modification in T2DM reported a 5% LDL-c reduction which equals an 8.5% reduced CVD risk, assuming 1% LDL-c reduction equals 1.7% reduced cardiovascular events (Jialal & Singh, 2019). Other lifestyle modifications like calorie restriction, increased fruit and vegetable intake and reduced sedentary time also significantly reduced disease risk. A combination of pharmacotherapy and lifestyle intervention is required for the effective reduction of cardiovascular events among T2DM (Sharma et al., 2020).

4   Conclusion

Type 2 diabetes mellitus (T2DM) is rising at a dangerous rate and is associated with several complications. Cardiovascular disease (CVD) is the most frequent complication associated with T2DM and is linked with increased mortality and morbidity among T2DM individuals. Elevated glucose and insulin levels can promote atherosclerosis by being pro-atherogenic. Epidemiological study has proven a number of risk factors other than glycaemic controls strongly affect cardiovascular events in patients with T2DM (Eckel et al., 2021). Mechanisms like hyperglycemia, insulin resistance, dyslipidemia, oxidative stress, hypercoagulability, vascular calcification, obesity and genetic polymorphism were found to be associated with vascularization in patients with DM thereby causing CVD. However, the role of these mechanisms and pathways as a diagnostic biomarker for early detection of CVD risk especially among asymptomatic diabetic individuals has not yet been explored. Pre-diabetes individuals were also found to have higher CVD risk; however, the roles of these mechanisms in pre-diabetic individuals are scarce and need further research. Several pharmacological interventions hold a promising novel targets for delaying cardiovascular events in DM. On the other hand, lowering body weight, increased fruit and vegetable intake, diet modification, smoking and alcoholism cessation, etc. were also found to significantly reduce the CVD risk among T2DM individuals. However, the pleiotropic effect of glucose-lowering drugs in CVD, drug safety, the long-term side effects of several drugs, and the efficacy of combined pharmacotherapy and lifestyle intervention is yet to be established. The effect of patient-centered interventions on macrovascular complications among T2DM needs to be studied. The roles of gene therapy, microRNAs, PKC signalling, etc. as a pharmacological targets also require further detailed research.

Table 2 List of current pharmacological drugs and its effect on the cardiovascular events among type 2 diabetes mellitus individuals (Schmidt, 2019; Wang et al., 2016)

S. no	Drug	TRIAL	Effect
1.	GLP-1 (glucagon-like peptide 1)	NA	Inhibit glucagon secretion, enhances insulin secretion
2.	Liraglutide: GLP-1 RA (GLP-1 receptor agonists)	LEADER (Liraglutide Effect and Action in Diabetes: Evaluation of Cardiovascular Outcome Results)	Significant cardiovascular benefits were observed
3.	Semaglutide: GLP-1 RA	SUSTAIN-6(Trial to Evaluate Cardiovascular and Other Long-Term Outcomes with Semaglutide in Subjects with Type 2 Diabetes)	Rate of CVD death, nonfatal myocardial infarction or stroke was significantly lowered
4.	DPP-4 (dipeptidyl peptidase-4) inhibitors	NA	DPP-4 inhibits GLP-1 degradation and increases GLP-1 availability. Reduced gain of weight
5.	SGLT-2 (sodium-glucose cotransporter-2) inhibitors	NA	Reduce hyperglycemia by decreasing plasma glucose levels by prevention of renal glucose resorption
6.	Empagliflozin: SGLT-2 inhibitor	EMPA-REG OUTCOME trial (Empagliflozin, Cardiovascular Outcomes, and Mortality in Type 2 Diabetes)	Decreased CVD and all-cause mortality, reduced hospitalization for heart failure among T2DM
7.	Canagliflozin: SGLT-2 inhibitor	CANVAS (Canagliflozin Cardiovascular Assessment Study)	Reduced deaths due to CVD, MI, stroke among T2DM
8.	Dapagliflozin: SGLT-2 inhibitor	DECLARE-TIMI 58 trial (Dapagliflozin Effect on Cardiovascular Events–Thrombolysis in Myocardial Infarction 58)	Lower deaths and hospitalization due to heart failure among T2DM
9.	Alirocumab: PCSK9 inhibitor	ODYSSEY COMBO-II trial (Efficacy and Safety of Alirocumab Versus Ezetimibe on Top of Statin in High Cardiovascular Risk Patients with Hypercholesterolaemia)	Reduced LDL-c among diabetes patients and non-diabetic individuals
10.	Evolocumab: PCSK9 inhibitor	FOURIER trial (Further Cardiovascular Outcomes Research with PCSK9 Inhibition in Subjects with Elevated Risk)	Reduced CVD risk among diabetes patients and non-diabetic individuals
11.	Canakinumab: IL-1β inhibitor	CANTOS (Canakinumab Anti-inflammatory Thrombosis Outcomes Study)	Reduced MI, stroke and all-cause deaths among diabetes

(continued)

Table 2 (continued)

S. no	Drug	TRIAL	Effect
			patients with inflammation and diabetes with high CVD risk
12.	Colchicine	NA	Reduced ACS, non-cardio metabolic ischemic stroke. Decreased inflammatory levels and infract size
13.	Icosapent ethyl	REDUCE-IT (Reduction of Cardiovascular Events Trial)	Decreased CVD, MI, stroke, coronary revascularization or unstable angina among diabetes with hypertriglyceridaemia
14.	Atorvastatin: statin treatment	CARDS (Collaborative Atorvastatin Diabetes Study) trial	37% reduced CVD risk among T2DM individuals
15.	Ezetimibe: non-statin lipid-lowering agent		Diminished LDL-c level among diabetic and non-diabetic individuals
16.	Ramipril: ACE inhibitor	HOPE (Heart Outcomes Prevention Evaluation) study	Diminished MI, stroke and CVD outcome in T2DM patient
17.	Losartan: RAAS inhibitor	LIFE (Losartan Intervention For Endpoint reduction in hypertension)	24% reduced risk for cardiovascular events

In conclusion, several mechanisms and pathways are involved in the development of CVD, especially among T2DM patients. Recent advances in understanding the mechanism of vascular complications have helped in developing certain promising novel targets and drugs to delay the development of vascular complications. However further research is required to improve the efficacy and suppress CVD risk which is essential to maintain long-term health and leading a better-quality life.

References

Adeva-Andany, M. M., Martínez-Rodríguez, J., Gonzalez-Lucan, M., et al. (2019). Insulin resistance is a cardiovascular risk factor in humans. *Diabetes and Metabolic Syndrome: Clinical Research & Reviews, 13*, 1449–1455.

Arnold, S. V., Bhatt, D. L., Barsness, G. W., et al. (2020). Clinical management of stable coronary artery disease in patients with type 2 diabetes mellitus: A scientific statement from the American heart association. *Circulation, 141*(19), e779–e806.

Bornfeldt, K. E., & Tabas, I. (2011). Insulin resistance, hyperglycemia, and atherosclerosis. *Cell Metabolism, 14*(5), 575–585.

Chen, S. C., & Tseng, C. H. (2013). Dyslipidemia, kidney disease, and cardiovascular disease in diabetic patients. *Review of Diabetic Studies, 10*(2–3), 88.

De Rosa, S., Arcidiacono, B., Chiefari, E., et al. (2018). Type 2 diabetes mellitus and cardiovascular disease: Genetic and epigenetic links. *Frontiers in Endocrinology, 9*, 2.

Doehner, W., Erdmann, E., Cairns, R., et al. (2012). Inverse relation of body weight and weight change with mortality and morbidity in patients with type 2 diabetes and cardiovascular co-morbidity: An analysis of the PROactive study population. *International Journal of Cardiology, 162*(1), 20–26.

Domingueti, C. P., Dusse, L. M., Mdas, C., et al. (2016). Diabetes mellitus: The linkage between oxidative stress, inflammation, hypercoagulability and vascular complications. *Journal of Diabetes and its Complications, 30*(4), 738–745.

Eckel, R. H., Bornfeldt, K. E., & Goldberg, I. J. (2021). Cardiovascular disease in diabetes, beyond glucose. *Cell Metabolism, 33*(8), 1519–1545.

Gajos, G. (2018). Diabetes and cardiovascular disease: From new mechanisms to new therapies. *Polish Archives of Internal Medicine., 128*(3), 178–186.

Gast, K. B., Tjeerdema, N., Stijnen, T., et al. (2012). Insulin resistance and risk of incident cardiovascular events in adults without diabetes: Meta-analysis. *PLoS One, 7*(12), e52036.

Glovaci, D., Fan, W., & Wong, N. D. (2019). Epidemiology of diabetes mellitus and cardiovascular disease. *Current Cardiology Reports, 21*(4), 21.

Guo, J., Erqou, S. A., Miller, R. G., et al. (2019). The role of coronary artery calcification testing in incident coronary artery disease risk prediction in type 1 diabetes. *Diabetologia, 62*, 259–268.

Haas, A. V., & McDonnell, M. E. (2018). Pathogenesis of cardiovascular disease in diabetes. *Endocrinology and Metabolism Clinics of North America, 47*(1), 51–63.

Harper, E., Forde, H., Davenport, C., et al. (2016). Vascular calcification in type-2 diabetes and cardiovascular disease: Integrative roles for OPG, RANKL and TRAIL. *Vascular Pharmacology, 82*, 30–40.

Hasheminasabgorji, E., & Jha, J. C. (2021). Dyslipidemia, diabetes and atherosclerosis: Role of inflammation and ROS-redox-sensitive factors. *Biomedicine, 9*, 1602.

IDF Diabetes Atlas. (2019). *IDF Diabetes Atlas* (9th ed.). International Diabetes Federation.

IDF Diabetes Atlas. (2021). *IDF Diabetes Atlas* (10th ed.). International Diabetes Federation.

Jialal, I., & Singh, G. (2019). Management of diabetic dyslipidemia: An update. *World Journal of Diabetes, 10*(5), 280–290.

Kannel, W. B., & McGee, D. L. (1979). Diabetes and cardiovascular disease. The Framingham study. *JAMA, 241*(19), 2035–2038.

Kaur, S., Bhatti, G. K., Vijayvergiya, R., et al. (2018). Paraoxonase 1 gene polymorphisms (Q192R and L55M) are associated with coronary artery disease susceptibility in Asian Indians. *Dubai Diabetes and Endocrinology Journal, 24*(1–4), 38–47.

Kayama, Y., Raaz, U., Jagger, A., et al. (2015). Diabetic cardiovascular disease induced by oxidative stress. *International Journal of Molecular Sciences, 16*, 25234–25263.

Kolb, H., Kempf, K., Röhling, M., et al. (2020). Insulin: Too much of a good thing is bad. *BMC Medicine, 18*(1), 224.

Król, M., & Kepinska, M. (2020). Human nitric oxide synthase-its functions, polymorphisms, and inhibitors in the context of inflammation, diabetes and cardiovascular diseases. *International Journal of Molecular Sciences, 22*(1), 56.

Mach, F., Baigent, C., Catapano, A. L., et al. (2020). ESC scientific document group. 2019 ESC/EAS guidelines for the Management of dyslipidemias: Lipid modification to reduce cardiovascular risk. *European Heart Journal., 41*, 111–188.

Mapanga, R. F., & Essop, M. F. (2016). Damaging effects of hyperglycemia on cardiovascular function: Spotlight on glucose metabolic pathways. *American Journal of Physiology. Heart and Circulatory Physiology, 310*, H153–H173.

Navarro-Pérez, J., Orozco-Beltran, D., Gil-Guillen, V., et al. (2018). Mortality and cardiovascular disease burden of uncontrolled diabetes in a registry-based cohort: The ESCARVAL-risk study. *BMC Cardiovascular Disorders, 18*, 180.

Nicholls, S. J., Tuzcu, E. M., Kalidindi, S., et al. (2008). Effect of diabetes on progression of coronary atherosclerosis and arterial remodeling: A pooled analysis of 5 intravascular ultrasound trials. *Journal of the American College of Cardiology, 52*, 255–262.

Ormazabal, V., Nair, S., Elfeky, O., et al. (2018). Association between insulin resistance and the development of cardiovascular disease. *Cardiovascular Diabetology, 17*, 122.

Pagidipati, N. J., Zheng, Y., Green, J. B., et al. (2020). Association of obesity with cardiovascular outcomes in patients with type 2 diabetes and cardiovascular disease: Insights from TECOS. Am. *Heart, J.219*, 47–57.

Paneni, F., Beckman, J. A., Creager, M. A., et al. (2013). Diabetes and vascular disease: Pathophysiology, clinical consequences, and medical therapy: Part I. *European Heart Journal, 34*(31), 2436–2443.

Patoulias, D., Stavropoulos, K., Imprialos, K., et al. (2020). Pharmacological management of cardiac disease in patients with type 2 diabetes: Insights into clinical practice. *Current Vascular Pharmacology, 18*, 125–138.

Pérez-Hernández, N., Vargas-Alarcón, G., Posadas-Sánchez, R., et al. (2016). PHACTR1 gene polymorphism is associated with increased risk of developing premature coronary artery disease in Mexican population. *International Journal of Environmental Research and Public Health, 13*(8), 803.

Petrie, J. R., Guzik, T. J., & Touyz, R. M. (2018). Diabetes, hypertension, and cardiovascular disease: Clinical insights and vascular mechanisms. *The Canadian Journal of Cardiology, 34*, 575–584.

Piché, M.-E., Tchernof, A., & Després, J.-P. (2020). Obesity phenotypes, diabetes, and cardiovascular diseases. *Circulation Research, 126*(11), 1477–1500.

Poznyak, A., Grechko, A. V., Poggio, P., et al. (2020). The diabetes mellitus–atherosclerosis connection: The role of lipid and glucose metabolism and chronic inflammation. *International Journal of Molecular Sciences, 21*(5), 1835.

Pretorius, L., Thomson, G. J. A., Adams, R. C. M., et al. (2018). Platelet activity and hypercoagulation in type 2 diabetes. *Cardiovascular Diabetology, 17*, 141.

Rao Kondapally Seshasai, S., Kaptoge, S., Thompson, A., et al. (2011). Diabetes mellitus, fasting glucose, and risk of cause-specific death. *The New England Journal of Medicine, 364*(9), 829–841.

Rattanatham, R., Settasatian, N., Komanasin, N., et al. (2021). Association of combined TCF7L2 and KCNQ1 gene polymorphisms with diabetic micro- and macrovascular complications in type 2 diabetes mellitus. *Diabetes and Metabolism Journal, 45*(4), 578–593.

Rawshani, A., Rawshani, A., Franzen, S., et al. (2018). Risk factors, mortality, and cardiovascular outcomes in patients with type 2 diabetes. *The New England Journal of Medicine, 379*, 633–644.

Regassa, L. D., Tola, A., & Yohanes, A. (2021). Prevalence of cardiovascular disease and associated factors among type 2 diabetes patients in selected hospitals of Harari region, Eastern Ethiopia. *Frontiers in Public Health, 8*, 532719.

Robson, R., Kundur, A. R., & Singh, I. (2018). Oxidative stress biomarkers in type 2 diabetes mellitus for assessment of cardiovascular disease risk. *Diabetes and Metabolic Syndrome: Clinical Research & Reviews, 12*, 455–462.

Roumeliotis, S., Dounousi, E., Eleftheriadis, T., et al. (2019). Association of the inactive circulating matrix Gla protein with vitamin K intake, calcification, mortality, and cardiovascular disease: A review. *International Journal of Molecular Sciences, 20*(3), 628.

Samanta, S. (2021). Glycated hemoglobin and subsequent risk of microvascular and macrovascular complications. *Indian Journal of Medical Sciences, 83*(2), 230–238.

Sarwar, N., Gao, P., Emerging Risk Factors Collaboration, et al. (2010). Diabetes mellitus, fasting blood glucose concentration, and risk of vascular disease: A collaborative meta-analysis of 102 prospective studies. *Lancet, 375*(9733), 2215–2222.

Schmidt, A. M. (2019). Diabetes mellitus and cardiovascular disease: Emerging therapeutic approaches. *Arteriosclerosis, Thrombosis, and Vascular Biology, 39*(4), 558–568.

Schwarz, P. E. H., Timpel, P., Harst, L., et al. (2018). Blood sugar regulation as a key focus for cardiovascular health promotion and prevention: An umbrella review. *Journal of the American College of Cardiology, 72*(15), 1829–1844.

Sharma, A., Mittal, S., Aggarwal, R., et al. (2020). Diabetes and cardiovascular disease: Interrelation of risk factors and treatment. *Future Journal of Pharmaceutical Sciences, 6*(1), 130.

Sobha, S. P., & Ebenezar, K. (2022). Susceptibility of glutathione-S-transferase polymorphism to CVD development in type 2 diabetes mellitus: A review. *Endocrine, Metabolic & Immune Disorders Drug Targets, 22*(2), 225–234.

Stahel, P., Xiao, C., Hegele, R. A., et al. (2018). The Atherogenic dyslipidemia complex and novel approaches to cardiovascular disease prevention in diabetes. *The Canadian Journal of Cardiology, 34*(5), 595–604.

Stamler, J., Vaccaro, O., Neaton, J. D., et al. (1993). Diabetes, other risk factors, and 12-yr cardiovascular mortality for men screened in the multiple risk factor intervention trial. *Diabetes Care, 16*(2), 434–444.

Strain, W. D., & Paldánius, P. M. (2018). Diabetes, cardiovascular disease and the microcirculation. *Cardiovascular Diabetology, 17*(1), 57.

Sunkara, N., & Ahsan, C. H. (2017). Hypertension in diabetes and the risk of cardiovascular disease. *Cardiovascular Endocrinology, 6*(1), 33–38.

Tang, W. H. W., McGee, P., Lachin, J. M., et al. (2018). Oxidative stress and cardiovascular risk in type 1 diabetes mellitus: Insights from the DCCT/EDIC study. *Journal of the American Heart Association, 7*, e008368.

Verges, B. (2015). Pathophysiology of diabetic dyslipidemia: Where are we? *Diabetologia, 58*, 886–899.

Wang, C. C. L., Hess, C. N., Hiatt, W. R., et al. (2016). Atherosclerotic cardiovascular disease and heart failure in type 2 diabetes–mechanisms, management, and clinical considerations. *Circulation, 133*(24), 2459–2502.

Wang, Y.-T., Li, Y., Ma, Y.-T., et al. (2018). Association between apolipoprotein B genetic polymorphism and the risk of calcific aortic stenosis in Chinese subjects, in Xinjiang, China. *Lipids in Health and Disease, 17*(1).

Wang, J., Chen, Y., Xu, W., et al. (2019). Effects of intensive blood pressure lowering on mortality and cardiovascular and renal outcomes in type 2 diabetic patients: A meta-analysis. *PLoS One, 14*(4), e0215362.

Yamagishi, S., & Matsui, T. (2018). Role of hyperglycemia-induced advanced glycation end product (AGE) accumulation in atherosclerosis. *Annals of Vascular Diseases, 11*(3), 253–258.

Yang, Y., Qiu, W., Meng, Q., et al. (2021). GRB10 rs 1800504 polymorphism is associated with the risk of coronary heart disease in patients with type 2 diabetes mellitus. *Frontiers in Cardiovascular Medicine, 8*, 728976.

Yang, Y., Peng, N., Chen, G., et al. (2022). Interaction between smoking and diabetes in relation to subsequent risk of cardiovascular events. *Cardiovascular Diabetology, 21*, 14.

Yen, F. S., Wei, J. C. C., Chiu, L. T., et al. (2022). Diabetes, hypertension, and cardiovascular disease development. *Journal of Translational Medicine, 20*, 9.

Yun, J. S., & Ko, S. H. (2021). Current trends in epidemiology of cardiovascular disease and cardiovascular risk management in type 2 diabetes. *Metabolism, Clinical and Experimental, 123*, 154838.

Diabetes and Neurological Disorder

Iyshwarya Bhaskar Kalarani and Ramakrishnan Veerabathiran

Abstract

Neurologic issues usually accompany diabetes mellitus, which can be incapacitating. Diabetic patients may have discomfort and sensory anomalies, weakness and paralysis, and symptoms of autonomic dysfunction. Neurological problems have been linked to diabetes mellitus for many years, but our understanding of this group of ailments is exceedingly restricted. Diabetes type 2 is a metabolic disorder characterized by hyperglycemia, dyslipidemia, and insulin resistance. Strokes, vascular dementia, and Alzheimer's have been associated with it. Pathologic examinations are ambiguous and incomplete. It is challenging to develop a diagnosis and interpret a pathophysiologic pattern given the vast range of clinical symptoms. Many studies have looked at the potential causes of various diseases, but none have addressed the core problem. Despite significant advancements in our knowledge of neurological problems and diabetes over the past 10 years, it is still unclear which specific pathways cause type 1 and type 2 diabetes-related neurological abnormalities. Future research on disease causation will be essential for dealing with diabetic neuropathy on all fronts, from prevention to therapy.

Keywords

Diabetes mellitus · Neurological disorders · Hyperglycemia · Insulin resistance · Pathogenesis · Physiological factors

I. B. Kalarani · R. Veerabathiran (✉)
Human Cytogenetics and Genomics Laboratory, Faculty of Allied Health Sciences, Chettinad Hospital and Research Institute, Chettinad Academy of Research and Education, Kelambakkam, Tamil Nadu, India

R. Noor (ed.), *Advances in Diabetes Research and Management*, https://doi.org/10.1007/978-981-19-0027-3_4

1   Introduction

Diabetes mellitus is commonly referred to as a set of metabolic diseases which is consistently characterized by the elevated the blood glucose level over the normal range. It happens either when there is not enough insulin present or when there are substances present that interfere with insulin's ability to work. An inadequate response to insulin causes hyperglycemia (Watkins & Thomas, 1998). The ramifications of chronic hyperglycemia include organ damage such as retinopathy, which may cause blindness; nephropathy, which causes renal failure; and peripheral neuropathy, which may cause significant consequences, including amputation. There are other related metabolic problems, most notably the onset of hyperketonemia in cases of acute insulin deficiency and changes in the turnover of fatty acids, lipids, and protein. Except for a few rare circumstances, diabetes is a chronic disorder that cannot be temporarily reversed. In individuals with diabetes mellitus, a wide range of abnormalities that either directly or indirectly affect the central and peripheral nervous systems may be present. Atherosclerosis, myocardial infarction, and cerebrovascular disorders are also more common among diabetics (Madden, 2013). It affected most people worldwide by diabetes, which has become one of the pandemic diseases (Dardano et al., 2014). According to Awotidebe et al., it is linked to a posh lifestyle, unhealthy eating practices, age, sedentary behavior, and hereditary predisposition (Awotidebe TO et al., 2016; Kaveeshwar & Cornwall, 2014). It now affects public health more than any other issue on the planet and has serious long-term consequences. It is becoming more prevalent, especially affecting young and middle-aged people in both developing and developed countries (Innes & Selfe, 2016). DM places a considerable burden on the global economy. It costs money both directly and indirectly since chronic impairment results in lost workdays and poorer productivity at work and home, and greater medical costs for things like medications, insulin, lab tests, diets, and hospital stays (Hogan et al., 2003; American Diabetes Association, 2008). This chapter focuses on recent developments in the description of clinical characteristics of neurological diseases associated with diabetes.

2   Types of Diabetes Mellitus

Diabetics have higher blood glucose levels. They classify it into many types based on its origin: the most common types are type 1 diabetes mellitus (T1DM), type 2 diabetes mellitus (T2DM), and gestational diabetes. It can also be caused by genetic abnormalities with insulin action, beta-cell function, pancreatic illness, endocrinopathies, or infections, which vary based on the causation.

2.1   Type 1 Diabetes Mellitus

The immune system-mediated loss of pancreatic beta cells results in T1DM. Environmental factors may bring it on, like a viral infection and genetic markers like human leucocyte antigen (HLA) (American Diabetes Association, 2009). The usual early years for insulin-dependent diabetes mellitus include childhood and adolescence (IDDM). The patients are not insulin-dependent in the early stages of the condition. Based on this need for insulin, the American Diabetes Association (ADA) further separated idiopathic DM into types 1 A and 1 B. The production of autoantibodies is characterized by an immune-mediated syndrome known as a type against islet cells, insulitis, and the selective degeneration of beta cells in the islets. A significant insulin deficit associated with type 1 B diabetes is the absence of beta-cell autoimmunity (Madden, 2013). Lymphocytic cells invade islet cells during the initiation of type 1 diabetes. The area around the beta cells is inflamed. The inflammation destroys beta cells, resulting in a total absence of insulin. When pancreatic beta cells are killed, insulin production is reduced, resulting in metabolic inefficiency. As a result, uncontrolled IDDM might cause excessive glucagon secretion. Muscles and adipose tissues receive glucose from insulin. In the absence of these tissues, glucose cannot be absorbed, causing high blood glucose levels. Free fatty acids (FFAs) in plasma are excessive due to uncontrolled lipolysis caused by insulin deficiency. Peripheral organs' glucose metabolism is suppressed, further hindering glucose utilization. Certain genes must be expressed in target tissues for insulin to be effective. In the liver, glucokinase is one of these genes, while in adipose tissue, GLUT4 is one of them. As a result, instead of being oxidized, hepatocytes transform into ketone bodies in the tricarboxylic acid cycle. It results in hyperglycemia because the body cannot use glucose effectively because of the abundance of FFA and ketone bodies (Ozougwu et al., 2013).

2.2   Diabetes Mellitus Type 2

Non-insulin-dependent diabetes mellitus, or T2DM, is the most common kind of diabetes. Insulin resistance and relative insulin deficiency distinguished it (American Diabetes Association, 2014). We can classify diabetes as having normal glucose tolerance, impaired glucose tolerance, diabetes with minimum fasting hyperglycemia (FPG) of 140 mg/dl, or diabetes with minimal fasting hyperglycemia (> 140 mg/dl). In the early stages of T2DM, the number of beta cells is normally or marginally reduced. A decrease in receptor sensitivity occurs at this stage of T2DM, which is compensated by an increase in insulin production and the resulting hyperinsulinemia. Diabetes occurs when beta cells begin to degenerate and insulin production diminishes in an environment characterized by insulin resistance. The beta-cell activity wanes around 10 years before the onset of diabetes mellitus; by this time, it has dropped to 30% or less (Sacks et al., 2011).

3   Neurological Disorders Associated with Diabetes Mellitus

Several genetic disorders increase the risk of diabetes. In addition, severe early-onset obesity is associated with genetic disorders such as Prader-Willi syndrome, Alstrom syndrome, and a variety of genetically characterized Bardet-Biedl syndrome. The second group of chromosomal anomalies includes Down's, Klinefelter's, and Turner's syndromes. There is also a group of neurological problems to consider, which will be discussed in more detail below (Kalra et al., 2018).

3.1   Diabetes and Depression

Because of short-term insulin-related hypoglycemia and long-term cardiovascular disease, neuropathy, nephropathy, and vision problems, diabetes has negative effects on mental health, especially depression (Kalra et al., 2018). Depressed patients have a persistent depressed mood for 2 weeks and have at least five of the following symptoms: decreased interest, changes in sleep patterns with insomnia and hypersomnia, weight loss with increased appetite, feelings of guilt and worthlessness, lack of energy, loss of concentration, suicidal ideation, excitement, and delay in psychomotor activity. They estimated that developed nations have a depression rate of 15%, while industrialized nations have a depression rate of 11% (Bromet et al., 2011). Most countries throughout the world have a high frequency and low emphasis on mental health issues among the elderly, with depression being the most common treatable condition (Kumar et al., 2021). We have found that diabetes and depression share similar symptoms as fatigue, weight gain, altered appetite, and impaired focus (deJoode et al., 2019). Insufficient depression treatment results in nonadherence to lifestyle changes, noncompliance with medications, and poor glycemic control, which leads to mortality (Hermanns et al., 2013). According to disability-adjusted life years (DALYs), depression and anxiety are the fourth and eighth leading causes of disability in industrialized nations (Hagstrom et al., 2015). Lack of social support and depression are psychological variables that can cause or worsen diabetes (Strodl & Kenardy, 2006). Neuroimmunology, neuroendocrine, and microvascular abnormalities can produce metabolic dysfunction, which can be long-term and play a significant role in the progression of diabetes and depression (Patterson et al., 2013).

Physiological Factors

Hypothalamic-pituitary-adrenal (HPA) axis dysregulation, autonomic nervous system (ANS) hyperactivity, and inflammatory processes are shared biological causes of depression and T2DM (Champaneri et al., 2010). Stress triggers the HPA axis, which secretes corticotrophin-releasing hormone (CRH), which activates adrenocorticotropic hormone (ACTH) release from the anterior pituitary gland. ACTH then stimulated the production of cortisol from the adrenal gland (Golden et al., 2011). Similarly to this, prolonged stress causes the sympathetic nervous system (SNS) to become overactive and increases the release of catecholamines

(Kyrou & Tsigos, 2009). High amounts of catecholamines and cortisol make it harder for insulin to attach to its receptor, which causes insulin resistance and the emergence of hyperglycemia (Weber et al., 2000).

Hyperglycemia might cause hippocampal atrophy. The level of glycated hemoglobin (HbA1c) shows an inverse relationship between blood sugar levels and hippocampal volume (Gold et al., 2007). Persistent hyperglycemia activates the polyol pathway or fluctuating glucose, which damages neurons by inducing oxidative stress and the production of advanced glycation end products (AGEs) (Chen et al., 2012). It has linked depression to neurodegenerative processes in the prefrontal cortex and hippocampal regions (Sapolsky, 2001). According to several studies, prolonged stress impairs the immune system and increases the production of inflammatory cytokines. This can happen either directly or indirectly via the HPA axis or the SNS. Inflammatory cytokines at high doses impair insulin sensitivity by interacting with pancreatic cells, hastening the development of T2DM (Pickup & Crook, 1998). Inflammatory cytokines have the potential to alter pathophysiological areas allied with depression, including neurotransmitter signaling, neuroendocrine function, and neurosynaptic transmission (Raison et al., 2006). In addition, microvascular dysfunction and arterial stiffness contribute to the cause of the association between diabetes and depression (Mitchell et al., 2011). Microvascular dysfunction may contribute to insulin resistance by reducing cellular glucose and insulin absorption (Liu et al., 2009). There are several behavioral factors that may contribute to depression, including poor eating habits, insufficient exercise, and inconsistent sleeping patterns (Bădescu et al., 2016). Because of these behavioral changes, which influence glucose metabolism and are independent of food and lifestyle, individuals with T2DM are at a higher risk of developing T2DM (Beydoun & Wang, 2010). We illustrate pathophysiological causes in Fig. 1.

3.2   Diabetes and Epilepsy

Epileptic seizures represent a group of neurological diseases known as epilepsy. People with epilepsy have a greater death rate than the general population, which is especially true for various types of fatalities, such as accidents, neoplastic disorders, and cerebrovascular disease (CVD) (Chang et al., 2012). Many problems associated with type 2 diabetes, such as stroke and dementia, have been related to epilepsy (Ferlazzo et al., 2016), besides other known risk factors for epilepsy, such as degenerative brain diseases and head trauma (Olafsson et al., 2005). Most of the evidence supporting the association between diabetes and epilepsy came from studies of T1DM (Ramakrishnan & Appleton, 2012); T2DM and the risk of epilepsy are less well-established. Several illnesses, including hypoglycemia and epilepsy, are common among people with type 2 diabetes. Some hypoglycemia-induced changes in electrical activity might persist for a long time despite the onset of seizures, unconsciousness, and even death at high levels of hypoglycemia (Frier, 2014).

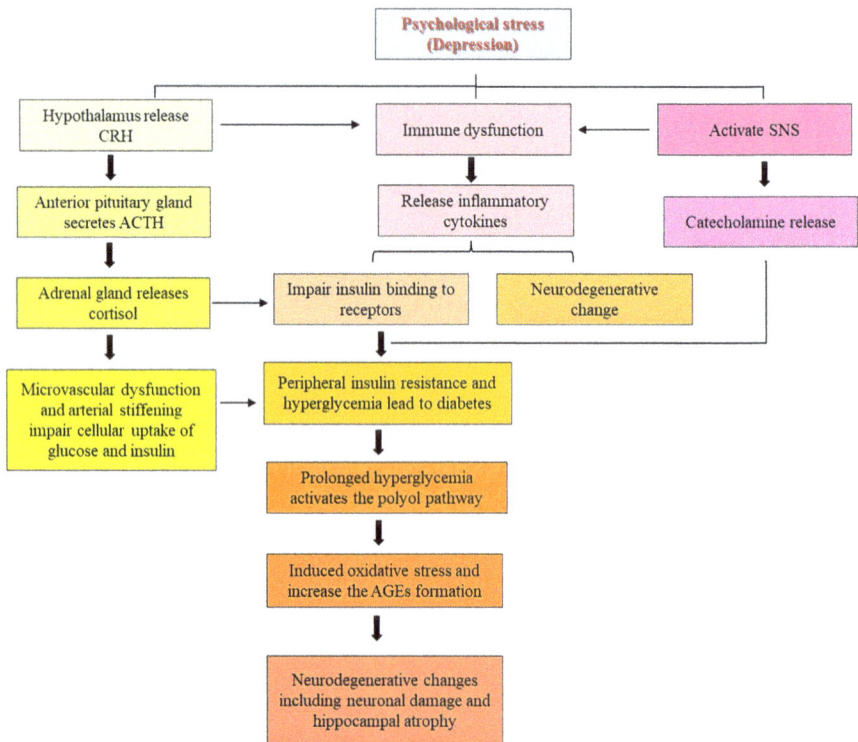

Fig. 1 Pathophysiology of diabetes and depression

Children and young adults are far more prone to suffering from autoimmune diseases. The most extensive population-based retrospective cohort study on epilepsy in autoimmune disorders found that among people with T1DM under 65, the risk of receiving an epilepsy diagnosis is over five times higher than in individuals without DM (Ong et al., 2014). Healthcare practitioners who treat people with epilepsy will only become more effective in their work if they are aware of the linkages between T1DM and epilepsy. The research they discovered in the two disorders connection in 1952 in a group of two infants who had both DM and epilepsy (Keezer et al., 2015). However, due to methodological issues, the prevalence of diabetes in epileptic patients remains variable. Despite their distinct pathologies, few studies distinguish T1DM from T2DM, confounding their respective relationships with seizures. Because patients with T1DM typically present with seizures provoked by glucose imbalances, identifying spontaneous seizures can be challenging and may be misdiagnosed. A growing number of population-based studies show that people with T1DM have a higher incidence of seizures than those without DM.

Pathophysiology of Epilepsy in Type 1 Diabetes Mellitus

Many theories explain the connection between T1DM and epilepsy. Although some genetic abnormalities cause both disorders, their rarity makes them unlikely to account for the bulk of this association. According to reports, there may be genetic factors that are common to both epilepsy and T1DM, where epilepsy usually develops first in some people and DM is usually the first symptom. Strong pathophysiological evidence supports the assumption that there is a shared underlying autoimmune mechanism, which helps to explain at least some links. There are two distinct but uncommon conditions that cause epilepsy and DM: persistent newborn, neonatal diabetic syndrome (PDNS), developmental delays, and persistent newborn diabetes mellitus (PNDM). It is possible that less harmful mutations that are more common may also interfere with neurons and pancreatic beta cells in the same way as these rare monogenic entities do in population-based studies that find epilepsy and T1DM associated (Slingerland et al., 2009).

In the pathophysiology of epilepsy, anti-glutamic acid decarboxylase (GAD) antibodies may be involved. Anti-GAD antibodies are hypothesized to cause seizures primarily through a change in GABA transmission, which raises neuronal excitability and lowers the seizure threshold. GAD is an enzyme that contributes to the production of GABA, and research investigations have shown that anti-GAD antibodies prevent hippocampus neurons in culture from functioning properly when GABA levels are adequate (Bien & Scheffer, 2011). Since GAD antibodies cannot pass the blood-brain barrier, intrathecal generation of these antibodies has been documented and appears to be crucial to the pathophysiology of the disease (Yoshimoto et al., 2005). Even though neurons contain GAD and it is not readily accessible to the plasma, it has shown neurons to internalize circulating antibodies, providing antibodies with intracellular access (Manto et al., 2007).

The two most likely manifestations of epilepsy with positive anti-GAD titers are limbic encephalitis or chronic epilepsy without active encephalitis. This condition, which is defined by temporal lobe hyperexcitability, may have two clinical manifestations, one chronic and one acute (Vincent et al., 2011). The presence of positive titers in epilepsy without encephalitis does not seem to correlate with the intensity of seizures. Anti-GAD titers were high in 2.8 percent of epileptics, with temporal lobe epilepsy accounting for 86 percent (Liimatainen et al., 2010). Studies of type 1 diabetes have provided most of the evidence for the link between diabetes and epilepsy (Chou et al., 2016; Dafoulas et al., 2017); type 2 diabetes and the risk of epilepsy is less well established. Hypoglycemia and other conditions that co-occur with epilepsy can develop in T2DM patients. While the electrical activity abnormalities caused by severe hypoglycemia may last for a very long time (Frier, 2014), severe hypoglycemia can also cause seizures, comas, and even life-threatening conditions quickly.

3.3   Diabetes and Dementia

It linked cognitive deficits to diabetes mellitus. In T1DM, there is a little to moderate deterioration in mental speed and a decrease in mental flexibility. The three cognitive alterations that influence learning and memory, mental flexibility, and mental speed most often take place in T2DM patients (Awad et al., 2004). According to several sizable longitudinal population-based studies (Allen et al., 2004), elderly people with T2DM have a faster rate of cognitive loss. The causes of this rapid cognitive deterioration, however, are less obvious; some studies have shown links with glycemic control (Kanaya et al., 2004), while others have suggested that hypertension may be a key mediator. There are still some disputes regarding the relationship between diabetes and dementia, although diabetes is well-known to cause mild cognitive deficits. According to a recent study, up to 80% of Alzheimer's patients may have T2DM or impaired fasting glucose levels (Janson et al., 2004). These contradictory findings show categorically that studies looking at the frequency of dementia in diabetics who have already had the condition are unlikely to produce information on the risk of dementia in diabetics. These disparities might be the consequence of technical inaccuracies, such as the survival bias of Alzheimer's patients who do not have diabetes and the effects of the illness on glucose metabolism, which could hide the association between DM and a common form of dementia (Watson & Craft, 2004). Research on dementia in people is less trustworthy than population-based studies on dementia incidence in patients with and without diabetes. In the coming decades, there will probably be an increase in both the total and relative numbers of older people with diabetes (Zimmet et al., 2001). Therefore, it is crucial to comprehend the nature and scope of the link between diabetes and dementia.

Diabetes may affect the development and progression of the many underlying disorders linked with dementia via a range of pathophysiological pathways (Gispen & Biessels, 2000). These pathways include those that aging shares with Alzheimer's disease and vascular dementia. It is becoming more well-understood that dementia patients' brains, particularly those of the very elderly, are more prone to having many disorders, including vascular abnormalities and Alzheimer's type pathologies (Esiri et al., 2001). A schematic diagram is a simplified view of some of the endocrinological, metabolic, and vascular problems linked to diabetes, and capable of causing these many diseases is shown in Fig. 2. These linked mechanisms may cause mixed pathology. This type of dementia, which is characterized by predominating vascular disease, affected some diabetics. Some individuals may display a "pure Alzheimer's disease" clinical profile, with a predominance of amyloid-related processes. Most people will first exhibit intermediate phases of these two dementia types. Several studies (Luchsinger et al., 2004) have linked hyperinsulinemia to dementia and sped-up cognitive decline. Because insulin has vasoactive effects, vascular disease is likely to have a role in this link. In many population-based investigations of people without diabetes, it linked the highest insulin concentrations to an increased risk of stroke; It maintained this effect even

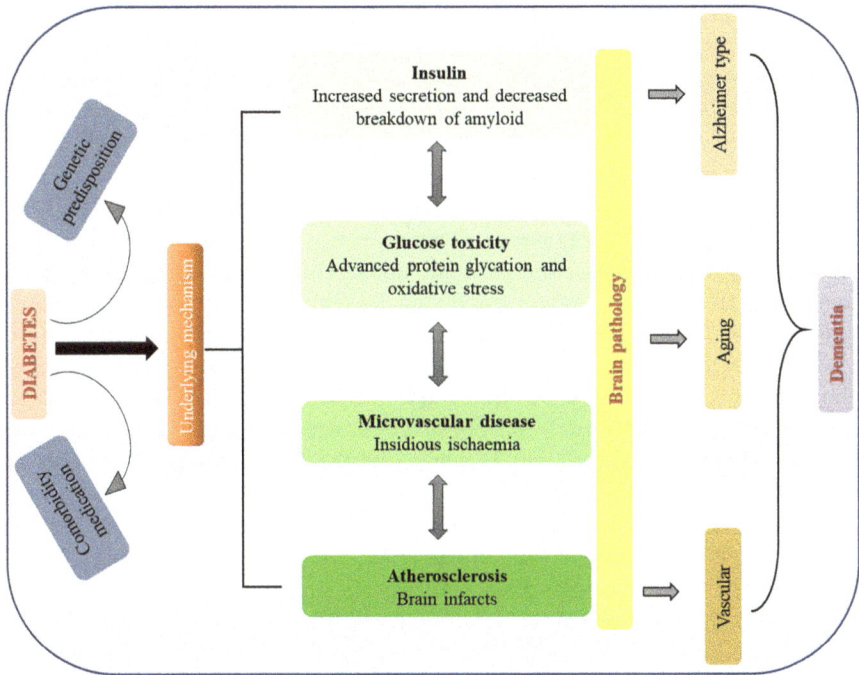

Fig. 2 Schematic of dementia associated with diabetes

after accounting for additional vascular risk factors from the metabolic syndrome that were also present (Kernan et al., 2002).

3.4 Diabetes and Stress

T2DM is a complex disease that causes dysregulation of multiple biological functions, including the immune, circulatory, metabolic, and neuroendocrine systems (Galicia-Garcia et al., 2020; Schwartz et al., 2017). Stress has a major impact on these physical systems, so it has increasingly identified chronic stress as a risk factor for T2DM (Hackett & Steptoe, 2017). It commonly defined chronic stress as exposure to stress that lasts for weeks to years, such as when one is unemployed or lives in a tough socioeconomic situation (Epel et al., 2018). The intensity of reported connections for chronic stress was comparable to that for more traditional health variables such as obesity or inactivity, showing that chronic stress increases the risk of developing T2DM. This is because multiple longitudinal studies have now shown this (Hackett & Steptoe, 2017; Akter et al., 2017; Kelly & Ismail, 2015; Maddatu et al., 2017). It did not explain the associations between behavioral risk factors that may have biased the results, suggesting that it directly implicated physiologic stress responses in the genesis of T2DM (Steptoe et al., 2014).

After the sympathetic nervous system (SNS) is activated, the hypothalamic-pituitary-adrenal (HPA) axis reacts later and more slowly, resulting in a biphasic plasma hormonal response in noradrenaline and glucocorticoids (GCs) (Romero & Butler, 2007; Ulrich-Lai & Herman, 2009). These hormones interact together to control a variety of neuronal, neuroendocrine, cardiovascular, immunological, and metabolic system aspects important in T2DM. They also aid in the maintenance of homeostasis, which improves behavioral responses to stress (de Kloet et al., 2005).

Initially, the stress response encompasses cognitive and neurobiological responses. There are substantial linkages among these three dimensions, even though they are normally examined individually (Epel et al., 2018). For instance, it has been discovered that uncontrolled stress and psychological states brought on by threats are linked to various neuroendocrine and autonomic reactions that may have variable degrees of negative effects on health (Dickerson & Kemeny, 2004). A higher incidence of T2D has been linked to particular stress paradigms that are characterized by unpredictability. Examples of these include being exposed to challenging working circumstances, a disparity between effort and reward at work, traumatic experiences, poor socioeconomic status (SES), or a greater frequency of racial or ethnic discrimination (Nordentoft et al., 2020; Whitaker et al., 2017). By combining the cognitive and biological aspects of stress reactions with their interconnections, it may be feasible to comprehend the connections between chronic stress and T2D better. The second difference is that stress-related neurotransmitters and hormones act not only on the SNS and the HPA axis but also on a wide range of brain and peripheral regions (Joëls & Baram, 2009; Stone et al., 2020). Besides regulating the release of other molecules, such as glucose, insulin, or cytokines, complicated dynamics often govern their physiological functions, such as the negative feedback they impose on their release (Dumbell et al., 2016; Koch et al., 2017; Oster, 2020).

A multitude of behavioral and physiological factors (Kudielka et al., 2009; Stenvers et al., 2019) influenced the physiology of the HPA axis. Sleep disruptions or nighttime light exposure can interfere with the HPA axis' ability to regulate the cortisol cycle because these factors affect how the HPA axis functions. This could then affect how tissue insulin sensitivity and pancreatic cell insulin production are regulated, leading to insulin resistance and T2D (Hackett et al., 2020; Vetter et al., 2018). Physical stressors, such as disease or pollution exposure, can activate or change the HPA axis (Snow et al., 2018; Thomson, 2019; Ulrich-Lai & Herman, 2009). Acute ozone exposure, for example, increased blood GCs and changed lipid metabolism in both rats and humans, showing that the HPA axis was activated (Miller et al., 2016). It has also linked the occurrence of T2D to prolonged exposure to ozone and air pollution (Li et al., 2021; Renzi et al., 2018). According to researchers, chronic psychological and physical stress may worsen one another and increase the risk of T2D in a reinforcing rather than additive way (Kodavanti, 2016; Wright & Schreier, 2013; Zänkert et al., 2019). In Fig. 3, the role of neuroendocrine, inflammatory, and autonomic pathways in diabetes-related processes is illustrated.

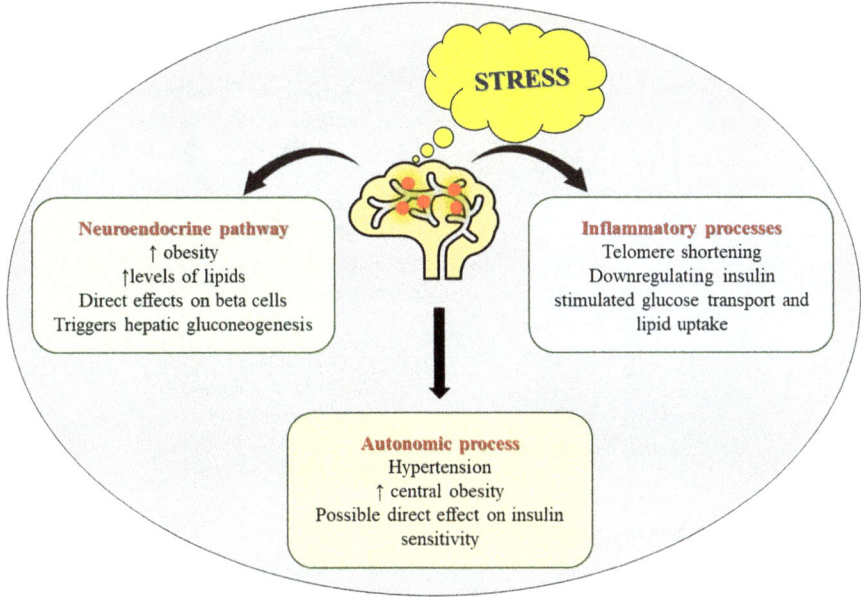

Fig. 3 Biological pathways related to stress and diabetes mellitus

3.5 Diabetes and Stroke

Strokes are associated with diabetes. Several sites, including the brain, can be affected by it and cause pathologic changes in blood vessels, which can cause strokes if cerebral vessels are directly damaged. Individuals with uncontrolled glucose levels also had a higher mortality rate and worse post-stroke outcomes. In order to prevent strokes and recurrences, diabetes and other risk factors must be controlled. Epidemiological studies well-established the risk of strokes, including ischemic and hemorrhagic strokes. People with diabetes who are young are at an increased risk of stroke.

Diabetes can induce strokes in a variety of ways. A few examples include vascular endothelial dysfunction, arterial stiffness at an early age, systemic inflammation, and capillary basal membrane thickening. Diabetes type 2 is associated with abnormalities of left ventricular diastolic filling. Rising insulin levels in most type 2 diabetics are a byproduct of increasing glucose load rather than the cause (Chockalingam et al., 2021). Microvascular disease, metabolic abnormalities, interstitial fibrosis, hypertension, and autonomic dysfunction are among the hypothesized causes of type 2 diabetes and congestive heart failure (Fig. 4). Keeping vessel walls structurally and functionally intact, as well as moderating vasomotor activity, is critically important for maintaining vessel wall integrity. In addition to inducing vasodilation, nitric oxide (NO) can also cause endothelial dysfunction and lead to atherosclerosis. Diabetes reduces NO-mediated vasodilation, due to increase NO inactivation or decreased smooth muscle sensitivity to NO. Compared to people with

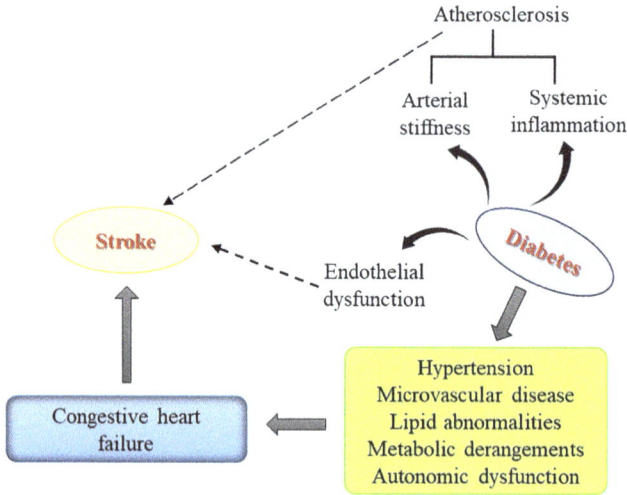

Fig. 4 Possible mechanisms of stroke in individuals with diabetes

normal glucose levels, diabetic patients had stiffer arteries and less flexibility. As a result of type 1 diabetes, the common carotid artery is often structurally damaged early, manifested as an increasing in intima-medial thickness (Chen et al., 2016).

4 Conclusion

The two categories are type 1 insulin-dependent diabetes and type 2 non-insulin-dependent diabetes. There are various neurological disorders where this type of syndrome can be seen, including Friedreich's ataxia, Wolfram syndrome, and mitochondrial diseases. Because of diabetes, a wide range of neurological symptoms are experienced. There are several reasons these symptoms may occur, including metabolic disorders or secondary symptoms following the disorder. Diabetes-related conditions might show acute metabolic decompensation, such as diabetic ketoacidosis, which is common in type 1 cases and causes diffuse cerebral edema, which is especially common in youngsters. Type 2 instances are more likely to have hyperosmolar nonketotic coma. The most significant are peripheral neuropathy and cerebrovascular illness because they affect the neurological system. A distal predominant sensory polyneuropathy is a common symptom of a variety of diabetes syndromes rather than a single diabetic neuropathy. It connected diabetes to a greater risk of macrovascular disease because diabetics are more likely to experience a stroke than non-diabetics. Despite significant progress in our knowledge of the intricacies of diabetic neuropathy over the last decade, the specific processes driving neuropathy in T1 and T2 DM remain unexplained. Future findings on disease pathophysiology will be critical in addressing all aspects of diabetic neuropathy, from prevention to therapy.

Acknowledgments The authors thank the Chettinad Academy of Research and Education for their constant support and encouragement.

References

Akter, S., Goto, A., & Mizoue, T. (2017). Smoking and the risk of type 2 diabetes in Japan: A systematic review and meta-analysis. *Journal of Epidemiology, 27*(12), 553–561.

Allen, K. V., Frier, B. M., & Strachan, M. W. J. (2004). The relationship between type 2 diabetes and cognitive dysfunction: Longitudinal studies and their methodological limitations. *European Journal of Pharmacology, 490*, 169–175.

American Diabetes Association. (2008). Economic costs of diabetes in the U.S. in 2007. *Diabetes Care, 31*(3), 596–615.

American Diabetes Association. (2009). Diagnosis and classification of diabetes mellitus. *Diabetes Care, 32*(Suppl 1), S62–S67.

American Diabetes Association. (2014). Diagnosis and classification of diabetes mellitus. *Diabetes Care, 37*(SUPPL.1), S81–S90.

Awad, N., Gagnon, M., & Messier, C. (2004). The relationship between impaired glucose tolerances, type 2 diabetes, and cognitive function. *Journal of Clinical and Experimental Neuropsychology, 26*, 1044–1080.

Awotidebe TO, Ativie, R. N., Oke, K. I., et al. (2016). Relationships among exercise capacity, dynamic balance and gait characteristics of Nigerian patients with type - 2 diabetes: An indication for fall prevention. *Journal of Exercise Rehabilitation, 12*(6), 581–588.

Bădescu, S., Tătaru, C., Kobylinska, L., Georgescu, E. L., Zahiu, D. M., Zăgrean, A. M., & Zăgrean, L. (2016). The association between diabetes mellitus and depression. *Journal of Medicine and Life, 9*, 120–125.

Beydoun, M. A., & Wang, Y. (2010). Pathways linking socioeconomic status to obesity through depression and lifestyle factors among young US adults. *Journal of Affective Disorders, 123*, 52–63. https://doi.org/10.1016/j.jad.2009.09.021

Bien, C. G., & Scheffer, I. E. (2011). Autoantibodies and epilepsy. *Epilepsia, 52*(Suppl 3), 18–22.

Bromet, E., Andrade, L. H., Hwang, I., et al. (2011). Cross-national epidemiology of DSM-IV major depressive episode. *BMC Medicine, 9*, 90. https://doi.org/10.1186/1741-7015-9-90

Champaneri, S., Wand, G. S., Malhotra, S. S., Casagrande, S. S., & Golden, S. H. (2010). Biological basis of depression in adults with diabetes. *Current Diabetes Reports, 10*, 396–405. https://doi.org/10.1007/s11892-010-0148-9

Chang, Y. H., Ho, W. C., Tsai, J. J., Li, C. Y., & Lu, T. H. (2012). Risk of mortality among patients with epilepsy in southern Taiwan. *Seizure, 21*, 254–259.

Chen, R., Ovbiagele, B., & Feng, W. (2016 Apr 1). Diabetes and stroke: Epidemiology, pathophysiology, pharmaceuticals and outcomes. *The American Journal of the Medical Sciences, 351*(4), 380–386.

Chen, G., Xu, R., Wang, Y., et al. (2012). Genetic disruption of soluble epoxide hydrolase is protective against streptozotocin-induced diabetic nephropathy. *American Journal of Physiology - Endocrinology and Metabolism, 303*, E563–E575. https://doi.org/10.1152/ajpendo.00591.2011

Chockalingam, A., Natarajan, P., Thanikachalam, P., & Pandiyan, R. (2021 Mar). Insulin resistance: The inconvenient truth. *Missouri Medicine, 118*(2), 119.

Chou, I. C., Wang, C. H., Lin, W. D., Tsai, F. J., Lin, C. C., & Kao, C. H. (2016). Risk of epilepsy in type 1 diabetes mellitus: A population-based cohort study. *Diabetologia, 59*, 1196–1203.

Dafoulas, G. E., Toulis, K. A., McCorry, D., Kumarendran, B., Thomas, G. N., Willis, B. H., et al. (2017). Type 1 diabetes mellitus and risk of incident epilepsy: A population-based, open-cohort study. *Diabetologia, 60*, 258–261.

Dardano, A., Penno, G., Prato, S. D., & Miccoli, R. (2014). Optimal therapy of type 2 diabetes: A controversial challenge. *Aging, 6*(3), 187–206.

de Kloet, E. R., Jöels, M., & Holsboer, F. (2005). Stress and the brain: From adaptation to disease. *Nature Reviews. Neuroscience, 6*(6), 463–475.

deJoode, J. W., van Dijk, S. E., Walburg, F. S., et al. (2019). Diagnostic accuracy of depression questionnaires in adult patients with diabetes: A systematic review and meta-analysis. *PLoS One, 14*, e0218512. https://doi.org/10.1371/journal.pone.0218512

Dickerson, S., & Kemeny, M. (2004). Acute stressors and cortisol responses: A theoretical integration and synthesis of laboratory research. *Psychological Bulletin, 130*, 355–391.

Dumbell, R., Matveeva, O., & Oster, H. (2016). Circadian clocks, stress, and immunity. *Frontiers in Endocrinology, 7*, 37.

Epel, E. S., Crosswell, A. D., Mayer, S. E., Prather, A. A., Slavich, G. M., Puterman, E., & Mendes, W. B. (2018). More than a feeling: A unified view of stress measurement for population science. *Frontiers in Neuroendocrinology Stress Brain, 49*, 146–169.

Esiri, M. M., Matthews, F., Brayne, C., & Neuropathology Group. Medical Research Council Cognitive Function and Aging Study; Neuropathology Group of the Medical Research Council Cognitive Function and Ageing Study (MRC CFAS). (2001). Pathological correlates of late-onset dementia in a multicentre, community-based population in England and Wales. *Lancet, 357*(9251), 169–175.

Ferlazzo, E., Gasparini, S., Beghi, E., Sueri, C., Russo, E., Leo, A., et al. (2016). Epilepsy in cerebrovascular diseases: Review of experimental and clinical data with meta-analysis of risk factors. *Epilepsia, 57*, 1205–1214.

Frier, B. M. (2014). Hypoglycaemia in diabetes mellitus: Epidemiology and clinical implications. *Nature Reviews. Endocrinology, 10*, 711–722.

Galicia-Garcia, U., Benito-Vicente, A., Jebari, S., Larrea-Sebal, A., Siddiqi, H., Uribe, K. B., Ostolaza, H., & Martín, C. (2020). Pathophysiology of type 2 diabetes mellitus. *International Journal of Molecular Sciences, 21*, 6275.

Gispen, W. H., & Biessels, G. J. (2000). Cognition and synaptic plasticity in diabetes mellitus. *Trends in Neurosciences, 23*, 542–549.

Gold, S. M., Dziobek, I., Sweat, V., et al. (2007). Hippocampal damage and memory impairments as possible early brain complications of type 2 diabetes. *Diabetologia, 50*, 711–719. https://doi.org/10.1007/s00125-007-0602-7

Golden, S. H., Wand, G. S., Malhotra, S., Kamel, I., & Horton, K. (2011). Reliability of hypothalamic-pituitary-adrenal axis assessment methods for use in population-based studies. *European Journal of Epidemiology, 26*, 511–525. https://doi.org/10.1007/s10654-011-9585-2

Hackett, R. A., Dal, Z., & Steptoe, A. (2020). The relationship between sleep problems and cortisol in people with type 2 diabetes. *Psychoneuroendocrinology, 117*, 104688.

Hackett, R. A., & Steptoe, A. (2017). Type 2 diabetes mellitus and psychological stress – A modifiable risk factor. *Nature Reviews. Endocrinology, 13*(9), 547–560.

Hagstrom, E. L., Patel, S., Karimkhani, C., et al. (2015). Comparing cutaneous research funded by the US National Institutes of Health (NIH) with the US skin disease burden. *Journal of the American Academy of Dermatology, 73*, 383–91.e1. https://doi.org/10.1016/j.jaad.2015.04.039

Hermanns, N., Caputo, S., Dzida, G., Khunti, K., Meneghini, L. F., & Snoek, F. (2013). Screening, evaluation and management of depression in people with diabetes in primary care. *Primary Care Diabetes, 7*, 1–10. https://doi.org/10.1016/j.pcd.2012.11.002

Hogan, P., Dall, T., & Nikolov, P. (2003). Economic costs of diabetes in the US in 2002. *Diabetes Care, 26*(3), 917–932.

Innes, K. E., & Selfe, T. K. (2016). Yoga for adults with type 2 diabetes: A systematic review of controlled trials. *Journal Diabetes Research, 2016*, 6979370.

Janson, J., Laedtke, T., Parisi, J. E., O'Brien, P., Petersen, R. C., & Butler, P. C. (2004). Increased risk of type 2 diabetes in Alzheimer disease. *Diabetes, 53*, 474–481.

Joëls, M., & Baram, T. Z. (2009). The neuro-symphony of stress. *Nature Reviews. Neuroscience, 10*(6), 459–466.

Kalra, S., Jena, B. N., & Yeravdekar, R. (2018). Emotional and psychological needs of people with diabetes. *Indian Journal of Endocrinology and Metabolism, 22*, 696–704. https://doi.org/10.4103/ijem.IJEM_579_17

Kanaya, A. M., Barrett-Connor, E., Gildengorin, G., & Yaffe, K. (2004). Change in cognitive function by glucose tolerance status in older adults: A 4-year prospective study of the rancho Bernardo study cohort. *Archives of Internal Medicine, 164*, 1327–1333.

Kaveeshwar, S. A., & Cornwall, J. (2014). The current state of diabetes mellitus in India. *The Australasian Medical Journal, 7*(1), 45–48.

Keezer, M. R., Novy, J., & Sander, J. W. (2015). Type 1 diabetes mellitus in people with pharmacoresistant epilepsy: Prevalence and clinical characteristics. *Epilepsy Research, 115*, 55–57.

Kelly, S. J., & Ismail, M. (2015). Stress and type 2 diabetes: A review of how stress contributes to the development of type 2 diabetes. *Annual Review of Public Health, 36*(1), 441–462.

Kernan, W. N., Inzucchi, S. E., Viscoli, C. M., Brass, L. M., Bravata, D. M., & Horwitz, R. I. (2002). Insulin resistance and risk for stroke. *Neurology, 59*, 809–815.

Koch, C. E., Leinweber, B., Drengberg, B. C., Blaum, C., & Oster, H. (2017). Interaction between circadian rhythms and stress. *Neurobiol. Stress, SI: Stressors in animals, 6*, 57–67.

Kodavanti, U. P. (2016). Stretching the stress boundary: Linking air pollution health effects to a neurohormonal stress response. Biochim. Biophys. *Acta BBA - Gen. Subj., SI: Air Pollution, 1860*, 2880–2890.

Kudielka, B. M., Hellhammer, D. H., & Wüst, S. (2009). Why do we respond so differently? Reviewing determinants of human salivary cortisol responses to challenge. *Psychoneuroendocrinology, 34*(1), 2–18.

Kumar, B. M., Raja, T. K., Liaquathali, F., Maruthupandian, J., & Raja, P. V. (2021 Apr). A study on prevalence and factors associated with depression among elderly residing in tenements under resettlement scheme, Kancheepuram District, Tamil Nadu. *Journal of Mid-Life Health, 12*(2), 137.

Kyrou, I., & Tsigos, C. (2009). Stress hormones: Physiological stress and regulation of metabolism. *Current Opinion in Pharmacology, 9*, 787–793. https://doi.org/10.1016/j.coph.2009.08.007

Li, Y.-L., Chuang, T.-W., Chang, P.-y., Lin, L.-Y., Su, C.-T., Chien, L.-N., & Chiou, H.-Y. (2021). Long-term exposure to ozone and sulfur dioxide increases the incidence of type 2 diabetes mellitus among aged 30 to 50 adult population. *Environmental Research, 194*, 110624.

Liimatainen, S., Peltola, M., Sabater, L., Fallah, M., Kharazmi, E., Haapala, A. M., et al. (2010). Clinical significance of glutamic acid decarboxylase antibodies in patients with epilepsy. *Epilepsia, 51*(5), 760–767.

Liu, Z., Liu, J., Jahn, L. A., Fowler, D. E., & Barrett, E. J. (2009). Infusing lipid raises plasma free fatty acids and induces insulin resistance in muscle microvasculature. *The Journal of Clinical Endocrinology and Metabolism, 94*, 3543–3549. https://doi.org/10.1210/jc.2009-0027

Luchsinger, J. A., Tang, M. X., Shea, S., & Mayeux, R. (2004). Hyperinsulinemia and risk of Alzheimer disease. *Neurology, 63*, 1187–1192.

Maddatu, J., Anderson-Baucum, E., & Evans-Molina, C. (2017). Smoking and the risk of type 2 diabetes. *Translational Research: The Journal of Laboratory and Clinical Medicine, 184*, 101–107.

Madden, K. M. (2013). Evidence for the benefit of exercise therapy in patients with type 2 diabetes. *Diabetes, Metabolic Syndrome and Obesity: Targets and Therapy, 6*(233), 39.

Manto, M. U., Laute, M. A., Aguera, M., Rogemond, V., Pandolfo, M., & Honnorat, J. (2007). Effects of anti-glutamic acid decarboxylase antibodies associated with neurological diseases. *Annals of Neurology, 61*(6), 544–551.

Miller, D. B., Ghio, A. J., Karoly, E. D., Bell, L. N., Snow, S. J., Madden, M. C., Soukup, J., Cascio, W. E., Gilmour, M. I., & Kodavanti, U. P. (2016). Ozone exposure increases circulating stress hormones and lipid metabolites in humans. *American Journal of Respiratory and Critical Care Medicine, 193*(12), 1382–1391.

Mitchell, G. F., van Buchem, M. A., Sigurdsson, S., et al. (2011). Arterial stiffness, pressure and flow pulsatility and brain structure and function: The age, Gene/Environment Susceptibility--Reykjavik study. *Brain, 134*, 3398–3407. https://doi.org/10.1093/brain/awr253

Nordentoft, M., Rod, N. H., Bonde, J. P., Bjorner, J. B., Madsen, I. E. H., Pedersen, L. R. M., Cleal, B., Hanson, M., Linda, L., Nexo, M. A., Pentti, J., Stenholm, S., Sterud, T., Vahtera, J., & Rugulies, R. (2020). Effort-reward imbalance at work and risk of type 2 diabetes in a national sample of 50,552 workers in Denmark: A prospective study linking survey and register data. *Journal of Psychosomatic Research, 128*, 109867.

Olafsson, E., Ludvigsson, P., Gudmundsson, G., Hesdorffer, D., Kjartansson, O., & Hauser, W. A. (2005). Incidence of unprovoked seizures and epilepsy in Iceland and assessment of the epilepsy syndrome classification: A prospective study. *Lancet Neurology, 4*, 627–634.

Ong, M. S., Kohane, I. S., Cai, T., Gorman, M. P., & Mandl, K. D. (2014). Population-level evidence for an autoimmune etiology of epilepsy. *JAMA Neurology, 71*(5), 569–574.

Oster, H. (2020). The interplay between stress, circadian clocks, and energy metabolism. *The Journal of Endocrinology, 247*, R13–R25.

Ozougwu, J., Obimba, K., Belonwu, C., & Unakalamba. (2013). The pathogenesis and pathophysiology of type 1 and type 2 diabetes mellitus. *Journal of Physiology and Pathophysiology., 4*(4), 46–57.

Patterson, Z. R., Khazall, R., Mackay, H., Anisman, H., & Abizaid, A. (2013). Central ghrelin signaling mediates the metabolic response of C57BL/6 male mice to chronic social defeat stress. *Endocrinology, 154*, 1080–1091. https://doi.org/10.1210/en.2012-1834

Pickup, J. C., & Crook, M. A. (1998). Is type II diabetes mellitus a disease of the innate immune system? *Diabetologia, 41*, 1241–1248. https://doi.org/10.1007/s001250051058

Raison, C. L., Capuron, L., & Miller, A. H. (2006). Cytokines sing the blues: Inflammation and the pathogenesis of depression. *Trends in Immunology, 27*, 24–31. https://doi.org/10.1016/j.it.2005.11.006

Ramakrishnan, R., & Appleton, R. (2012). Study of prevalence of epilepsy in children with type 1 diabetes mellitus. *Seizure, 21*, 292–294.

Renzi, M., Cerza, F., Gariazzo, C., Agabiti, N., Cascini, S., Di Domenicantonio, R., Davoli, M., Forastiere, F., & Cesaroni, G. (2018). Air pollution and occurrence of type 2 diabetes in a large cohort study. *Environment International, 112*, 68–76.

Romero, M. L., & Butler, L. K. (2007). Endocrinology of stress. *International Journal of Comparative Psychology, 20*, 89–95.

Sacks, D. B., Arnold, M., Bakris, G. L., et al. (2011). Guidelines and recommendations for laboratory analysis in the diagnosis and Management of Diabetes Mellitus. *Clinical Chemistry, 57*(6), e1 47.

Sapolsky, R. M. (2001). Depression, antidepressants, and the shrinking hippocampus. *The Proceedings of the National Academy of Sciences U S A, 98*, 12320–12322. https://doi.org/10.1073/pnas.231475998

Schwartz, S. S., Epstein, S., Corkey, B. E., Grant, S. F. A., Gavin, J. R., III, Aguilar, R. B., & Herman, M. E. (2017). A unified pathophysiological construct of diabetes and its complications. *Trends in Endocrinology and Metabolism, 28*, 645–655.

Slingerland, A. S., Shields, B. M., Flanagan, S. E., Bruining, G. J., Noordam, K., Gach, A., et al. (2009). Referral rates for diagnostic testing support an incidence of permanent neonatal diabetes in three European countries of at least 1 in 260,000 live births. *Diabetologia, 52*(8), 1683–1685.

Snow, S. J., Henriquez, A. R., Costa, D. L., & Kodavanti, U. P. (2018). Neuroendocrine regulation of air pollution health effects: Emerging insights. *Toxicological Sciences, 164*, 9–20.

Stenvers, D. J., Scheer, F. A. J. L., Schrauwen, P., la Fleur, S. E., & Kalsbeek, A. (2019). Circadian clocks and insulin resistance. *Nature Reviews. Endocrinology, 15*(2), 75–89.

Steptoe, A., Hackett, R. A., Lazzarino, A. I., Bostock, S., La Marca, R., Carvalho, L. A., & Hamer, M. (2014). Disruption of multisystem responses to stress in type 2 diabetes: Investigating the dynamics of allostatic load. *Proceedings of the National Academy of Sciences, 111*(44), 15693–15698.

Stone, L. A., Harmatz, E. S., & Goosens, K. A. (2020). Ghrelin as a stress hormone: Implications for psychiatric illness. *Biological Psychiatry, 88*(7), 531–540.

Strodl, E., & Kenardy, J. (2006). Psychosocial and non-psychosocial risk factors for the new diagnosis of diabetes in elderly women. *Diabetes Research and Clinical Practice, 74*, 57–65. https://doi.org/10.1016/j.diabres.2006.02.011

Thomson, E. M. (2019). Air pollution, stress, and allostatic load: Linking systemic and central nervous system impacts. *Journal of Alzheimer's Disease, 69*(3), 597–614.

Ulrich-Lai, Y. M., & Herman, J. P. (2009). Neural regulation of endocrine and autonomic stress responses. *Nature Reviews. Neuroscience, 10*(6), 397–409.

Vetter, C., Dashti, H. S., Lane, J. M., Anderson, S. G., Schernhammer, E. S., Rutter, M. K., Saxena, R., & Scheer, F. A. J. L. (2018). Night shift work, genetic risk, and type 2 diabetes in the UK biobank. *Diabetes Care, 41*(4), 762–769.

Vincent, A., Bien, C. G., Irani, S. R., & Waters, P. (2011). Autoantibodies associated with diseases of the CNS: New developments and future challenges. *Lancet Neurology, 10*(8), 759–772.

Watkins, P. J., & Thomas, P. K. (1998 Nov 1). Diabetes mellitus and the nervous system. *Journal of Neurology, Neurosurgery & Psychiatry., 65*(5), 620–632.

Watson, G. S., & Craft, S. (2004). Modulation of memory by insulin and glucose: Neuropsychological observations in Alzheimer's disease. *European Journal of Pharmacology, 490*, 97–113.

Weber, B., Schweiger, U., Deuschle, M., & Heuser, I. (2000). Major depression and impaired glucose tolerance. *Experimental and Clinical Endocrinology and Diabetes, 108*, 187–190. https://doi.org/10.1055/s-2000-7742

Whitaker, K. M., Everson-Rose, S. A., Pankow, J. S., Rodriguez, C. J., Lewis, T. T., Kershaw, K. N., Diez Roux, A. V., & Lutsey, P. L. (2017). Experiences of discrimination and incident type 2 diabetes mellitus: The multi-ethnic study of atherosclerosis (MESA). *American Journal of Epidemiology, 186*, 445–455.

Wright, R. J., & Schreier, H. M. C. (2013). Seeking an integrated approach to assessing stress mechanisms related to asthma. *American Journal of Respiratory and Critical Care Medicine, 187*(2), 115–116.

Yoshimoto, T., Doi, M., Fukai, N., Izumiyama, H., Wago, T., Minami, I., et al. (2005). Type 1 diabetes mellitus and drug-resistant epilepsy: Presence of high titer of anti-glutamic acid decarboxylase autoantibodies in serum and cerebrospinal fluid. *Internal Medicine, 44*(11), 1174–1177.

Zänkert, S., Bellingrath, S., Wüst, S., & Kudielka, B. M. (2019). HPA axis responses to psychological challenge linking stress and disease: What do we know on sources of intra- and interindividual variability? *Psychoneuroendocrinology, 105*, 86–97.

Zimmet, P., Alberti, K. G., & Shaw, J. (2001). Global and societal implications of the diabetes epidemic. *Nature, 414*, 782–787.

Diabetic and Nephropathy

Langeswaran Kulanthaivel, Geevaprabhakaran Ganesan, Chandrashekar Kirubhanand, and Gowtham Kumar Subbaraj

Abstract

Recently, several genes that predispose to type 2 diabetes have been discovered. There is ample evidence to indicate a genetic predisposition to the microvascular complication of nephropathy in people with both type 1 and type 2 diabetes, in addition to the well-known and potent effects of environmental variables. In populations all over the world, familial aggregation of phenotypes such as end-stage renal disease, albuminuria, and chronic kidney disease has frequently been recorded. Heritability estimations for albuminuria and glomerular filtration rate also show considerable influences from inherited variables. Recent genome-wide linkage analyses have examined positional candidate genes under numerous chromosomal areas that are more likely to contain genes that increase the risk of developing diabetic nephropathy. The hereditary elements of diabetic kidney disease are reviewed in this book chapter, with a focus on recently identified genes and pathways. It appears likely that inheriting risk alleles at numerous susceptibility loci, in the presence of hyperglycemia, increases the risk for diabetes-associated kidney damage. In contrast to the molecular genetic studies,

L. Kulanthaivel
Molecular Cancer Biology Lab, Department of Biotechnology, Science Campus, Alagappa University, Karaikudi, Tamil Nadu, India

G. Ganesan
Chettinad Hospital & Research Institute, Chettinad Academy of Research and Education (Deemed to be University), Kelambakkam, Chennai, Tamil Nadu, India

C. Kirubhanand
Department of Anatomy, All India Institute of Medical Sciences, Nagpur, Maharashtra, India

G. K. Subbaraj (✉)
Faculty of Allied Health Sciences, Chettinad Hospital and Research Institute, Chettinad Academy of Research and Education (Deemed to be University), Kelambakkam, Chennai, Tamil Nadu, India

R. Noor (ed.), *Advances in Diabetes Research and Management*,
https://doi.org/10.1007/978-981-19-0027-3_5

which have already been fully reviewed elsewhere, this book chapter focuses on the gathered data on hereditary factors from family studies in order to assess the role of genetic vulnerability in diabetic nephropathy.

Keywords

Diabetic nephropathy · Genetic factors · End-stage renal disease · Microalbuminuria · Gene polymorphism

1   Introduction

Diabetes mellitus (DM) is recognised to have a wide range of consequences, including microvascular and macrovascular issues as well as other metabolic issues. These have been chosen as a group to increase their global influence. The frequency of diabetes is rapidly increasing, having significant negative socioeconomic impacts in economically developed and underdeveloped countries (Tandon et al., 2018). Additionally, the rise in DM-related issues, specifically diabetic diabetes-related retinopathy, diabetic neuropathy, diabetic nephropathy, and other related cardiovascular conditions are placing enormous demands on healthcare. According to the International Diabetes Federation (IDF), around 463 million persons worldwide had diabetes as of 2019, and by 2030, the population could reach 578 million worldwide (Saeedi et al., 2019). According to estimates, 30 to 40 percent of people with diabetes mellitus go on to develop the disease, later developing nephropathy into end-stage renal disease (ESRD) (Wei et al., 2018). There are fundamental mechanisms underlying type 1 (T1DM) and type 2 diabetic kidney disease. Despite the overall clinical appearance of both diseases is remarkably similar, and T2DM histology findings. The primary cause of the increase in hyperglycemic levels is acknowledged to be the development of renal lesions that precisely detect DN progression.

Diabetic nephropathy is rarely seen in type 2 diabetes in the first 10 years after diagnosis; however, roughly about 3 percent of patients with type 2 diabetes, recently diagnosed with diabetes, will be already have overt nephropathy. The variance in this part reveals the role of comorbidities and the link between age-related effects of hypertension as well as coexisting atherosclerosis. Although the major reason for DN is hemodynamic and metabolic factors, it has been indicated that DN is an inflammatory process including immune cells (Duran-Salgado & Rubio-Guerra, 2014). The key pathology variations of diabetic nephropathy are factors such as hypertension, proteinuria, and aggregation of advanced glycation end products (AGEs) as well growth factors along with a perpetual worsening in the function of the kidney. The etiology of DN is multifactorial, accompanied by genetic and ecological factors (Magee et al., 2017). Albumin/creatinine ratios of more than 300 mg/g creatinine or abnormally high levels of albuminuria, which are found in at least two out of three samples and show an obvious result, as well as the occurrence of diabetic retinopathy condition and the non-appearance of other renal diseases, are characteristics of diabetic nephropathy (Forbes & Cooper, 2013). According to studies by Parving et al. (2011), Alter et al. (2012), and Bonventre (2012), it is also a clinical disease characterised by extreme extracellular matrix protein deposition, reduced glomerular filtration rate (GFR), lower creatinine clearance, reduced excretion of albumin, tubulointerstitial fibrosis, increased thickness of glomerular basement membrane, and glomerular hypertrophy. Through repeated measurement and rising values of albuminuria within the usual and unusual ranges are linked with an increased threat in the progression of cardiovascular and renal disease, the normal level of albuminuria is defined as being 30 mg per gram or 30 mg in 24 h, whereas the unusual range is beyond 30 mg per gram.

The most frequent cause of ESRD progression is diabetic nephropathy. Although T2DM is a preventable and treatable foundation for DN to ESRD, it is known that T2DM-related ESRD cases account for more than 50% of all cases globally (American Diabetes Association, 2017). High morbidity and mortality in cardiovascular disease are also associated with DN, most often in the initial stages of the disease. Although the pathophysiology of DN is very well recognised, several causes of the disease still need to be found. A number of previously published studies have shown that some non-modifiable risk variables, such as family history, higher GFR following the diagnosis of type 2 diabetes mellitus, and ethnicity, are associated with DN. Obesity, hypertension, smoking, excessive blood sugar, and dyslipidemia are the modifiable causal factors for the development of DN (Shaza et al., 2005) (Fig. 1).

Family clustering and the fact that not all diabetes patients proceed to obvious DN despite long-standing poor glycaemic management reveal the genetic component of DN vulnerability. The family clustering studies provide more evidence for the major role that inherited variables play in ESRD and diabetic nephropathy. A strategy to distinguish between the mechanisms of DN initiation and advancement is the investigation of predisposed genetic loci or genes. Due to this, multiple determinations have been undertaken to recognise the genetic relationships between DN and various traits using an assortment of techniques, together with genome-wide association studies (GWAS), and candidate gene linkage analysis (Liu et al., 2015).

2 Epidemiology

Diabetes renal disease is well documented to have a substantial part in the progression of ESRD worldwide, particularly in nations like the USA, Japan, and Europe. Numerous population-based studies have found that Blacks and Asians with diabetes are more likely than White people to have the end-stage renal disease (ESRD) (Karter et al., 2002). When compared to White people, Hispanic people with and without diabetes have a higher incidence of renal illness (Young et al., 2003). Despite having low body mass indices, diabetic patients in India frequently have high levels of inflammatory markers, low levels of adiponectin, and insulin resistance (Pugh et al., 1988). While overt diabetes is very common among older Indian patients in other places as well, intolerance is well-established within the Indian people. However, the dangerously high prevalence of prediabetes and obesity in

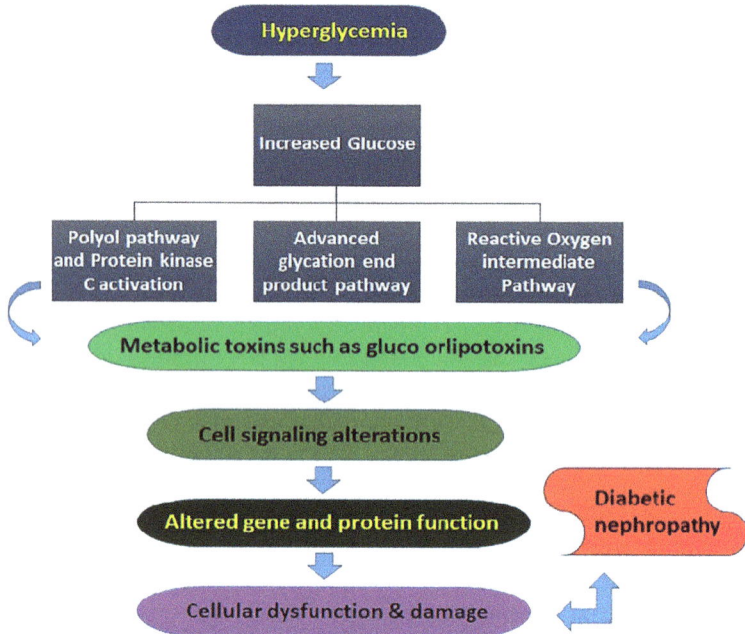

Fig. 1 Pathophysiology of diabetic nephropathy

young individuals raises serious concerns in the current environment (Mohan et al., 2003, 2007; Ravikumar et al., 2011).

Along with the rise in diabetes, there was also a marked rise in the frequency of DN, which has appeared as the leading cause of ESRD. Currently, chronic renal disease in elderly persons accounts for no less than 46% of cases of diabetic nephropathy. Overt nephropathy and microalbuminuria were found to be prevalent in urban residents with diabetes, with respective incidences of 2.1% and 26.9%, according to the Urban-Rural Epidemiology Study (Goyal et al., 2010). It is quite concerning that Southeast Asians have a considerably higher risk of developing impaired glucose tolerance and low fasting glucose levels than the local inhabitants of European ancestry (Unnikrishnan et al., 2007). Additionally, it was noted in the UK and Canada that people of Asian descent had a higher prevalence of chronic kidney-related issues, notably DN, which was reportedly caused by differing lifestyles and genetics (Gray et al., 2010).

When present in a diabetic patient, proteinuria of more than 300 mg/day was a clear indicator of diabetic nephropathy (Barbour et al., 2010). Studies have shown that DN patients exhibit a variety of clinical manifestations as ESRD progresses, despite the fact that the hallmark of DN is a steady progression of micro- to macroalbuminuria, and hyperfiltration in the initial stage leads to the impairment of kidney function. There are several phases involved in the progression of DN in diabetic patients, which leads to end-stage renal disease.

3   Stage I: Early Hypertrophy-Hyperfiltration

Past investigations have mostly revealed that several renal alterations are involved from the clinical onset of diabetes mellitus. As well as the early stage of hypertrophy and hyperfiltration, these modifications mostly depend on the function, biochemistry, and structure providing the definition of a new entity (American Diabetes Association, 2004). The majority of individuals with newly diagnosed diabetes experience this stage, which is characterised by an upsurge in glomerular filtration (GF) rate and a rise in capillary permeability that causes microalbuminuria (MAU) and, ultimately, renal hypertrophy. Glomerular filtration and microalbuminuria may worsen as a result of insulin therapy (Satchell & Tooke, 2008).

4   Stage II: Latent Diabetic Nephropathy

This is a stage that advances slowly and manifests structural lesions in biopsy after many years without any obvious clinical indications of renal disease. While an obvious glomerular impairment will be present, the glomerular filtration rate either stays high or rebounds to normal. Some diabetics will remain in stage II for the rest of their lives. In this stage as well, improved glucose management can result in a reduction in glomerular filtration rate (MacIsaac et al., 2017).

5   Stage III: Incipient Diabetic Nephropathy

The glomerular basement membrane (GBM) dilatation, increased albuminuria of between 30–300 mg/day (20–200 g/min), and a practically unchanged glomerular filtration rate are the hallmarks of this stage. The progression of albuminuria in the setting of declining renal function is significantly influenced by blood pressure. Another symptom is microalbuminuria; thus screening for it once a year in all diabetic patients can help identify the potential renal disease and cardiovascular risk (Satchell & Tooke, 2008; Clarke et al., 2003).

6   Stage IV: Overt Nephropathy

Irreversible proteinuria, ongoing hypertension, and GFR under 60 ml/min/1.73 m^2 are the hallmarks of this stage. It has been determined that between 20 and 40 percent of T2DM patients who have microalbuminuria are more likely to develop overt nephropathy (Clarke et al., 2003). When albumin levels in the urine exceed 300 mg per day or 200 g/min after 10 to 15 years, hypertension is also known to worsen in almost all patients. On the other hand, when glomerular damage worsens, the kidneys' ability to filter out albumin is increasingly diminished, causing a rise in urine albumin excretion of roughly 10–20 percent annually (Molitch et al., 2004). This stage typically leads to renal insufficiency and nephrotic syndrome with concurrent advancement of additional diabetic-related complications like diabetic nephropathy, diabetic neuropathy, retinopathy, or diabetic foot as well as increased hypertension if blood sugar and blood pressure are not strictly controlled (Drummond & Mauer, 2002).

7   Stage V: End-Stage Renal Disease

The glomerular filtration rate (2–20 ml/min/year), which is known to be extremely varied in individuals, steadily decreases with the occurrence of overt nephropathy, necessitating kidney replacement therapy (peritoneal dialysis, hemodialysis, and kidney transplantation). The variability of diseases, such as glomerular filtration barrier malfunction and loss in kidney function among individuals at this stage, points to a variety of disease pathways associated with a variety of hereditary determinants (Kramer et al., 2003; Ellis et al., 2012). The research of relative risk alleles and low-frequency alleles of the diabetic nephropathy phenotype provided experimental study support for this interpretation (Placha et al., 2005).

Variations in structural and functional characteristics identify diabetic nephropathy. Glomerular hyperfiltration, glomerular basement membrane dilation, mesangial cell growth, and nodular glomerulosclerosis are typical symptoms in the early stages of DN. Diabetes induces significant histopathological changes in the kidney structure, according to Kimmelstiel-Wilson (Chan et al., 2014). It is well established that DN's pathophysiological mechanisms are complex. In the early stages of DN, tubular hypertrophy is seen to exist, and this gradually gives rise to interstitial fibrosis, tubular atrophy, and arteriolar hyalinosis (Kimmelstiel & Wilson, 1936). Hyperglycemia is the initiating cause that leads to structural and functional alterations, which progress to mesangial matrix buildup, overt proteinuria, glomerular hyperfiltration, tubular epithelial hypertrophy, and ultimately end-stage renal disease (Weil et al., 2012).

The development and progression of DN, which affect a variety of cell types including podocytes, inflammatory cells, endothelial, mesangial kidney cells, as well as renal tubular and accumulation duct systems, are known to be significantly influenced by high glucose levels. These procedures involve a number of different paths. Transforming growth factor (TGF) cellular expression and strong reactive oxygen species (ROS) generation in the mitochondria are important effects of hyperglycemia (Vallon & Komers, 2011). In addition to increasing the expression of growth factors, cytokines, extracellular matrix proteins (ECM), and leucocyte adhesion molecules in the kidneys, reactive oxygen species also stimulate the signalling and transcription factors (Fig. 2).

TGF stimulates the production of the ECM, which results in cellular hypertrophy and better collagen synthesis (Weil et al., 2012). The diffuse mesangial enlargement is a recognisable sign of diabetic nephropathy (Forbes & Cooper, 2013). Wilson nodules, or areas of strong mesangial deposition, are visible in 40 to 50 percent of patients who later develop proteinuria (Chan et al., 2014). The mesangial matrix expansion also has a relationship to proteinuria and declining kidney function. The

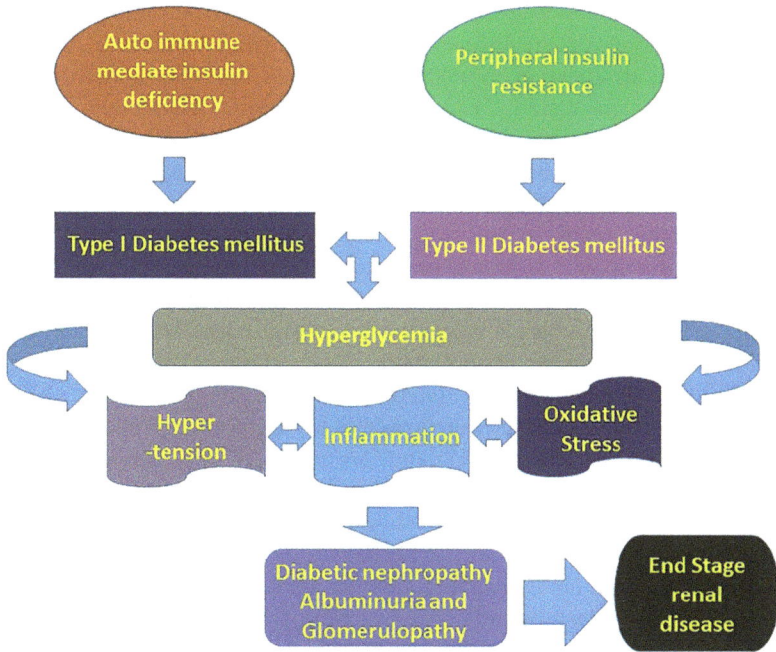

Fig. 2 Schematic representation of the consequences of diabetes mellitus

accumulation of matrix in the mesangial area, which lowers the capillary surface area accessible for the filtration process, causes advanced loss of renal function (Bilous, 1997). The glomerular basement membrane dilatation caused by a change in the structure and composition of the membrane is another distinct characteristic of diabetic glomerulopathy; however, it is not strictly related because it occurs regardless of nephropathy (Fioretto & Mauer, 2007). Although it is more permeable to proteins, the thick GBM contributes as an efficient barrier in the filtering of proteins (Arif & Nihalani, 2013). The lack of charge selectivity by the GBM, which permits the passage of positively charged proteins like albumin, is shown by a decrease in negatively charged proteoglycans. The excretion of proteinuria helps to explain this effect in part. The glomerular visceral epithelial cell is the primary barrier separating plasma proteins from the vasculature (slit diaphragm of podocytes). The podocytes' variants are adaptable. When microalbuminuria first develops, there is a noticeable drop in the number of podocytes as well as a broadening of the pedicels, but these changes are unrelated to the degree of proteinuria. The majority of the podocytes will be lost as DN progresses, and the remaining podocytes should compensate for this loss (Lin & Susztak, 2016).

The fall in proteinuria and GFR is a result of all the aforementioned alterations in the glomerular filtration barrier, which is made up of glomerular endothelium, podocytes, and GBM. The tubules that have interstitial fibrosis and nephron atrophy are observed to be negatively impacted by proteinuria (Williams, 2005). As a result, tubular transport will be impaired by ongoing hyperglycemia, resulting in hemodynamic imbalance and a decline in renal function. Hyaline arteriosclerosis, which also causes glomerular dysfunction, is another early sign of DN. This syndrome is made worse by hyperlipidemia and hypertension, which constrict the arteriole lumen and increase renal vascular hypertension (Nishi et al., 2000).

In the initial phases of diabetic nephropathy, the glomerulus experiences hemodynamic alterations, including hyperfiltration and hyperperfusion damage. This stage is thus multifaceted. A rise in glomerular capillary pressure, which in turn enhances the transcapillary hydraulic pressure gradient and intensifies the glomerular plasma flow, results from a reduction in the arteriolar resistance (Hostetter, 2003). There is a vast range of probable processes and intermediates. There are sufficiently clear indications that diabetic hyperfiltration is caused by an imbalance among the vasoactive factors that regulate the afferent and efferent arteriolar tonus, in addition to the fact that metabolic variables are linked to hyperglycemia (Vallon & Komers, 2011). Hyperperfusion and hyperfiltration are produced as a result of variables including growth hormone, prostanoids, insulin, atrial natriuretic factor, nitric oxide, glucagon, and angiotensin-II (ANG-II) (Chawla et al., 2010). On the other hand, glomerular basement membrane thickening increased mesangial cell matrix synthesis, and elevated intraglomerular pressure are recognised to be the causes of glomerulosclerosis. Hyperglycemia causes effects on the renal cells that are hemodynamic, inflammatory, and profibrogenic, which enhance the production of angiotensin-II. Cytokines like TGF and vascular endothelial growth factors (VEGF) are the main mediators of hyperfiltration (Wolf et al., 2003).

It is well-recognised that beta plays a significant role in diabetic vascular instability. Thus, this process increases arginine re-synthesise and endothelial NOS (nitric oxide synthase) mRNA expression, both of which lead to an increase in nitric oxide production (Sharma et al., 2003). The enhanced glomerular filtration will be a result of the elevated nitric oxide levels in the kidneys. According to recent research, endothelial nitric oxide (NOS3, eNOS) is a causal enzyme that contributes to the excess nitric oxide generation in diabetic kidneys. Additionally, this leads to an increase in the amount of ROS in the diabetic kidneys (Veelken et al., 2000; Sugimoto et al., 1998). Other research has also indicated that neuronal NOS plays a significant role in hyperfiltration, modulating the negative tubuloglomerular feedback loop (Ito et al., 2001).

Additional variables that are connected to the pathophysiology of diabetic nephropathy are also regulated by nitric oxide. Growth factors and cytokines are released via the autocrine and paracrine pathways when glomerular hemodynamics are altered as a result of mechanical strain and shear stress (Chawla et al., 2010). Further structural changes to DN are caused by this hemodynamic stress and are initiated locally by growth factors and cytokines. Because of the macula densa mechanism and tubular hypertrophy, the increase in sodium chloride reabsorption in the loops of Henle increases the frequency of the glomerular filtration rate. As a result, it facilitates sodium chloride's accelerated reabsorption, which may be crucial in this phase and once again links structural alterations to hemodynamic variation in diabetic nephropathy (Thomson et al., 2004).

Hyperglycemia, elevated glucagon levels, and growth factors that characterise diabetes mellitus are considered metabolic variables. It is well-established that changes in insulin levels have a clear effect on microvascular problems (Vithian & Hurel, 2010). Even after many histological defects in their kidneys, the majority of diabetes people will not advance to renal failure. Thus, it is clear from this that hyperglycemia is a necessary but insufficient component to trigger renal damage that ultimately results in kidney failure. The glucose transport activity is a known important regulator in the extracellular growth of mesangial cells. The key regulator that facilitates the entry of glucose into kidney cells is the glucose transporter 1 (GLUT1). Additional metabolic pathways are stimulated by glucose, which results in the growth and formation of mesangial cell matrix (Mishra et al., 2005). Even though the glucose levels are normal, changes in the renal cells will result from GLUT1 overexpression (Heilig et al., 1995). The mesangial cells' presence of GLUT1, a brain-type glucose transporter, and GLUT4, an insulin-sensitive extracellular glucose transporter, makes it simple for a high level of glucose to enter the cell in an insulin-independent fashion (Haneda et al., 2003; Heilig et al., 1995).

Numerous experimental investigations show that diabetes significantly increases the rate of matrix and glomerular basement membrane components, which improves collagen synthesis. As a result of higher glycaemic levels, GLUT1 is typically overexpressed; however, the pathophysiological role of glucose transporter 1 in DN is not yet known (Heilig et al., 1995). It is also known that glucose interacts with structural proteins, causing a process known as non-enzymatic glycosylation that leads to the formation of advanced glycosylation end products (Nishikawa et al., 2000). The end products of this reaction distress the glomerular basement membrane components, leading to the formation of cross-links that are prominent to diminished sensitivity, which results in the buildup and deposit of proteases (Brownlee et al., 1986). AGEs alter the macrophage removal system by impeding the mechanisms of mesangial clearance, which causes an expansion of the mesangial cell and eventually results in glomerular blockage (Forbes et al., 2001). The final result of a series of as-yet-unknown metabolic processes is advanced glycation end products (AGEs). Although there is a clear difference between regulated glycaemic levels as expressed by HbA1c (glycated haemoglobin), AGE formation is high in diabetes induced by persistent hyperglycemia (Cohen et al., 2003). There is also known to be a strong association between the decrease in kidney function and the accumulation of AGE in kidney disorders because filtration of the kidney plays a critical role in the clearance of circulating AGEs (Nitta et al., 2013).

Due to hemodynamic changes brought on by non-enzymatic glycosylation, protein kinase C (PKC) is also activated, accelerating the polyol pathway and stimulating the production of TGF, VEGF, interleukin-6, interleukin-1, interleukin-18, and tumour necrosis factor-alpha (TNF). Oxidative stress and the formation of ROS both destroy DNA and proteins and act as signalling amplifiers for stress pathways like mitogen-activated protein kinase (MAPK), PKC, and nuclear factor kappa B (NF-KB) (Ha & Lee, 2000). Aldose reductase converts extra glucose into sorbitol, activating the polyol pathways. Sorbitol dehydrogenase then converts fructose, causing oxidative stress due to an increase in the nicotinamide adenine dinucleotide phosphate (NADPH) ratio (Srivastava et al., 2005). Protein structural and functional variation, the expression of inflammatory cytokines and growth factors, and oxidative stress are all consequences of AGE formation, which is caused by the non-enzymatic binding of lipids, glucose, and nucleic acids to proteins as well as nucleic acids (Sheetz & King, 2002). All of these lead to increased proteinuria, glomerulosclerosis, and undoubtedly tubulointerstitial fibrosis by generating the glomerular basement membrane's high albumin permeability and extracellular matrix buildup.

The generation of ROS by oxidative stress is believed to be a causative factor in the development of diabetes complications (Brownlee, 2001). The cytotoxic nature of ROS is known to promote fibrosis and inflammation. In addition, molecular disruption is caused by the oxidation of macromolecules such as carbohydrates, proteins, lipids, and DNA (Nishikawa et al., 2000). According to studies, oxidative stress is significantly influenced by insulin resistance because free glucose and low NADPH ratios activate the polyol and aldose reductase pathways. Through the de novo synthesis of diacylglycerol (DAG), the elevated intracellular glucose level activates the PKC (Geraldes & King, 2010). The transition of metal-catalysed Fenton reactions, peroxidases, deficits in the mitochondrial respiratory chain, xanthine oxidase activity, NADPH oxidase coupled with NO synthase, and autooxidation of glucose is known to contribute to the formation of ROS in diabetes (Fakhruddin et al., 2017). The production of ROS in diabetes is significantly influenced by the activity and expression of NADPH oxidase (Wautier et al., 2001). The membranous p22phox and gp91phos, also known as nox-1 and nox-4 homologues, the cytosolic p47phox and p67phox subunits, and the regulatory G protein rac-1 make up the five subunits that make up vascular NADPH (Harrison et al., 2003). Due to the stimulation, the cytosolic and membranous subunits will come together to create a functional oxidase that will use NADPH as an electron donor, resulting in a finely defined oxidative burst. In experimental diabetes, there has been a rise in the cytosolic NADPH oxidase subunit, p47phox, in the kidney and vascular system (Christ et al., 2002).

It is well established that a rise in blood sugar levels can boost the mitochondria's production of ROS. The increased glucose absorption also stimulates the citric acid cycle and glycolysis, which results in an excess synthesis of electron donors such as NADH and flavin adenine dinucleotide (FADH2) (Brownlee, 2001). The inner membrane of the mitochondria contains a localised electron transport chain, and when the number of electron donors increases, this pumping of protons across the inner membrane raises the membrane potential. As a result, the transit of electrons at complex III is restricted by lengthening the half-life of the coenzyme Q's free radical intermediates, which further reduces O2 to superoxide. Furthermore, although type 2 diabetic people can also develop renal lesions without diabetes, the distinguishing features of DN are recognised to be brought on by diabetes. All of these judgments reach the conclusion that detecting the accumulation of AGEs significantly indicates hyperglycemia and also advances knowledge of other metabolic variables. The excessive production of superoxides and cellular pseudohypoxia also contributes to inflammatory processes and oxidative stress. In diabetic individuals, the accumulation of AGE is well recognised as an early indicator of renal problems (Amore et al., 2004). Numerous familial studies, as well as epidemiological studies, have shown that the main factors that show an important role in the progression of nephropathy as well hyperglycemia, obesity, insulin resistance, dyslipidemia, and genetic susceptibility along with other environmental factors and dietary habits (Adler et al., 2003). Age, race, ethnicity, and family history are non-modifiable characteristics. Modifiable factors typically have to do with lifestyle choices and environment-related elements that can be changed.

Individual differences in the risk of developing DN in diabetes patients bring the genetic predisposition relationship to a close. Long-term effects of DN include increased blood pressure, declining renal function, and ESRD, according to knowledge. A combination of metabolic and hemodynamic disturbances increased expression of inflammatory cytokines, growth hormones, and genetic susceptibility works together to activate these causes (Jacobsen, 2005). It is yet unknown how nephropathy is genetically transmitted. Its multifactorial character makes polygenic transmission more likely than other modes of transmission. Several familial investigations have produced ample data which confirms the abovementioned information (Adler et al., 2003).

It is well-recognised that ethnicity can also affect the likelihood of developing diabetic nephropathy. Numerous studies have shown that Asians with diabetes are more likely than Caucasians to develop ESRD, which raises their likelihood of receiving kidney replacement therapy by 5.8 times (Burden et al., 1992). According to a cross-sectional study done in the UK, South Asians are further probable than Europeans to have microalbuminuria. Another study that supported this finding showed that South Asians had a proteinuria prevalence rate of 21%, while White people in the UK only had a prevalence rate of 14% (Dreyer et al., 2009). In a follow-up study, it was shown that South Asians had a 1.45 times larger drop in eGFR and were four times more likely to develop albuminuria than Europeans (Shaw et al., 2006). Hence, all these findings emphasise the notion that an increased understanding of the risk factors which lead to the pathogenesis of DN specifically in the South Asian population is crucial. Learning more about the pathogenesis of DN in South Asians will be essential for preventing the disease's start and delaying its progression.

One of the main complications of diabetes is diabetic nephropathy, which is characterised by elevated levels of albuminuria that accelerate the loss of renal function. Because elevated levels of albuminuria do not manifest until there has been severe kidney damage, clinical identification of nephropathy is difficult (Parving et al., 2011). Heavy albuminuria rate is currently the gold standard for DN detection. Typically, after 15 to 20 years of uncontrolled diabetes, DN incidence is noted (Nazar, 2014). Another fact that not all diabetes patients with microalbuminuria proceed to DN is still not fully understood. This emphasises the necessity of developing genetic indicators for the prognosis of DN at the onset of the disease. The fact that not all patients with diabetes progress to diabetic nephropathy (DN) can shed light on the role of hereditary factors in the development of diabetes and its progression to DN. According to this assertion, inherited genetic characteristics in the family may have an impact on a person who has both diabetes and diabetic nephropathy. However, each person experiences renal functional impairment at the same rate. When compared to patients with a normal amount of albumin excretion, studies have shown that those with nephropathy had a higher incidence of diabetes risk factors (Brennan et al., 2013). Comparing first-degree relatives of diabetes patients with and without DN, familial investigations have shown that these features are more common in DN patients (Fogarty et al., 2000). As a result, the development of diabetic nephropathy may have been influenced by genetic factors, which further suggest a vulnerability to cardiovascular illnesses. The impact of genetic factors, in addition to the link to single-gene changes, can likely be complicated and multivariate. Consequently, it is also known that gene-gene interactions and gene-environment interactions show a substantial role in the development of disease (Liu et al., 2015).

Numerous genetic investigations have already been conducted to categorise potential candidate genes in large sample size studies, which also adds to the study of diabetic nephropathy pathogenesis. The identification of DN-related genes will also aid in identifying those people who are more likely to experience this renal problem. Additionally, it increases understanding of the illness progression mechanisms. Due to their small sample sizes and weak statistical power, the majority of earlier studies about the genes of DN are often constrained. Recent observations on diversity in genotype expression brought on by intraglomerular pressure and high glucose levels make it more difficult to make sense of the results. The prediction of diabetic people who are likely to develop a high risk of DN will benefit from the discovery of vulnerable genes and causal variations. It will also help to identify potential prognostic indicators for early nephropathy detection and inhibition, as well as for delaying the progression to the subsequent advanced stages. The major goal is to raise the patients with DN's survival rate. Single-nucleotide polymorphisms (SNPs), extensive, structural chromosomal disruptions, and other variations of genetic deviation can all be found in the human genome. The SNPs linked to gene expression may also be involved in a number of disorders brought on by disruptions in the homeostasis of the genes. This could include transcriptional mechanisms such as transcription factor binding, atypical gene splicing, and messenger RNA degradation (Cheung & Spielman, 2009). The human genome has been sequenced, and it shows that 99.9% of each person's DNA is the same. Only 0.1 percent of people differ from one another. This distinction is what determines how each individual's phenotype is unique. Single base deviations in the human genome are known as single-nucleotide polymorphisms, which are small genetic changes. Despite the relatively low danger that the normal genetic variants display, bigger populations indicate increased risk due to their larger frequencies. A genetic locus is often considered polymorphic if the frequency of the sporadic allele exceeds 1% and results in a heterozygotic frequency of at least 2%. It is generally recognised that a combination of hereditary genetic polymorphisms and numerous environmental circumstances can predispose someone to specific diseases or change how a disease develops. Additionally, recent research has demonstrated that genetic polymorphisms might influence a person's susceptibility to conditions like diabetes mellitus, osteoporosis, hypertension, and Alzheimer's disease. Comparatively to the detection of single gene mutations or other significant chromosomal aberrations, the discovery of such vulnerable genes is difficult because a single polymorphism seems to display only a minimal effect on the phenotype. Since the occurrence of many sensitive alleles at a multiplex locus may not always result in overt clinical symptoms, it is hypothesised that the change in environmental factors will be the cause (Holtzman & Marteau, 2000). With the completion of the Human Genome Project (HGP), opportunities for identifying the genes linked to a variety of multi-faceted diseases have also quickly advanced. This has improved our understanding of the haplotype structure of the genome along with recent developments in epide-miological studies relating to DN. Genome-wide association studies, genetic linkage studies, association analysis studies/case-control studies, and candidate gene approach studies are a basic list of genetic techniques (Smith et al., 2004). The application of GWAS in the realm of medical genetics is well established. Investigating the relationship between particular genetic variations and particular diseases is done using this method. The described method further analyses the genomes of numerous people to find genetic indicators that can be used for disease prognosis. Once these genetic markers have been discovered, they may be utilised to research the role of genes in the development of a disease and the creation of superior preventive and therapeutic approaches. One benefit of GWAS is that it can find the genes regardless of whether their function has previously been established. How-ever, problems including low statistical power, high costs, ambiguous genetic effects, and unreliable phenotypic definitions are some of the causes of partial success (Cooke et al., 2008). However, SNP-based genome-wide association studies are utilised in humans to significantly identify the regions that are important for the onset of some diseases, including diabetes, Crohn's disease, and even other heredi-tary problems (Barrett et al., 2008; Sladek et al., 2007). Additionally, this approach has been utilised to identify specific genetic loci, namely, for diabetic kidney disease (DKD), eGFR, and albuminuria in people with diabetes mellitus. This method pinpoints the chromosomal location of genes that are naturally associated with the disease. It simply works by keeping an eye on the genes that remain close to one another on the same chromosome and are recognised as being linked throughout the meiotic stage (Altshuler et al., 2008). The description of a specific phenotype associated with each gene is essential. The two main categories of linkage analysis investigations include parametric and nonparametric test methods. In the conven-tional linkage analysis, the method of inheritance, phenocopies, as well as gene frequencies must be identified (Lin et al., 2008). For linkage analysis research, choosing suitable genetic markers is also essential. Microsatellite markers were used in the initial linkage studies; however, SNPs are now more frequently used due to their equivalent importance and low cost (Table 1).

The comprehensive information from the entire genome is known to be delivered through genome-wide linkage analysis. The downside of this strategy, though, is that it is less sensitive than association studies. However, it also has the capacity to identify genomic regions that include a variety of genes and gene variations. Because it is typically undertaken on a large number of instructional families, it is more

Table 1 Gene polymorphisms associated with diabetes nephropathy

Gene	SNP	Study type	Ethnicity	Sample size (cases/control)	Association	References
TNF-α	308 G/A	Meta	Asian and Caucasian	1724/1714	Associated	Tiongco et al. (2020)
COL4A1	(rs614282 and rs679062)	CC CC	Caucasian	115	Associated	Ewens et al. (2005)
LAM	(rs3734287 and rs20557)	CC	Caucasian	115	Associated	Ewens et al. (2005)
ACE I/D	rs1799752	CC CC	Caucasian Asian	1365 432	Associated	Boright et al. (2005)
eNOS	rs1799983	CC	Asian	280	Associated	Varghese and Kumar (2022)
MnSOD gene	(V16A, rs4880)	CC	Asian	478/261	Associated	Nomiyama et al. (2003)
ADIPOQ	rs2241766	Cohort	Asian	708 556	Associated	Chung et al. (2014)
APOE	E2, E3, and E4	Meta	Asian, Caucasian	1257/1555	Associated	Choe et al. (2013)
ALR2	Z + 2	Meta	Asian, Caucasian	1481/1374	Associated	Xu et al. (2008)
GLUT1	rs841853	CC	Caucasian	227	Associated	Ramadan et al. (2016)
PPARG2		CC	Asian	432	Associated	Satirapoj et al. (2019)
TCF7L2	rs7903146 and rs111962	CC	Caucasian Asian	1355 898	Associated	Buraczynska et al. (2014), Fu et al. (2012)
Glucose metabolism-related genes GCKR	rs1260326	CC	Caucasian	2097	Associated	Deshmukh et al. (2013)
VEGF-A	rs833061	Meta	Caucasian	543	Associated	Mooyaart et al. (2011)
FRMD3	rs10868025 rs1888747	Meta	Caucasian	1705 1705	Associated	Pezzolesi et al. (2009)
SHROOM3	rs17319721	Meta	Caucasian	3028	Associated	Deshmukh et al. (2013)

TGF-β1	rs1800470 and 915 G/C rs1800471	CC CC	Caucasian Asian	1818 4073 280	Associated	Valladares-Salgado et al. (2010), Varghese and Kumar (2022)
AGTR1	rs5186	Meta	Asian, Caucasian	ND	Associated	Möllsten et al. (2011)
SLC12A3	rs11643718	CC CC CC	Asian Asian Asian	1417 372 358	Associated	Seman et al. (2014), Zhang et al. (2018), Kim et al. (2006)
RAB38/CTSC	rs649529	GWAS	Caucasian	7787	Associated	Teumer et al. (2016)

ND no data, *GWAS* genome-wide association study, *CC* case-control study, *cohort* prospective cohort, *SNP* single-nucleotide polymorphism difficult, expensive, and time-consuming. Pima Indians' entire genome linkage investigation resulted in the identification of DN-susceptible loci on chromosomes 3, 7, and 20 (Imperatore et al., 1998). This method entails the identification of genetic variants in particular candidate genes that may alter or alter the structure and function of proteins, causing the advancement of specific diseases. In case-control studies, participant selection is the key element. Case-control association studies are used to investigate candidate genetic polymorphisms by comparing the frequency of polymorphisms in certain genes between patients and control participants. To avoid unfavourable outcomes, it is necessary to properly recruit a sizable number of carefully characterised cases and control individuals. It is necessary to enroll a high-enough percentage of qualified patients to guarantee that there is no chance of developing a significant association.

To achieve this, uniform diagnostic tests for the illness should be investigated, especially if subgroup analysis is being done with varied disease stages and severity. Given that occupational characteristics and changes in lifestyle have a significant impact on susceptibility to various illnesses, it is imperative to develop a relevant questionnaire that accounts for all of these variables and to ensure that all participants complete it (Daly & Day, 2001). The family-based association analysis is another common technique used in genetic association investigations. The transmission disequilibrium test (TDT) is a different approach to investigating the frequency of diseased alleles that are transmitted or not to the affected offspring from the diseased heterozygous parents (Feng et al., 2007). There is no need for a control group because the genotypes of the patient's parents are already known. It is thought that the illness allele is connected with the trait if allele transmission occurs in more than half of the cases. It is well-known that looking into the relationships between alleles in these groups can provide light on the population association and assist localise the vulnerable locus within a 2 Mb-wide area. These chromosomal regions may be incredibly small, yet they will be big enough to allow for physical mapping. Further discovery of causative polymorphisms and thorough genetic alterations will also be possible. In case-control studies, the candidate gene method is essential for identifying the genes that are predisposed to diabetic nephropathy. The candidate gene study approach can be well understood in comparison to genome-wide studies because DN-associated genes will be selectively chosen for the study. Around the world, candidate gene studies for DN are being carried out in an effort to find genetic indicators that may help predict the likelihood that a diabetic patient would eventually develop diabetic nephropathy. Single-nucleotide polymorphisms in one or more genes that suggest a potential role in the pathology of a certain disease are studied while examining the candidate genes. Therefore, even when a gene's influence on a disease is minimal, this approach appears to be helpful. It is thought to be an appropriate investigation for complicated genetic transmission. Additionally, this can be particularly helpful in situations when the genetic influence is very low but the prevalence of disease-associated alleles is high (Patnala et al., 2013). Although candidate gene studies are known to have strong statistical power, no novel genes will be found. Linkage analyses, case-control association studies, and genome-wide association studies, among others, have demonstrated the role of genetic vulnerability in the development of DN in people with type 2 diabetes mellitus.

However, as is the case with the majority of complicated diseases, it has been demonstrated that it is difficult to pinpoint the precise genetic variants that cause nephropathy. Therefore, a deeper comprehension of the genetic heritability underpinning DN may even aid in the development of novel biomarkers to identify illness vulnerability and support novel therapy modalities. The combination of multiple information sources tends to be more beneficial to improve the understanding of the pathophysiology of illnesses with diverse phenotypes, such as diabetic nephropathy (Brennan et al., 2013). When compared to external environmental influences, genes have a significant impact on the development of nephritis, which is already a well-known and established fact. Single-nucleotide polymorphism analysis is used to study the impact of these polymorphisms on parameters such as illness susceptibility, healthcare to change treatment trajectories, and drug resistance.

In light of this, a wide variety of genes have been examined to determine their relationship to DN and the contribution of multiple SNPs found in the DN-associated genes. The fact that different ethnic groups exhibit different susceptibility patterns is another obsession. Positive associations of DN with several genes that have already undergone investigation are evident. Profibrotic growth factors genes like TGF-1, connective tissue growth factor (CTGF), VEGF, intercellular adhesion molecule 1 (ICAM1), the receptor for advanced glycation end products (RAGE), monocyte chemoattractant protein-1 (MCP-1) and antagonists, as well as genes linked to dyslipidemia, are a few of the genes. Predisposed maps for diseases will benefit from the Indian Genome Variation Consortium (IGVC) data as well as additional epidemiology and related phenotypic information (Narang et al., 2010).

This will raise awareness for those who are more susceptible to developing DN. Genetic association studies are widely employed and sensitive when contrasted to the scant information on genetic linkage research. Fatty acid-binding protein-2 (FABP2), angiotensin-converting enzyme (ACE), GLUT1, and ectonucleotide pyrophosphatase (ENPP1) are a few other DN-associated genes that have been discovered using the candidate genetic association technique (Tang et al., 2015). It has been established that the genetic variants chosen for the current study play a substantial role in the many signs and symptoms of diabetic nephropathy, including renal hypertrophy, mesangial cell growth, and endothelial dysfunction.

TGF-1 is a multifunctional cytokine that is linked to the pathogenesis of many renal issues, including diabetic nephropathy, by advancing renal hypertrophy and increasing extracellular matrix buildup (Reeves & Andreoli, 2000). The TGF-1 gene, which is at the chromosomal location 19q13.1–q13.3, is known to contain seven exons and six big introns. Currently, information on ten polymorphic loci that are visible between introns and exons as well as in various flanking regions is being acknowledged (Watanabe et al., 2002). The transforming growth factor is known to promote renal hypertrophy and fibrosis, while other cytokines, such as TNF and monocyte chemoattractant protein-1 (MCP-1), interfere with the CCR2 and CCR5 chemokine receptors to further regulate the influx of macrophages into the kidney (Prasad et al., 2007). High glycaemic levels seen in people with type 2 diabetes mellitus are linked to the activation of glucose transport 1, which prompts the renal and mesangial tubular cells to overexpress TGF. Additionally, an increase in intraglomerular pressure, activation of the renin-angiotensin system, reactive oxygen species, and advanced glycation end products cause the renal system to produce TGF (Qian et al., 2008). Numerous studies have looked at significant individual differences and shown that genetic variables may regulate the amounts of TGF-1 in the serum and mRNA (Ahluwalia et al., 2009). The two genetic polymorphisms, 915G > C and 869 T > C, were examined in the German population, and the results revealed a substantial connection between 869 T > C with DN but not with the other SNP, 915 G > C. (Babel et al., 2006). In a study conducted by Valladares-Salgado et al. (2010), it was important to explore the relationship between cholesterol levels and the TGF-1 gene's T869C polymorphism. Results showed a significant correlation between total cholesterol levels and the CC + CT genotype. Although many different variations of this gene are expected to exist and be deposited in the dbSNP database, there is very little information regarding the function and significance of these variants in relation to renal problems. Numerous investigations in various populations revealed an enhanced correlation between the 915G > C polymorphism and a high risk of ESRD and chronic renal failure. However, it is still unclear how exactly this polymorphism affects diabetes patients whose condition worsens and eventually leads to DN. There hasn't been any research on the association between the TGF-1915G > C polymorphism and diabetic nephropathy in the South Indian population. The genetic polymorphism was included in the proposed study since it was found to play a significant role in the pathophysiology of DN and was associated with greater plasma levels in diabetic patients in other groups. Translocation of a variant plasmacytoma one gene is well-known for being involved in frequent translocations between the area of chromosome 2 that codes for it and the region of chromosome 22 (Graham & Adams, 1986). Chromosome 8q24 is where it is situated (Millis et al., 2007). Although it has long been known that the kidney contains significant quantities of PVT1, its function is not yet understood. Through the use of a genome-wide SNP association research among American Indians with diabetes mellitus, the gene for plasmacytoma variant translocation 1 is identified as a candidate gene for end-stage renal disease (ESRD) (Nazar, 2014). PVT1 has been found to have an impact on unchecked cell proliferation, especially mesangial cell expansion, which is a well-known key characteristic of diabetic kidney disorders. PVT1 has been found to affect the formation and progression of DN by a variety of processes, including the aggregation of extracellular matrix proteins in the glomerular mesangium (Alvarez & DiStefano, 2011). It has been demonstrated that PVT1 increases TGFB1 and plasminogen activator inhibitor 1 in human mesangial cells, which are two of the main causes of extracellular matrix buildup in the glomeruli under high glucose circumstances (Alvarez et al., 2013). The PVT1 gene cluster was found to contain the markers that were found to be strongly linked with ESRD. The expression of the PVT1 gene results in a number of transcript variations (Nelson et al., 1988). PVT1 is a 1.9 kb long non-coding RNA that further encodes a number of transcripts but is unable to synthesise proteins. According to Alvarez et al. (2013), it is the first and first non-coding RNA to have demonstrated a relationship with renal disorders. Breast and ovarian cancers are also known to be linked to the overexpression of PVT1 (Guan et al., 2007). A genome-wide single-nucleotide polymorphism association study among Pima Indians demonstrated a connection between the rs2648875 gene polymorphism of the PVT1 gene and ESRD. It is well-known that a rise in glucose and TGF-1 levels causes PVT1 to be increased (Hanson et al., 2007). The genetic variation rs2648875 has been linked to diabetic mellitus, which further contributes to DN.

To learn more about the association between PVT1 and DN and to research the precise role of the gene causing ESRD, additional studies in various groups are needed. It is known that pentraxin 3 belongs to the long pentraxin group and shares structural similarities with short pentraxins like C-reactive protein and serum amy-loid P component (SAP) (CRP). But when it comes to cell formation and stimulus activation, it shows a variety of variances. While IL-6 is known to trigger the formation of CRP and SAP in liver cells, PTX3 is known to be produced by a variety of cell types as a result of activation brought on by specific stimuli. Additionally, it has been discovered that PTX3 is stimulated by inflammatory signals and changed lipoproteins in cells such vascular endothelium and smooth muscle cells. Additionally, epithelial cells, neurons, fibroblasts, glial cells, and adipocytes have all been discovered to contribute to its formation (Bottazzi et al., 2016). According to certain research, acute ischemia-reperfusion renal injuries and chronic kidney disorders both cause higher-than-normal plasma levels of PTX3. PTX3 is a proinflammatory cytokine with a documented propensity for having a pathogenic effect (Yilmaz et al., 2009). A biomarker of immune-inflammatory response and a proven conservative defence mechanism, high plasma levels of PTX3 are likely to protect the kidneys as well (Speeckaert et al., 2013).

As established by prior investigations, DN is an inflammatory disease with changed plasma levels that may also play a role in the disease's progression. Similar findings demonstrated that plasma levels of PTX3 were negatively correlated with eGFR before positively correlating with the levels of proteinuria in DN patients. This led to the further suggestion of PTX3's involvement in the development of DN (Uzun et al., 2016). Additionally, it has been discovered that the SNPs in this gene affect the amounts of circulating PTX3, which is once again connected to several additional illnesses, including hepatocellular carcinoma, primary graft malfunction, and pulmonary tuberculosis (Carmo et al., 2016; Diamond et al., 2012; Olesen et al., 2007), while only one study, conducted in the Chinese population, found a link between DN and genetic variations that increase susceptibility to diabetes (Zhu et al., 2017). According to linkage studies, the chromosome 3q is linked to diabetes and nephropathy in a number of populations. Additionally, the gene that codes for PTX3 is situated in the region of 3q25.32 that is too close to the linkage region (Vionnet et al., 2006). This strengthens the connection between PTX3 and DN that may exist. Further research on the genetic variation of PTX3 will pave the path for the creation of a potential biomarker for the early diagnosis of DN in diabetic patients. Endothe-lial dysfunction is another defining feature of DN. Nitric oxide synthase activity in endothelial cells is decreased as a result of endothelial dysfunction, a risk factor. Additionally, this action is critical to the pathophysiology of DN (Nakagawa et al.,

2007). The nitric oxide synthase 3 gene is located on chromosome 7 in the 7q35–7q36 region. With 26 exons and 25 introns, it has a total length of 21 kilobases (Zanchi et al., 2000).

Endothelial nitric oxide synthase, an enzyme, identifies nitric oxide as a crucial endothelial cell by product that is from L-arginine (eNOS). It has been demonstrated that the NO plays a variety of physiological and regulatory activities as well as taking part in processes like blocking platelet aggregation, neurotransmission, controlling blood pressure, and relaxing smooth muscle (Syed et al., 2011). Inducible NOS (iNOS), endothelial NOS (eNOS), and neuronal NOS (nNOS) are the three isoforms of nitric oxide synthase (Noiri et al., 2002). The likelihood of developing issues connected to diabetes increases when ROS generation increases, particularly superoxide anion (O_2-) (Mclennan et al., 2000). Peroxynitrite ($ONOO-$), which is known as a weak agonist necessary to activate the cyclic guanosine monophosphate, is created when the superoxide anion combines with nitric oxide (cGMP). O_2- effectively deactivates the NO as a result. Nitric oxide activity may be reduced by the NADPH oxidase induction seen in diabetes mellitus. By reducing the expression of matrix metalloproteinases and boosting the expression of tissue inhibitor of metalloproteinases in the kidney, this further accelerates the mRNA expression required for the fibronectin and transforming growth factor-1 (TGF-1) present in the glomerulus (Chiarelli et al., 2009).

It has been established that the eNOS gene variations both reduce nitric oxide production and contribute to endothelial dysfunction. The 786 T > C alteration found in the promoter region of the eNOS gene is one of the most significant single-nucleotide polymorphisms (b). Glu is changed to Asp at codon 298 by the 894 G > T alteration found in exon 7 (c). Intron 4 contains a 27-bp variable tandem repeat (VNTR) polymorphism that has been shown to exhibit an insertion and deletion with two alleles (Asakimori et al., 2002). The 894G > T polymorphism among these has been found to continue to be significantly related with DN. Studies have shown that the eNOS 894G > T polymorphism T allele and TT genotype significantly increase the chance of developing DN. Diabetic individuals from Tunisia and Japan have a higher prevalence of this polymorphism, which has further contributed to renal failure (Santos et al., 2011).

However, other studies have also claimed to have found no appreciable effects in the specific population. This variation can be explained by ethnic and geographic factors that may affect the connection of polymorphisms and hence affect whether DN will proceed or not. Other illnesses such acute myocardial infarction, coronary artery disease, hypertension, and atherosclerosis are associated with an increased frequency of this polymorphism (Colombo et al., 2002; Cilingir et al., 2019). The polymorphism can be proposed as a primary factor for the progression of diabetic nephropathy specifically in individuals with diabetes, based on the results of previous studies, it can be concluded; furthermore, the majority of the Caucasian and Asian populations were associated with the 894G > T polymorphism's significant risk of DN among diabetic patients as opposed to in participants who were not diabetic (Chalasova et al., 2014).

8 Conclusion

DN and ESRD continue to be major issues despite greatest attempts to reduce the disease's effect on such end-organ damage. A multifactorial strategy is still the most rational one in the extremely complicated environment of diabetes, where no one medication can stop the progression of DN. This should include single RAS inhibition for hypertension or albuminuria, as well as appropriate glycaemic management. The general health around the world is seriously threatened by diabetes mellitus (DM). One of the most frequent side effects of diabetes and the main contributor to end-stage renal disease is diabetic nephropathy (DN) (ESRD). The progression of DM patients to ESRD is estimated to be between 30 and 40 percent worldwide, highlighting the influence of hereditary variables on DN. Family clustering further supports the crucial part that inherited variables play in DN and ESRD. In order to find susceptibility genes in various diabetic cohorts, numerous genetic investigations have been conducted. It has only recently been discovered that DN and ESRD have extensive susceptibility genes.

References

Adler, A. I., Stevens, R. J., Manley, S. E., Bilous, R. W., Cull, C. A., Holman, R. R., & UKPDS Group. (2003). Development and progression of nephropathy in type 2 diabetes: The United Kingdom prospective diabetes study (UKPDS 64). *Kidney International, 63*(1), 225–232.

Ahluwalia, T. S., Khullar, M., Ahuja, M., Kohli, H. S., Bhansali, A., Mohan, V., Venkatesan, R., Rai, T. S., Sud, K., & Singal, P. K. (2009). Common variants of inflammatory cytokine genes are associated with risk of nephropathy in type 2 diabetes among Asian Indians. *PLoS One, 4*(4), e5168.

Alter, M. L., Ott, I. M., Von Websky, K., Tsuprykov, O., Sharkovska, Y., Krause-Relle, K., Raila, J., Henze, A., Klein, T., & Hocher, B. (2012). DPP-4 inhibition on top of angiotensin receptor blockade offers a new therapeutic approach for diabetic nephropathy. *Kidney and Blood Pressure Research, 36*(1), 119–130.

Altshuler, D., Daly, M. J., & Lander, E. S. (2008). Genetic mapping in human disease. *Science, 322*(5903), 881–888.

Alvarez, M. L., & DiStefano, J. K. (2011). Functional characterization of the plasmacytoma variant translocation 1 gene (PVT1) in diabetic nephropathy. *PLoS One, 6*(4), e18671.

Alvarez, M. L., Khosroheidari, M., Eddy, E., & Kiefer, J. (2013). Role of microRNA 1207-5P and its host gene, the long non-coding RNA Pvt1, as mediators of extracellular matrix accumulation in the kidney: Implications for diabetic nephropathy. *PLoS One, 8*(10), e77468.

American Diabetes Association. (2004). Nephropathy in diabetes (position statement). *Diabetes Care, 27*, 79–83.

American Diabetes Association. (2017). 10. Microvascular complications and foot care. *Diabetes Care, 40*, S88–S98.

Amore, A., Cirina, P., Conti, G., Cerutti, F., Bagheri, N., Emancipator, S. N., & Coppo, R. (2004). Amadori-configurated albumin induces nitric oxide-dependent apoptosis of endothelial cells: A possible mechanism of diabetic vasculopathy. *Nephrology Dialysis Transplantation, 19*(1), 53–60.

Arif, E., & Nihalani, D. (2013). Glomerular filtration barrier assembly: An insight. *Postdoc Journal: A Journal of Postdoctoral Research and Postdoctoral Affairs, 1*(4), 33.

Asakimori, Y., Yorioka, N., Taniguchi, Y., Ito, T., Ogata, S., Kyuden, Y., & Kohno, N. (2002). T-786→C polymorphism of the endothelial nitric oxide synthase gene influences the progression of renal disease. *Nephron, 91*(4), 747–751.

Babel, N., Gabdrakhmanova, L., Hammer, M. H., et al. (2006). Predictive value of cytokine gene polymorphisms for the development of end-stage renal disease. *Journal of Nephrology, 19*(6), 802–807.

Barbour, S. J., Er, L., Djurdjev, O., Karim, M., & Levin, A. (2010). Differences in progression of CKD and mortality amongst Caucasian, oriental Asian and South Asian CKD patients. *Nephrology Dialysis Transplantation, 25*(11), 3663–3672.

Barrett, J. C., Hansoul, S., Nicolae, D. L., Cho, J. H., Duerr, R. H., Rioux, J. D., Brant, S. R., Silverberg, M. S., Taylor, K. D., Barmada, M. M., & Bitton, A. (2008). Genome-wide association defines more than 30 distinct susceptibility loci for Crohn's disease. *Nature Genetics, 40*(8), 955–962.

Bilous, R. W. (1997). The pathology of diabetic nephropathy. In K. Alberti, P. Zimmet, R. A. DeFronzo, & H. Keen (Eds.), *International textbook of diabetes mellitus* (pp. 1349–1362). Wiley.

Bonventre, J. V. (2012). Can we target tubular damage to prevent renal function decline in diabetes. *Seminars in Nephrology, 32*(5), 452–462.

Boright, A. P., Paterson, A. D., Mirea, L., Bull, S. B., Mowjoodi, A., Scherer, S. W., Zinman, B., & DCCT/EDIC Research Group. (2005). Genetic variation at the ACE gene is associated with persistent microalbuminuria and severe nephropathy in type 1 diabetes: The DCCT/EDIC Genetics Study. *Diabetes, 54*(4), 1238–1244.

Bottazzi, B., Inforzato, A., Messa, M., Barbagallo, M., Magrini, E., Garlanda, C., & Mantovani, A. (2016). The pentraxins PTX3 and SAP in innate immunity, regulation of inflammation and tissue remodelling. *Journal of Hepatology, 64*(6), 1416–1427.

Brennan, E., McEvoy, C., Sadlier, D., Godson, C., & Martin, F. (2013). The genetics of diabetic nephropathy. *Genes, 4*(4), 596–619.

Brownlee, M. (2001). Biochemistry and molecular cell biology of diabetic complications. *Nature, 414*(6865), 813–820.

Brownlee, M., Vlassara, H., Kooney, A., Ulrich, P., & Cerami, A. (1986). Aminoguanidine prevents diabetes-induced arterial wall protein cross-linking. *Science, 232*(4758), 1629–1632.

Buraczynska, M., Zukowski, P., Ksiazek, P., Kuczmaszewska, A., Janicka, J., & Zaluska, W. (2014). Transcription factor 7-like 2 (TCF7L2) gene polymorphism and clinical phenotype in end-stage renal disease patients. *Molecular Biology Reports, 41*(6), 4063–4068.

Burden, A. C., McNally, P. C., Feehally, J., & Walls, J. (1992). Increased incidence of end-stage renal failure secondary to diabetes mellitus in Asian ethnic groups in the United Kingdom. *Diabetic Medicine, 9*(7), 641–645.

Carmo, R. F., Aroucha, D., Vasconcelos, L. R., Pereira, L. M., Moura, P., & Cavalcanti, M. S. (2016). Genetic variation in PTX 3 and plasma levels associated with hepatocellular carcinoma in patients with HCV. *Journal of Viral Hepatitis, 23*(2), 116–122.

Chalasova, K., Dvorakova, V., Pacal, L., Bartakova, V., Brozova, L., Jarkovsky, J., & Kankova, K. (2014). NOS3 894G>T polymorphism is associated with progression of kidney disease and cardiovascular morbidity in type 2 diabetic patients: NOS3 as a modifier gene for diabetic nephropathy? *Kidney and Blood Pressure Research, 38*(1), 92–98.

Chan, Y., Lim, E. T., Sandholm, N., Wang, S. R., AJ, M. K., Ripke, S., Daly, M. J., Neale, B. M., Salem, R. M., Hirschhorn, J. N., & DIAGRAM Consortium. (2014). An excess of risk increasing low-frequency variants can be a signal of polygenic inheritance in complex diseases. *The American Journal of Human Genetics, 94*(3), 437–452.

Chawla, T., Sharma, D., & Singh, A. (2010). Role of the renin angiotensin system in diabetic nephropathy. *World Journal of Diabetes, 1*(5), 141.

Cheung, V. G., & Spielman, R. S. (2009). Genetics of human gene expression: Mapping DNA variants that influence gene expression. *Nature Reviews Genetics, 10*(9), 595–604.

Chiarelli, F., Gaspari, S., & Marcovecchio, M. L. (2009). Role of growth factors in diabetic kidney disease. *Hormone and Metabolic Research, 41*(08), 585–593.

Choe, E. Y., Wang, H. J., Kwon, O., Kim, K. J., Kim, B. S., Lee, B. W., Ahn, C. W., Cha, B. S., Lee, H. C., Kang, E. S., & Mantzoros, C. S. (2013). Variants of the adiponectin gene and diabetic microvascular complications in patients with type 2 diabetes. *Metabolism, 62*(5), 677–685.

Christ, M., Bauersachs, J., Liebetrau, C., Heck, M., Günther, A., & Wehling, M. (2002). Glucose increases endothelial-dependent superoxide formation in coronary arteries by NAD (P) H oxidase activation: Attenuation by the 3-hydroxy-3-methylglutaryl coenzyme A reductase inhibitor atorvastatin. *Diabetes, 51*(8), 2648–2652.

Chung, H. F., Long, K. Z., Hsu, C. C., Al Mamun, A., Chiu, Y. F., Tu, H. P., Chen, P. S., Jhang, H. R., Hwang, S. J., & Huang, M. C. (2014). Adiponectin gene (ADIPOQ) polymorphisms correlate with the progression of nephropathy in Taiwanese male patients with type 2 diabetes. *Diabetes Research and Clinical Practice, 105*(2), 261–270.

Cilingir, V., Donder, A., Milanlioğlu, A., Yilgör, A., & Tombul, T. (2019). Association between endothelial nitric oxide synthase polymorphisms T786C and G894T and ischaemic stroke. *Eastern Journal of Medicine, 24*(4), 472–477.

Clarke, P., Gray, A., Legood, R., Briggs, A., & Holman, R. (2003). The impact of diabetes-related complications on healthcare costs: Results from the United Kingdom prospective diabetes study (UKPDS Study No. 65). *Diabetic Medicine, 20*(6), 442–450.

Cohen, R. M., Holmes, Y. R., Chenier, T. C., & Joiner, C. H. (2003). Discordance between HbA1c and fructosamine: Evidence for a glycosylation gap and its relation to diabetic nephropathy. *Diabetes Care, 26*(1), 163–167.

Colombo, M. G., Andreassi, M. G., Paradossi, U., Botto, N., Manfredi, S., Masetti, S., Rossi, G., Clerico, A., & Biagini, A. (2002). Evidence for association of a common variant of the endothelial nitric oxide synthase gene (Glu298→Asp polymorphism) to the presence, extent, and severity of coronary artery disease. *Heart, 87*(6), 525–528.

Cooke, G. S., Campbell, S. J., Bennett, S., Lienhardt, C., McAdam, K. P., Sirugo, G., Sow, O., Gustafson, P., Mwangulu, F., van Helden, P., & Fine, P. (2008). Mapping of a novel susceptibility locus suggests a role for MC3R and CTSZ in human tuberculosis. *American Journal of Respiratory and Critical Care Medicine, 178*(2), 203–207.

Daly, A. K., & Day, C. P. (2001). Candidate gene case-control association studies: Advantages and potential pitfalls. *British Journal of Clinical Pharmacology, 52*(5), 489–499.

Deshmukh, H. A., Palmer, C. N., Morris, A. D., & Colhoun, H. M. (2013). Investigation of known estimated glomerular filtration rate loci in patients with type 2 diabetes. *Diabetic Medicine, 30*(10), 1230–1235.

Diamond, J. M., Meyer, N. J., Feng, R., Rushefski, M., Lederer, D. J., Kawut, S. M., Lee, J. C., Cantu, E., Shah, R. J., Lama, V. N., & Bhorade, S. (2012). Variation in PTX3 is associated with primary graft dysfunction after lung transplantation. *American Journal of Respiratory and Critical Care Medicine, 186*(6), 546–552.

Dreyer, G., Hull, S., Aitken, Z., Chesser, A., & Yaqoob, M. M. (2009). The effect of ethnicity on the prevalence of diabetes and associated chronic kidney disease. *QJM: An International Journal of Medicine, 102*(4), 261–269.

Drummond, K., & Mauer, M. (2002). The early natural history of nephropathy in type 1 diabetes: II. Early renal structural changes in type 1 diabetes. *Diabetes, 51*(5), 1580–1587.

Duran-Salgado, M. B., & Rubio-Guerra, A. F. (2014). Diabetic nephropathy and inflammation. *World Journal of Diabetes, 5*(3), 393.

Ellis, J. W., Chen, M. H., Foster, M. C., Liu, C. T., Larson, M. G., de Boer, I., Köttgen, A., Parsa, A., Bochud, M., Böger, C. A., & Kao, L. (2012). Validated SNPs for eGFR and their associations with albuminuria. *Human Molecular Genetics, 21*(14), 3293–3298.

Ewens, K. G., George, R. A., Sharma, K., Ziyadeh, F. N., & Spielman, R. S. (2005). Assessment of 115 candidate genes for diabetic nephropathy by transmission/disequilibrium test. *Diabetes, 54*(11), 3305–3318.

Fakhruddin, S., Alanazi, W., & Jackson, K. E. (2017). Diabetes-induced reactive oxygen species: Mechanism of their generation and role in renal injury. *Journal of Diabetes Research*.

Feng, B. J., Goldgar, D. E., & Corbex, M. (2007). Trend-TDT – A transmission/disequilibrium based association test on functional mini/microsatellites. *BMC Genetics, 8*(1), 1–8.

Fioretto, P., & Mauer, M. (2007). Histopathology of diabetic nephropathy. *Seminars in Nephrology*, 195–207.

Fogarty, D. G., Rich, S. S., Hanna, L., Warram, J. H., & Krolewski, A. S. (2000). Urinary albumin excretion in families with type 2 diabetes is heritable and genetically correlated to blood pressure. *Kidney International, 57*(1), 250–257.

Forbes, J. M., & Cooper, M. E. (2013). Mechanisms of diabetic complications. *Physiological Reviews, 93*(1), 137–188.

Forbes, J. M., Soulis, T., Thallas, V., Panagiotopoulos, S., Long, D. M., Vasan, S., Wagle, D., Jerums, G., & Cooper, M. E. (2001). Renoprotective effects of a novel inhibitor of advanced glycation. *Diabetologia, 44*(1), 108–114.

Fu, L. L., Lin, Y., Yang, Z. L., & Yin, Y. B. (2012). Association analysis of genetic polymorphisms of TCF7L2, CDKAL1, SLC30A8, HHEX genes and microvascular complications of type 2 diabetes mellitus. *Chinese Journal of Medical Genetics., 29*(2), 194–199.

Geraldes, P., & King, G. L. (2010). Activation of protein kinase C isoforms and its impact on diabetic complications. *Circulation Research, 106*(8), 1319–1331.

Goyal, R. K., Shah, V. N., Saboo, B. D., Phatak, S. R., Shah, N. N., Gohel, M. C., Raval, P. B., & Patel, S. S. (2010). Prevalence of overweight and obesity in Indian adolescent school going children: Its relationship with socioeconomic status and associated lifestyle factors. *The Journal of the Association of Physicians of India, 58*, 151–158.

Graham, M., & Adams, J. M. (1986). Chromosome 8 breakpoint far 3′ of the c-myc oncogene in a Burkitt's lymphoma 2; 8 variant translocation is equivalent to the murine pvt-1 locus. *The EMBO Journal, 5*(11), 2845–2851.

Gray, L. J., Tringham, J. R., Davies, M. J., Webb, D. R., Jarvis, J., Skinner, T. C., Farooqi, A. M., & Khunti, K. (2010). Screening for type 2 diabetes in a multiethnic setting using known risk factors to identify those at high risk: A cross-sectional study. *Vascular Health and Risk Management, 6*, 837.

Guan, Y., Kuo, W. L., Stilwell, J. L., Takano, H., Lapuk, A. V., Fridlyand, J., Mao, J. H., Yu, M., Miller, M. A., Santos, J. L., & Kalloger, S. E. (2007). Amplification of PVT1 contributes to the pathophysiology of ovarian and breast cancer. *Clinical Cancer Research, 13*(19), 5745–5755.

Ha, H., & Lee, H. B. (2000). Reactive oxygen species as glucose signalling molecules in mesangial cells cultured under high glucose. *Kidney International, 58*, 19–25.

Haneda, M., Koya, D., Isono, M., & Kikkawa, R. (2003). Overview of glucose signaling in mesangial cells in diabetic nephropathy. *Journal of the American Society of Nephrology, 14*(5), 1374–1382.

Hanson, R. L., Craig, D. W., Millis, M. P., Yeatts, K. A., Kobes, S., Pearson, J. V., Lee, A. M., Knowler, W. C., Nelson, R. G., & Wolford, J. K. (2007). Identification of PVT1 as a candidate gene for end-stage renal disease in type 2 diabetes using a pooling-based genome-wide single nucleotide polymorphism association study. *Diabetes, 56*(4), 975–983.

Harrison, D. G., Cai, H., Landmesser, U., & Griendling, K. K. (2003). Interactions of angiotensin II with NAD (P) H oxidase, oxidant stress and cardiovascular disease. *Journal of the Renin-Angiotensin-Aldosterone System: JRAAS, 4*(2), 51–61.

Heilig, C. W., Concepcion, L. A., Riser, B. L., Freytag, S. O., Zhu, M., & Cortes, P. (1995). Overexpression of glucose transporters in rat mesangial cells cultured in a normal glucose milieu mimics the diabetic phenotype. *The Journal of Clinical Investigation, 96*(4), 1802–1814.

Holtzman, N. A., & Marteau, T. M. (2000). Will genetics revolutionize medicine. *New England Journal of Medicine, 343*(2), 141–144.

Hostetter, T. H. (2003). Hyperfiltration and glomerulosclerosis. *Seminars in Nephrology, 23*(2), 194–199.

Imperatore, G., Hanson, R. L., Pettitt, D. J., Kobes, S., Bennett, P. H., & Knowler, W. C. (1998). Sib-pair linkage analysis for susceptibility genes for microvascular complications among Pima Indians with type 2 diabetes. Pima Diabetes Genes Group. *Diabetes, 47*(5), 821–830.

Ito, A., Uriu, K., Inada, Y., Qie, Y. L., Takagi, I., Ikeda, M., Hashimoto, O., Suzuka, K., Eto, S., Tanaka, Y., & Kaizu, K. (2001). Inhibition of neuronal nitric oxide synthase ameliorates renal hyper filtration in streptozotocin-induced diabetic rat. *Journal of Laboratory and Clinical Medicine, 138*(3), 177–185.

Jacobsen, P. K. (2005). Preventing end stage renal disease in diabetic patients—Genetic aspect (part I). *Journal of the Renin-Angiotensin-Aldosterone System, 6*(1), 1–14.

Karter, A. J., Ferrara, A., Liu, J. Y., Moffet, H. H., Ackerson, L. M., & Selby, J. V. (2002). Ethnic disparities in diabetic complications in an insured population. *Journal of the American Medical Association, 287*(19), 2519–2527.

Kim, J. H., Shin, H. D., Park, B. L., Moon, M. K., Cho, Y. M., Hwang, Y. H., Oh, K. W., Kim, S. Y., Lee, H. K., Ahn, C., & Park, K. S. (2006). SLC12A3 (solute carrier family 12 member [sodium/chloride] 3) polymorphisms are associated with end-stage renal disease in diabetic nephropathy. *Diabetes, 55*(3), 843–848.

Kimmelstiel, P., & Wilson, C. (1936). Intercapillary lesions in the glomeruli of the kidney. *The American Journal of Pathology, 12*(1), 83.

Kramer, H. J., Nguyen, Q. D., Curhan, G., & Hsu, C. Y. (2003). Renal insufficiency in the absence of albuminuria and retinopathy among adults with type 2 diabetes mellitus. *Journal of the American Medical Association, 289*(24), 3273–3277.

Lin, J. S., & Susztak, K. (2016). Podocytes: The weakest link in diabetic kidney disease. *Current Diabetes Reports, 16*(5), 1–9.

Lin, G., Wang, Z., Wang, L., Lau, Y. L., & Yang, W. (2008). Identification of linked regions using high-density SNP genotype data in linkage analysis. *Bioinformatics, 24*(1), 86–93.

Liu, R., Lee, K., & He, J. C. (2015). Genetics and epigenetics of diabetic nephropathy. *Kidney Diseases, 1*(1), 42–51.

MacIsaac, R. J., Jerums, G., & Ekinci, E. I. (2017). Effects of glycaemic management on diabetic kidney disease. *World Journal of Diabetes, 8*(5), 172.

Magee, C., Grieve, D. J., Watson, C. J., & Brazil, D. P. (2017). Diabetic nephropathy: A tangled web to unweave. *Cardiovascular Drugs and Therapy, 31*(5–6), 579–592.

Mclennan, S. V., Fisher, E., Martell, S. Y., Death, A. K., Williams, P. F., Lyons, J. G., & Yue, D. K. (2000). Effects of glucose on matrix metalloproteinase and plasmin activities in mesangial cells: Possible role in diabetic nephropathy. *Kidney International, 58*, 81–87.

Millis, M. P., Bowen, D., Kingsley, C., Watanabe, R. M., & Wolford, J. K. (2007). Variants in the plasmacytoma variant translocation gene (PVT1) are associated with end-stage renal disease attributed to type 1 diabetes. *Diabetes, 56*(12), 3027–3032.

Mishra, R., Emancipator, S. N., Kern, T., & Simonson, M. S. (2005). High glucose evokes an intrinsic proapoptotic signaling pathway in mesangial cells. *Kidney International, 67*(1), 82–93.

Mohan, V., Shanthirani, C. S., & Deepa, R. (2003). Glucose intolerance (diabetes and IGT) in a selected South Indian population with special reference to family history, obesity and lifestyle factors: The Chennai Urban Population Study (CUPS 14). *The Journal of the Association of Physicians of India., 51*, 771–777.

Mohan, V., Sandeep, S., Deepa, R., Shah, B., & Varghese, C. (2007). Epidemiology of type 2 diabetes: Indian scenario. *The Indian Journal of Medical Research, 125*(3), 217–230.

Molitch, M. E., DeFronzo, R. A., Franz, M. J., & Keane, W. F. (2004). Nephropathy in diabetes. *Diabetes Care, 27*, 79.

Möllsten, A., Vionnet, N., Forsblom, C., Parkkonen, M., Tarnow, L., Hadjadj, S., Marre, M., Parving, H. H., & Groop, P. H. (2011). A polymorphism in the angiotensin II type 1 receptor gene has different effects on the risk of diabetic nephropathy in men and women. *Molecular Genetics and Metabolism, 103*(1), 66–70.

Mooyaart, A. L., Valk, E. J., van Es, L. A., Bruijn, J. A., de Heer, E., Freedman, B. I., Dekkers, O. M., & Baelde, H. J. (2011). Genetic associations in diabetic nephropathy: A meta-analysis. *Diabetologia, 54*(3), 544–553.

Nakagawa, T., Sato, W., Glushakova, O., Heinig, M., Clarke, T., Campbell-Thompson, M., Yuzawa, Y., Atkinson, M. A., Johnson, R. J., & Croker, B. (2007). Diabetic endothelial nitric oxide synthase knockout mice develop advanced diabetic nephropathy. *Journal of the American Society of Nephrology, 18*(2), 539–550.

Narang, A., Roy, R. D., Chaurasia, A., Mukhopadhyay, A., Mukerji, M., Dash, D., & Indian Genome Variation Consortium. (2010). IGVBrowser–A genomic variation resource from diverse Indian populations. *Database*.

Nazar, C. M. (2014). Diabetic nephropathy; principles of diagnosis and treatment of diabetic kidney disease. *Journal of Nephropharmacology, 3*(1), 15.

Nelson, R. G., Newman, J. M., Knowler, W. C., Sievers, M. L., Kunzelman, C. L., Pettitt, D. J., Moffett, C. D., Teutsch, S. M., & Bennett, P. H. (1988). Incidence of end-stage renal disease in type 2 (non-insulin-dependent) diabetes mellitus in Pima Indians. *Diabetologia, 31*(10), 730–736.

Nishi, S., Ueno, M., Hisaki, S., et al. (2000). Ultrastructural characteristics of diabetic nephropathy. *Medical Electron Microscopy, 33*, 65–73.

Nishikawa, T., Edelstein, D., & Brownlee, M. (2000). The missing link: A single unifying mechanism for diabetic complications. *Kidney International, 58*, 26–30.

Nitta, K., Okada, K., Yanai, M., & Takahashi, S. (2013). Aging and chronic kidney disease. *Kidney and Blood Pressure Research, 38*(1), 109–120.

Noiri, E., Satoh, H., Taguchi, J. I., Brodsky, S. V., Nakao, A., Ogawa, Y., Nishijima, S., Yokomizo, T., Tokunaga, K., & Fujita, T. (2002). Association of eNOS Glu298Asp polymorphism with end-stage renal disease. *Hypertension, 40*(4), 535–540.

Nomiyama, T., Tanaka, Y., Piao, L., Nagasaka, K., Sakai, K., Ogihara, T., Nakajima, K., Watada, H., & Kawamori, R. (2003). The polymorphism of manganese superoxide dismutase is associated with diabetic nephropathy in Japanese type 2 diabetic patients. *Journal of Human Genetics, 48*(3), 138–141.

Olesen, R., Wejse, C., Velez, D. R., Bisseye, C., Sodemann, M., Aaby, P., Rabna, P., Worwui, A., Chapman, H., Diatta, M., & Adegbola, R. A. (2007). DC-SIGN (CD209), pentraxin 3 and vitamin D receptor gene variants associate with pulmonary tuberculosis risk in West Africans. *Genes & Immunity, 6*, 456–467.

Parving, H. H., Mauer, M., Fioretto, P., Rossing, P., & Ritz, E. (2011). Diabetic nephropathy. In *Brenner and Rector's the Kidney*. WB Saunders Company.

Patnala, R., Clements, J., & Batra, J. (2013). Candidate gene association studies: A comprehensive guide to useful in silico tools. *BMC Genetics, 14*(1), 1–1.

Pezzolesi, M. G., Poznik, G. D., Mychaleckyj, J. C., Paterson, A. D., Barati, M. T., Klein, J. B., Ng, D. P., Placha, G., Canani, L. H., Bochenski, J., & Waggott, D. (2009). Genome-wide association scan for diabetic nephropathy susceptibility genes in type 1 diabetes. *Diabetes, 58*(6), 1403–1410.

Placha, G., Canani, L. H., Warram, J. H., & Krolewski, A. S. (2005). Evidence for different susceptibility genes for proteinuria and ESRD in type 2 diabetes. *Advances in Chronic Kidney Disease, 12*(2), 155–169.

Prasad, P., Tiwari, A. K., Kumar, K. P., Ammini, A. C., Gupta, A., Gupta, R., & Thelma, B. K. (2007). Association of TGFβ1, TNFα, CCR2 and CCR5 gene polymorphisms in type-2 diabetes and renal insufficiency among Asian Indians. *BMC Medical Genetics, 8*(1), 20.

Pugh, J. A., Stern, M. P., Haffner, S. M., Eifler, C. W., & Zapata, M. (1988). Excess incidence of treatment of end-stage renal disease in Mexican Americans. *American Journal of Epidemiology, 127*(1), 135–144.

Qian, Y., Feldman, E., Pennathur, S., Kretzler, M., & Brosius, F. C. (2008). From fibrosis to sclerosis: Mechanisms of glomerulosclerosis in diabetic nephropathy. *Diabetes, 57*(6), 1439–1445.

Ramadan, R. A., Zaki, A. M., Magour, G. M., Zaki, M. A., Aglan, S. A., Madkour, M. A., & Shamseya, M. M. (2016). Association of XbaI GLUT1 polymorphism with susceptibility to type 2 diabetes mellitus and diabetic nephropathy. *American Journal of Molecular Biology, 6*, 71–78.

Ravikumar, P., Bhansali, A., Walia, R., Shanmugasundar, G., & Ravikiran, M. (2011). Alterations in HbA1c with advancing age in subjects with normal glucose tolerance: Chandigarh Urban Diabetes Study (CUDS). *Diabetic Medicine, 28*(5), 590–594.

Reeves, W. B., & Andreoli, T. E. (2000). Transforming growth factor β contributes to progressive diabetic nephropathy. *Proceedings of the National Academy of Sciences, 97*(14), 7667–7669.

Saeedi, P., Petersohn, I., Salpea, P., Malanda, B., Karuranga, S., Unwin, N., Colagiuri, S., Guariguata, L., Motala, A. A., Ogurtsova, K., & Shaw, J. E. (2019). Global and regional diabetes prevalence estimates for 2019 and projections for 2030 and 2045: Results from the international diabetes federation diabetes atlas. *Diabetes Research and Clinical Practice, 157*, 107843.

Santos, K. G., Crispim, D., Canani, L. H., Ferrugem, P. T., Gross, J. L., & Roisenberg, I. (2011). Association of eNOS gene polymorphisms with renal disease in Caucasians with type 2 diabetes. *Diabetes Research and Clinical Practice, 91*(3), 353–362.

Satchell, S. C., & Tooke, J. E. (2008). What is the mechanism of microalbuminuria in diabetes: A role for the glomerular endothelium. *Diabetologia, 51*(5), 714–725.

Satirapoj, B., Tasanavipas, P., & Supasyndh, O. (2019). Role of TCF7L2 and PPARG2 gene polymorphisms in renal and cardiovascular complications among patients with type 2 diabetes: A cohort study. *Kidney Diseases, 5*(4), 220–227.

Seman, N. A., He, B., Ojala, J. R., Mohamud, W. N., Östenson, C. G., Brismar, K., & Gu, H. F. (2014). Genetic and biological effects of sodium-chloride cotransporter (SLC12A3) in diabetic nephropathy. *American Journal of Nephrology, 40*(5), 408–416.

Sharma, K., Deelman, L., Madesh, M., Kurz, B., Ciccone, E., Siva, S., Hu, T., Zhu, Y., Wang, L., Henning, R., & Ma, X. (2003). Involvement of transforming growth factor-β in regulation of calcium transients in diabetic vascular smooth muscle cells. *American Journal of Physiology-Renal Physiology, 285*(6), 258–270.

Shaw, P. K., Baboe, F., van Es, L. A., van der Vijver, J. C., van de Ree, M. A., de Jonge, N., & Rabelink, T. J. (2006). South-Asian type 2 diabetic patients have higher incidence and faster progression of renal disease compared with Dutch-European diabetic patients. *Diabetes Care, 29*(6), 1383–1385.

Shaza, A. M., Rozina, G., Izham, M. M., & Azhar, S. S. (2005). Dialysis for end stage renal disease: A descriptive study in Penang Hospital. *Medical Journal of Malaysia, 60*(3), 320.

Sheetz, M. J., & King, G. L. (2002). Molecular understanding of hyperglycemia's adverse effects for diabetic complications. *Journal of the American Medical Association, 288*(20), 2579–2588.

Sladek, R., Rocheleau, G., Rung, J., Dina, C., Shen, L., Serre, D., Boutin, P., Vincent, D., Belisle, A., Hadjadj, S., & Balkau, B. (2007). A genome-wide association study identifies novel risk loci for type 2 diabetes. *Nature, 445*(7130), 881–885.

Smith, M. W., Patterson, N., Lautenberger, J. A., Truelove, A. L., McDonald, G. J., Waliszewska, A., Kessing, B. D., Malasky, M. J., Scafe, C., Le, E., & De Jager, P. L. (2004). A high density admixture map for disease gene discovery in African Americans. *The American Journal of Human Genetics, 74*(5), 1001–1013.

Speeckaert, M. M., Speeckaert, R., Carrero, J. J., Vanholder, R., & Delanghe, J. R. (2013). Biology of human pentraxin 3 (PTX3) in acute and chronic kidney disease. *Journal of Clinical Immunology, 33*(5), 881–890.

Srivastava, S. K., Ramana, K. V., & Bhatnagar, A. (2005). Role of aldose reductase and oxidative damage in diabetes and the consequent potential for therapeutic options. *Endocrine Reviews, 26*(3), 380–392.

Sugimoto, H., Shikata, K., Matsuda, M., Kushiro, M., Hayashi, Y., Hiragushi, K., Wada, J., & Makino, H. (1998). Increased expression of endothelial cell nitric oxide synthase (ecNOS) in afferent and glomerular endothelial cells is involved in glomerular hyper filtration of diabetic nephropathy. *Diabetologia, 41*(12), 1426–1434.

Syed, R., Biyabani, M. U., Prasad, S., Deeba, F., & Jamil, K. (2011). Evidence of association of a common variant of the endothelial nitric oxide synthase gene (Glu298 Asp polymorphism) to coronary artery disease in South Indian population. *Journal of Medical Genetics and Genomics, 3*(1), 13–18.

Tandon, N., Anjana, R. M., Mohan, V., Kaur, T., Afshin, A., Ong, K., Mukhopadhyay, S., Thomas, N., Bhatia, E., Krishnan, A., & Mathur, P. (2018). The increasing burden of diabetes and variations among the states of India: The Global Burden of Disease Study 1990–2016. *The Lancet Global Health, 6*(12), 1352–1362.

Tang, Z. H., Zeng, F., & Zhang, X. Z. (2015). Human genetics of diabetic nephropathy. *Renal Failure, 37*(3), 363–371.

Teumer, A., Tin, A., Sorice, R., Gorski, M., Yeo, N. C., Chu, A. Y., Li, M., Li, Y., Mijatovic, V., Ko, Y. A., & Taliun, D. (2016). Genome-wide association studies identify genetic loci associated with albuminuria in diabetes. *Diabetes, 65*(3), 803–817.

Thomson, S. C., Vallon, V., & Blantz, R. C. (2004). Kidney function in early diabetes: The tubular hypothesis of glomerular filtration. *American Journal of Physiology-Renal Physiology, 286*(1), F8–F15.

Tiongco, R. E., Aguas, I. S., Cabrera, F. J., Catacata, M., Flake, C. C., Manao, M. A., & Policarpio, A. (2020). The role of the TNF-α gene -308 G/A polymorphism in the development of diabetic nephropathy: An updated meta-analysis. *Diabetes and Metabolic Syndrome: Clinical Research and Reviews, 14*(6), 2123–2129. https://doi.org/10.1016/j.dsx.2020.10.032

Unnikrishnan, R., Rema, M., Pradeepa, R., Deepa, M., Shanthirani, C. S., Deepa, R., & Mohan, V. (2007). Prevalence and risk factors of diabetic nephropathy in an urban South Indian population: The Chennai Urban Rural Epidemiology Study (CURES 45). *Diabetes Care, 30*(8), 2019–2024.

Uzun, S., Ozari, M., Gursu, M., Karadag, S., Behlul, A., Sari, S., Koldas, M., Demir, S., Karaali, Z., & Ozturk, S. (2016). Changes in the inflammatory markers with advancing stages of diabetic nephropathy and the role of pentraxin-3. *Renal Failure, 38*(8), 1193–1198.

Valladares-Salgado, A. D., Angeles-Martínez, J. A., Rosas, M., García-Mena, J. A., Utrera-Barillas, D. O., Gómez-Díaz, R. I., Escobedo-De La Peña, J. O., Parra, E. J., & Cruz, M. (2010). Association of polymorphisms within the transforming growth factor-β1 gene with diabetic nephropathy and serum cholesterol and triglyceride concentrations. *Nephrology, 15*(6), 644–648.

Vallon, V., & Komers, R. (2011). Pathophysiology of the diabetic kidney. *Comprehensive Physiology, 1*(3), 1175–1232.

Varghese, S., & Kumar, S. G. (2022). Role of eNOS and TGFβ1 gene polymorphisms in the development of diabetic nephropathy in type 2 diabetic patients in South Indian population. *Egyptian Journal of Medical Human Genetics, 23*(1), 10.

Veelken, R., Hilgers, K. F., Hartner, A., Haas, A., & BÖHMER KP, Sterzel RB. (2000). Nitric oxide synthase isoforms and glomerular hyper filtration in early diabetic nephropathy. *Journal of the American Society of Nephrology, 11*(1), 71–79.

Vionnet, N., Tregouët, D., Kazeem, G., Gut, I., Groop, P. H., Tarnow, L., Parving, H. H., Hadjadj, S., Forsblom, C., Farrall, M., & Gauguier, D. (2006). Analysis of 14 candidate genes for diabetic nephropathy on chromosome 3q in European populations: Strongest evidence for association with a variant in the promoter region of the adiponectin gene. *Diabetes, 55*(11), 3166–3174.

Vithian, K., & Hurel, S. (2010). Microvascular complications: Pathophysiology and management. *Clinical Medicine, 10*(5), 505.

Watanabe, Y., Kinoshita, A., Yamada, T., Ohta, T., Kishino, T., Matsumoto, N., Ishikawa, M., Niikawa, N., & Yoshiura, K. I. (2002). A catalog of 106 single-nucleotide polymorphisms (SNPs) and 11 other types of variations in genes for transforming growth factor-β1 (TGF-β1) and its signaling pathway. *Journal of Human Genetics, 47*(9), 478–483.

Wautier, M. P., Chappey, O., Corda, S., Stern, D. M., Schmidt, A. M., & Wautier, J. L. (2001). Activation of NADPH oxidase by AGE links oxidant stress to altered gene expression via RAGE. *American Journal of Physiology-Endocrinology and Metabolism, 280*(5), 685–694.

Wei, L., Xiao, Y., Li, L., Xiong, X., Han, Y., Zhu, X., & Sun, L. (2018). The susceptibility genes in diabetic nephropathy. *Kidney Diseases, 4*(4), 226–237.

Weil, E. J., Lemley, K. V., Mason, C. C., Yee, B., Jones, L. I., Blouch, K., Lovato, T., Richardson, M., Myers, B. D., & Nelson, R. G. (2012). Podocyte detachment and reduced glomerular capillary endothelial fenestration promote kidney disease in type 2 diabetic nephropathy. *Kidney International, 82*(9), 1010–1017.

Williams, M. E. (2005). Diabetic nephropathy: The proteinuria hypothesis. *American Journal of Nephrology, 25*(2), 77–94.

Wolf, G., Butzmann, U., & Wenzel, U. O. (2003). The renin-angiotensin system and progression of renal disease: From hemodynamics to cell biology. *Nephron Physiology, 93*(1), 3–13.

Xu, M., Chen, X., Yan, L., Cheng, H., & Chen, W. (2008). Association between (AC) n dinucleotide repeat polymorphism at the 50-end of the aldose reductase gene and diabetic nephropathy: A meta-analysis. *Journal of Molecular Endocrinology, 40*, 243–251.

Yilmaz, M. I., Axelsson, J., Sonmez, A., Carrero, J. J., Saglam, M., Eyileten, T., Caglar, K., Kirkpantur, A., Celik, T., Oguz, Y., & Vural, A. (2009). Effect of renin angiotensin system blockade on pentraxin 3 levels in type-2 diabetic patients with proteinuria. *Clinical Journal of the American Society of Nephrology, 4*(3), 535–541.

Young, B. A., Maynard, C., & Boyko, E. J. (2003). Racial differences in diabetic nephropathy, cardiovascular disease, and mortality in a national population of veterans. *Diabetes Care, 26*(8), 2392–2399.

Zanchi, A., Moczulski, D. K., Hanna, L. S., Wantman, M., Warram, J. H., & Krolewski, A. S. (2000). Risk of advanced diabetic nephropathy in type 1 diabetes is associated with endothelial nitric oxide synthase gene polymorphism. *Kidney International, 57*(2), 405–413.

Zhang, R., Zhuang, L., Li, M., Zhao, W., Ge, X., Chen, Y., Wang, F., Wang, N., Bao, Y., Liu, L., & Liu, Y. (2018). Arg913Gln of SLC12A3 gene promotes development and progression of end-stage renal disease in Chinese type 2 diabetes mellitus. *Molecular and Cellular Biochemistry, 437*(1), 203–210.

Zhu, H., Yu, W., Xie, Y., Zhang, H., Bi, Y., & Zhu, D. (2017). Association of pentraxin 3 gene polymorphisms with susceptibility to diabetic nephropathy. *Medical Science Monitor: International Medical Journal of Experimental and Clinical Research, 23*, 428.

Technology in the Management of Type 1 and Type 2 Diabetes Mellitus: Recent Status and Future Prospects

Titas Biswas ⓘ, Biplab Kumar Behera, and Nithar Ranjan Madhu ⓘ

Abstract

Diabetes mellitus (DM), a metabolic condition with several causes, is characterized by persistent hyperglycemia and improper glucose, lipid, and protein metabolisms due to insulin deficiency. Type 1 diabetes mellitus (T1DM) treatment is improving. Most type 2 diabetes mellitus (T2DM) patients live in low- and middle-income countries, although most diabetes innovation research has been done in wealthier nations. Due to cost, health knowledge, and basic access, these folks cannot access most of these products. Glucose measurement and insulin delivery have improved results. Ensuring equal access to diabetes technologies across countries and socioeconomic groups is crucial. Data management tools will improve T1DM and T2DM services for all ages. T1DM and T2DM patients of all ages need more clinical trials. T2DM patients with severe insulin insufficiency who require exogenous insulin nevertheless struggle with glucose management. Safer, simpler insulin administration is needed clinically. Over the past decade, technology has made insulin treatment and disorder management safe. Highly customized insulin therapy for adult diabetics has shown significant results. New methods for the lower and middle classes are needed to increase device usage and improve the quality of life for those with inadequate technical skills or the elderly. This is crucial to promote treatment intensity and improve results. This must be tested in diabetic outpatients.

T. Biswas
Department of Chemistry, Gurudas College, Kolkata, West Bengal, India

B. K. Behera
Department of Zoology, Siliguri College, Siliguri, West Bengal, India

N. R. Madhu (✉)
Department of Zoology, Acharya Prafulla Chandra College, New Barrackpore, West Bengal, India

© The Author(s), under exclusive license to Springer Nature Singapore Pte Ltd. 2023
R. Noor (ed.), *Advances in Diabetes Research and Management*,
https://doi.org/10.1007/978-981-19-0027-3_6

Keywords

Diabetes mellitus · Glucose monitoring · Insulin · T1DM · T2DM

1   Introduction

Diabetes mellitus (DM), a metabolic disorder with numerous etiologies is defined by persistent hyperglycemia and abnormalities in glucose, fat, and protein metabolism as a result of deficiencies in insulin secretion (Ramadasa et al., 2011). Around 463 million people worldwide have diabetes mellitus, which is growing (Daly & Hovorka, 2021). Due to population growth, aging, urbanization, rising obesity rates, junk food consumption, and lack of exercise, diabetes is a severe global issue that is getting worse every day. Diabetic complications cost most of the worldwide cost, negatively impacting the global market and public health system (Fagherazzi & Ravaud, 2018). Type 1 diabetes mellitus (T1DM), type 2 diabetes mellitus (T2DM), and gestational diabetes are the three main kinds of diabetes mellitus (Garg, 2011).

Over the past 10 years, the idea of type 1 diabetes has transformed from being an illness mostly affecting children and teenagers to no longer being influenced by age. The death of insulin-producing pancreatic cells caused by an autoimmune disease, if not directly immune-mediated, is widely thought to hasten the progression of type 1 diabetes (Atkinson et al., 2014). T1DM accounts for 5 to 10% of diabetes patients globally (Tauschmann & Hovorka, 2018). T1DM is currently incurable despite cutting-edge research and technology. T1DM still has no known cure, even though patient health and survival have significantly improved over the previous 25 years. Additionally, despite technological advancements, most T1DM patients do not have optimal glycemic control, but most cannot obtain current medications due to the high expenses of even primary healthcare (DiMeglio et al., 2018).

Type 2 diabetes is brought on by the pancreatic beta cells not functioning properly. It is characterized by glucose intolerance and is connected to relative insulin shortage and insulin resistance. In addition to these complications, type 2 diabetes is linked to incretin deficiency in the gut, increased cardiovascular mortality, enhanced renal glucose reabsorption, increased glucagon pancreatic α-cell secretion, and increased lipolysis in fat cells.

Microvascular and macrovascular complications are also linked to type 2 diabetes (Kaissi & Sherbeeni, 2011). Diabetes type 2 accounts for around 90% of all occurrences of the disease (T2D), and the number is fast increasing worldwide. This has resulted in a significant burden for those affected by the disease and growing demand for the resources available for healthcare (Daly & Hovorka, 2021).

In developing countries, childhood obesity is becoming more common. Type 2 diabetes (T2DM) levels have risen in tandem with the rise in obesity, which is a key risk factor for the disease (Singhal & Kumar, 2021). Compared to people in the healthy weight range, obese men and women had a seven- and 12-fold increased risk of type 2 diabetes (T2DM), respectively, in a meta-analysis of studies that compared the US and Europe. If diagnosed and treated early enough, diabetes can be effectively managed, and one can avoid developing its complications. Poorly controlled T2D, however, is linked to life-threatening consequences. In the absence of quick and effective treatment, conditions such as chronic kidney disease (CKD), amputations, blindness, and cardiovascular disease (CVD) might develop (Seidu et al., 2021). Although it is obvious that many T2D patients need their treatment to be intensified, this does not always happen. This may be because of side effects, anxiety, patient preference, or a lack of health education. The precision and safety of insulin therapy can be improved, and glucose monitoring can be made simpler, thanks to recent technological advancements. This should be used to promote the right therapy intensification for better results.

2 What Is Diabetes?

The primary source of our body's energy, glucose, comes from the food we eat. The pancreas, one of the organs in our body, produces the hormone insulin. The pancreas releases the precise amount of the hormone insulin after determining the blood glucose level in order to ensure that the glucose is transported smoothly from the bloodstream into cells. In this way, insulin keeps the level of blood sugar in your body within a normal range. High blood glucose levels are a symptom of the condition known as diabetes. When the body is not able enough insulin or when the insulin produced does not function properly, glucose cannot be transported from the bloodstream into the cells, so it remains in the blood. This occurs when the body either does not produce enough insulin or when the insulin that is produced does not function properly. Long-term exposure to these high blood glucose levels harms our body, especially our eyes, kidneys, and heart.

Diabetes cannot be cured however it may be managed and one can maintain good health. Medication, physical activity, eating well, and leading a healthy lifestyle can help glycemic control. Glycemic control can be assessed using two main techniques in diabetes treatment. First up is the hemoglobin A1C blood test, which gauges the average blood sugar level over the past 3 months. The goal for many adults with diabetes is to achieve an A1C of 7 percent or below, or less than 7.5% for juvenile populations, even if treatment targets must be customized. Continuous testing of one's blood glucose levels every day. The second method that is commonly used for managing blood sugar levels involves using a blood glucose meter at home in conjunction with a finger stick to collect a little volume of blood. Individuals can adjust their treatment and lifestyle choices based on their blood glucose monitoring results. They can decide in virtually real-time whether or not their glycemic targets are being met through self-monitoring of their blood glucose levels (Hunter, 2016). Type 1 diabetes (T1D), type 2 diabetes (T2D), and gestational diabetes are the three different kinds of disease. The two most common kinds of diabetes out of these three are type 1 diabetes and type 2 diabetes.

2.1   What Is Type I Diabetes?

The immune system is responsible for destroying pancreatic beta cells in those with type 1 diabetes, which stops the body from making insulin. Exogenous insulin replacement, a lifetime therapy, is required to survive. As a result, type 1 diabetes burdens the patient, their family, the healthcare system, and the country (Diabetes Atlas Second Edition). Although the age at symptom start is no longer a limiting factor (Leslie, 2010). T1D is most frequently diagnosed in children and adolescents, with adults receiving fewer diagnoses.

T1D is a severe chronic disease requiring daily attention to a complex disease management regimen. This regimen includes paying attention to the type, quantity, and timing of food intake and physical activity, as well as regular monitoring of blood glucose levels and insulin administration. People with T1D, or their caregiver, must monitor their blood glucose regularly, generally six to ten times per day, because failing to control blood glucose levels can have dangerous short- and long-term implications.

A small amount of blood is drawn from the finger with a needle to be placed on a strip and analysed by a portable blood glucose monitoring device. Using a continuous glucose monitoring (CGM) device, which measures glucose in the fluid in the tissues and is connected to the body through an electrode beneath the skin, is another option. The CGM regularly collects blood glucose levels, providing real-time information that enables patients, family members, and their medical team to understand how glucose patterns change depending on behaviour or the time of day and follows trends over time. These recorded glucose readings frequently require some action on the part of the patient or caregiver, such as eating or supplying insulin through injection or a wearable insulin pump that provides continuous subcutaneous insulin infusion. For example, these recorded glucose readings are frequently required to bring blood glucose levels into a healthy range. These disease management practices must be customized to be suitable for people's varying demands throughout their lives.

In addition to the regular stress of life and raising children, the diagnosis of type 1 diabetes in a kid and the accompanying treatment requirements can strain the family's functioning and increase the likelihood that the parents will experience depressive and anxious symptoms. Diabetes in young children is extremely difficult to manage. Young children are more susceptible to hypoglycemic episodes, and smaller insulin dosages are more difficult to calculate; this is especially true at night. Due to nightly blood glucose readings or awakenings in reaction to a CGM alarm, it is typical for parents and their kids to experience sleep disruption.

The gap in life expectancy between those with T1D and the rest of the US population is getting smaller due to considerable improvements in treatment over the past few decades (Miller et al., 2012). Even though this is a tremendous achievement, it is still a relatively new concept. More extensive research is needed to understand how treatment barriers and facilitators change as people age in the psychosocial environment. Adults with T1D must juggle their employment, starting and raising a family, and/or other responsibilities with the demands of controlling their disease. It can be more challenging for people to manage their diabetes medical regimen and maintain their quality of life if cardiovascular illness, cognitive decline, musculoskeletal diseases, eyesight and hearing loss, end-stage renal disease, and unpleasant neuropathies are age-related comorbidities. Additionally, providing the medication regimen may need revisions or additional assistance when living circumstances change (e.g. nursing facilities). Individualized care may require an upward adjustment to the glycemic control targets to reduce the risk of hypoglycemia associated with excessively tight control, but this must be balanced against concerns that extremely high blood glucose levels may also be detrimental (ADA, 2015).

2.2   What Is Type 2 Diabetes?

T2DM complications include increased hepatic glucose production due to pancreatic beta-cell loss, insulin resistance, and a relative insulin deficit (Luna & Feinglos, 2001; El-Kaissi & Sherbeeni, 2011). T2DM is characterized by an increase in the production of glucose from pancreatic cells, increased glucose absorption in the kidney, and rapid lipolysis in fat cells (Defronzo, 2009). T2DM is becoming more and more common everywhere. It's anticipated that pandemic levels will spread across most regions worldwide (Wild et al., 2004). Given this knowledge, medicines used to treat this condition are created to correct one or more of these physiologic anomalies. If diet and exercise do not yield the target degree of glycemic control in a few months, then the appropriate pharmacologic treatment is required.

According to estimates, T2DM accounts for over 90 percent of all cases and is rising quickly (International Diabetes Federation, 2020). This puts a considerable burden on persons who have the disease, their families, and healthcare resources. T2D can be treated with anti-hyperglycemic medications, including long- and short-acting insulin analogs. However, many people have trouble reaching the advised glycemic levels, leading to long-term micro- and macrovascular problems (Khunti et al., 2013). Despite the urgent need to escalate T2D treatment, many people choose not to use the right insulin therapy, most likely out of personal preference, a lack of health literacy, or concern for adverse effects. T2DM in adolescents differs from T2DM in adults and is shown to start earlier in adolescents than in adults. Youths are found to have more complications at the time of type 2 diabetes diagnosis than adults (Katz et al., 2018; Magliano et al., 2020). The risk of cardiovascular disease is substantially increased in young people with T2DM due to additional risk factors such as hypertension and dyslipidaemia (Singhal & Kumar, 2021).

Obesity and T2DM are closely associated, and obese people have a higher chance of developing T2DM, according to various studies. Additionally, it has been shown that obese people are more likely to acquire cardiovascular disease (CVD) (Bogers et al., 2007), and this risk is considerably higher for obese people who have type 2 diabetes (Jonsson et al., 2002). In a meta-analysis conducted in the US and Europe, obese women had a 12-fold increased risk of T2DM compared to people in the healthy weight range, while obese males had a sevenfold high chance (Guh et al., 2009).

Globally, it is acknowledged that T2D carries a heavy cost, accounting for around 90% of all cases of diabetes (Blaslov et al., 2018; Seidu et al., 2021). Especially if recognized and treated early, diabetes can be successfully managed and its accompanying problems avoided (Khunti & Seidu, 2019).

3 Technology in the Management of T1D Mellitus

This section outlines five technological applications for treating T1D diabetes: "insulin administration", "glucose sensing", "glucose-responsive insulin delivery systems", "data management tools", and islet cell transplant.

3.1 Insulin Delivery

The field of treating diabetes underwent a sea change after the discovery of insulin. To establish glycemic control, all T1DM patients require insulin therapy based on frequent blood glucose monitoring (Shah et al., 2013) as- (1) an "insulin pen" or (2) an "insulin pump" (Shah et al., 2016).

Insulin Pens

Previously, insulin was administered using a vial and syringe, but this method has drawbacks due to inconvenience and inaccurate preparation of the insulin dose. These issues were almost resolved with the invention of insulin pens (Fig. 2). A fine replaceable needle and an insulin cartridge are included in insulin pens. In 1985, Novo Nordisk produced its first insulin pen (Novo Nordisk Blue sheet, 2010).

The more recent insulin pens are reusable and come with changeable, safer, and more accurate cartridges. The new insulin pens' auditory clicks after each dose increase accuracy and reduce the chance of human error (Penfornis et al., 2011). Recently, a lot of nations have approved Bluetooth-enabled smart pens. These smart pens keep track of dosages, wirelessly send information to smartphone diabetes management apps, and automatically upload information for sharing with healthcare specialists (Tauschmann & Hovorka, 2018).

Insulin Pumps

As an alternative to numerous daily injections, insulin pumps were first offered in 1970 (Tamborlane et al., 1979; Pickup et al., 1978). However, the widespread availability of insulin pump therapy didn't happen for another 20 years. The first models were basic; they only permitted a fixed rate for basal infusion and required manual entry of total units of insulin for extra boluses during meals or correction. Over the past 20 years, the use of insulin pump therapy has increased, leading to more advanced equipment that can be customized and makes some autonomous modifications based on actual or anticipated blood glucose levels. In addition, coverage by commercial insurance and public healthcare systems helped increase insulin pump use. As a result, large-scale diabetes registration statistics show that 40–60% of T1DM patients in Western nations are insulin pump users (Sherr et al., 2016).

According to a 2002 meta-analysis of randomized controlled studies, T1D patients using insulin pumps have significantly better glycemic control and a lower level of glycosylated hemoglobin. Compared to those who used conventional injections, it required less total daily insulin (Pickup et al., 2002). Compared to several daily injections, the use of insulin pumps resulted in significant long-term glycemic control for children with T1D for up to 6 years (Burckhardt et al., 2018).

The fundamental purpose of an insulin pump is to eliminate the requirement for separate basal insulin by continuously delivering rapid-acting insulin subcutaneously. Depending on the activity level, pumps can have various basal rates. Reducing insulin's basal rate, for instance, may help prevent a hypoglycemia episode since the perception of glucose straight into muscles reduces the requirement for insulin during activity. The amount of glucose utilized is quite low compared to when sleeping, but raising the basal rate can assist lower hyperglycemia caused by the dawn phenomena. In addition to the basal delivery, bolus insulin is given to treat high blood sugar levels and cover meals. By entering the number of scheduled carbohydrates, insulin pumps may automatically compute doses and predetermine insulin-to-carbohydrate ratios for each patient.

Programming insulin sensitivity variables and blood glucose targets also allow for properly correcting hyperglycemia. The pump also monitors the amount of active insulin on board to avoid hypoglycemia caused by insulin stacking if doses are given too soon apart in time. When the insulin supply needs to alter, insulin pumps also offer settings for other activities like exercise and sleep. Patients have fewer intrusive events than several daily injections because a single device controls all insulin. Furthermore, supplementary snacks, second helpings at a meal, or dessert can be covered by insulin via the pump without requiring an additional injection, aiding adherence. Finally, an insulin pump can be continuously utilized for up to 3 days before changing.

There are still some drawbacks to insulin pumps despite all of these amenities. The insulin pump patient is permanently connected to the delivery system while the insulin is being administered. The insulin pump hardware may malfunction for various reasons and is not always under our control. Tubing occlusion or unintentional disconnect can occur. These circumstances may prevent the administration of insulin. An alternative method of manually administering insulin, such as a pen or vial and syringe, should be available before employing an insulin pump. Additionally, an insulin pump simplifies diabetes management but cannot take the place of patient-guided control. Before using an insulin pump, it is always advised to go over its advantages and cons with the patient (Figs. 1, 2 and 3).

Fig. 1 Position of the pancreas gland

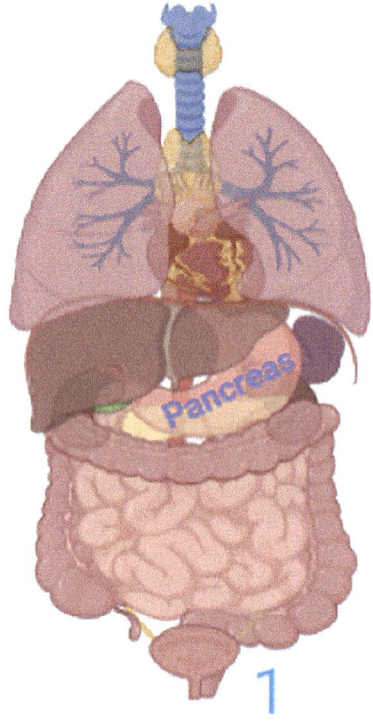

Fig. 2 Insulin pen for diabetes patient

Fig. 3 Process of insulin
pump therapy

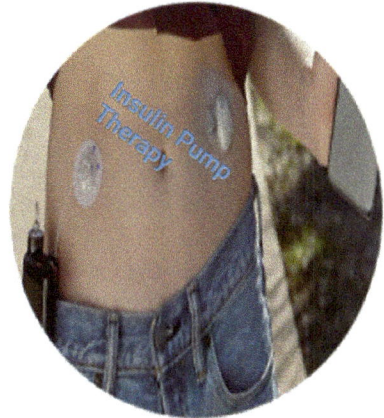

3.2   Glucose Monitoring

Measurements of Glucose in the Capillary Blood

The typical method of glucose monitoring is capillary blood glucose measurement. This technique uses a lancet and a hand-held, portable meter measuring glucose levels. Capillary testing must be performed more frequently to improve diabetes control—typically six to ten times per day (Rewers et al., 2014). Acute dysglycaemia rates are inversely correlated with the frequency of more capillary tests and higher HbA1c values. Similar to how a bolus calculator is used in an insulin pump, an integrated bolus adviser is used in an export meter to compute insulin dosage. According to RCTs, a bolus calculator is more practical than a control person. According to Vallejo Mora et al. (2017), utilizing a bolus calculator increases the proportion of patients who reach the HbA1c target.

Capillary blood glucose monitoring has significant drawbacks. Even though blood samples are often taken with this technique, they only provide snapshots of the glucose level. Missing instances of hyperglycemia and hypoglycemia could lead to treatment decisions not being taken into account (Tauschmann & Hovorka, 2018).

Constant Glucose Measuring

The most recent insulin pump systems that offer fully automated basal rate management rely mostly on continuous glucose monitoring (CGM). The new technology for CGM converts the interstitial glucose levels that are detected into an electrical signal that is then delivered to a receiving device and recorded. This signal is generated by the glucose-oxidation reaction (Cappon et al., 2019). Using this technique, subcutaneous tissue is given an injection of a very little filament. The glucose measurements and any relevant information regarding patterns in the patient's serum blood sugar levels that can assist in directing care are relayed to the patient. One advantage is reducing the number of finger-stick blood samples required to control type 1 diabetes thoroughly.

Although CGM has made significant progress since its inception, some significant restrictions still exist. The values that are displayed on the transmitter do not always provide an accurate representation of the real levels of glucose in the blood. For instance, the glucose measurement in the subcutaneous blood could be different in real-time from the value in the serum. Therefore, cybersecurity is one of the most important aspects (Klonoff, 2015). Because personal blood glucose data are subject to the Health Insurance Portability and Accountability Act (HIPAA), more sophisticated measures should be implemented to protect them.

Periodic Glucose Monitoring's Effectiveness

Early-generation device RCTs and meta-analyses raised concerns about the overall benefit of CGM systems, especially in T1DM-affected children and adolescents (Juvenile Diabetes Research Foundation Continuous Glucose Monitoring Study Group, 2008). However, evidence from the previous 10 years consistently shows that using CGM is linked to better HbA1c levels, a decrease in the frequency of mild to moderate hypoglycemia, and less variability in glucose levels. Although early studies and recommendations supported using CGM in conjunction with pump therapy, new research favours utilizing CGM as part of MDI therapy. The previous RTCs and meta-analyses have limited validity because technology is rapidly evolving (Tauschmann & Hovorka, 2018).

Flash Glucose Monitoring

The FreeStyle Libre flash glucose monitoring system was launched in the UK in 2013. It is compact and lightweight. Users of MDI therapy and insulin pumps both benefited from the FGM system. Flash glucose monitoring was still superior in this comparison even if traditional CGM more successfully decreased the amount of time spent in hypoglycemia in those with T1DM and poor awareness of hypoglycemia (Reddy et al., 2018). According to many studies, FGM is ineffective for the pediatric population. Flash glucose monitoring, a more economical option for CGM, could undoubtedly be viewed as a breakthrough in managing diabetes despite the scant evidence from RCTs (Tauschmann & Hovorka, 2018).

3.3   Delivery of Glucose-Sensitive Insulin

Automated suspension of insulin supply at low glucose levels or when low glucose levels are anticipated is an early example of glucose-responsive management of insulin delivery enabled by technology to solve the problem of hypoglycemia. Contrarily, closed-loop strategies are more intricate and deal with hypoglycemia and hyperglycemia as problems.

Threshold-Based Insulin Suspension

The improved version of threshold-based insulin suspension was approved in the USA in 2013 after the first product became available in 2009. Threshold-based insulin delivery devices stop when sensor glucose reaches a predefined low sensor threshold. Automated insulin suspension is safe and reduces the frequency and duration of overall and nocturnal hypoglycemia episodes when compared to insulin pump therapy alone or pump therapy with sensors (Bergenstal et al., 2013). In addition, threshold-based suspension reduces the incidence of severe hypoglycemia as well as the risk of moderate and severe hypoglycemia in people who are most at risk for developing these conditions.

Predictive Low-Glucose Insulin Suspension

The implementation of predictive low-glucose-suspend technology results in a reduction in the manifestation of nocturnal and overall hypoglycemia in RCTs. This reduction applies to adults (Maahs et al., 2014), children, and adolescents (Battelino et al., 2012), and it also reduces the prevalence of protracted nocturnal incidents. These benefits came at the expense of extended periods of usual hyperglycemia or slightly raised glucose levels throughout the night and in the morning. This was the case even though the benefits were greater.

Closed-Loop Systems

The present objective of research into insulin delivery technology is to develop a closed-loop system that can be referred to as an artificial pancreas or an automated insulin administration system. Essentially, this will involve the construction of an artificial pancreas. A glucose sensor transmits the glucose value and direction to a control algorithm in this automated insulin administration system. The control algorithm, having calculated the necessary amount, then informs the insulin pump to provide the necessary amount of insulin. The duration that a patient can spend within the controlled range for blood glucose is the primary focus of these programs. The systems that are now employed in clinical treatment are hybrid closed-loop systems. This is because automated basal insulin delivery, human entry of carbohydrate intake, and manual direction to provide correction insulin doses are all components of these systems. A fundamental shortcoming of the system is the time lag that exists between detecting glucose and adjusting the rate at which insulin is delivered (Christopher et al., 2020).

3.4   Data Management

Along with insulin administration and glucose monitoring improvements, distal technology for communication, instruction, intervention, and remote service delivery has also advanced. For the benefit of T1DM patients, numerous studies on distal diabetic technologies have been evaluated (Duke et al., 2018). These technologies include telemedicine, mobile health applications, game-based assistance, social platforms, and patient portals.

Data Download

Individuals can evaluate and view glucose levels and the ambulatory glucose profile using the cloud after downloading data from insulin pumps and CGM monitors to computers. These data make it easier for medical experts, T1DM patients, and caregivers to identify the key characteristics that need to be treated more effectively. Regularly downloading and evaluating glycemic data are associated with statistically significantly lower HbA1c values (Wong et al., 2015). A tiny proportion of T1DM patients and caregivers regularly download and analyse data due to certain impediments, including inconvenience, issues with the software and hardware, and a lack of training in data interpretation (Tauschmann & Hovorka, 2018).

Remote Monitoring

Continuous data transmission from CGM devices to the cloud is made possible by apps like Minimed Connect, LibreLink, and the Dexcom G5 Mobile. If enabled, third parties can view users' CGM traces and get notifications for low blood sugar or other events on their cellphones. Examples of these parties include spouses and carers. Beyond increased convenience, the benefits of remote monitoring in daily life have not yet been fully evaluated in an RCT. Surprisingly, the collaborative parent-led Nightscout Project anticipated and sparked the creation of remote monitoring features (Lee et al., 2017).

Mobile Diabetes Applications

Diabetes mobile apps are becoming more and more common. Over 1100 diabetes-specific applications and 165,000 apps relating to general health are available (Trawley et al., 2016). Diabetes apps may help with diabetes self-management since they include a variety of functions and activities, from straightforward logs and dosage reminders to bolus calculators and carbohydrate counting, as well as peer support and incentives to use boluses (Cafazzo et al., 2012). Users can keep a thorough track of parameters relating to their diabetes on their phones using apps like Bant, Glooko, mySugr, One Drop, and Tidepool (Tauschmann & Hovorka, 2018).

3.5   Islet Cell Transplant

One of the main study areas for the possibility of a T1D treatment is the transplant of islet cells. To achieve insulin independence, physiologically active islet cells are transplanted. However, the autoimmune nature of the disease and the scarcity of islet cells have presented significant obstacles to the application. In people with T1D who experience hypoglycemia unawareness and glycemic lability despite adequate insulin administration, islet cell transplant is currently accessible as an experimental therapy.

Isolated islet cells from deceased donors are pumped into the transplant recipient's portal vein during this procedure. It may take more than one infusion to provide the recipient with enough islet cells to become insulin-independent. To avoid transplant rejection, the recipient must take immunosuppressant medicine.

Other cell types have been considered potential implants due to the insufficient supply of islet cell donors. Recent investigations identified pancreatic nonendocrine cells and pluripotent stem cells as leading candidates (Dominguez-Bendala et al., 2016). Although this system is still in its infancy, future developments are expected to completely transform the way T1D is treated (Christopher et al., 2020).

4 Technology in the Management of T2D Mellitus

4.1 Technology-Enabled Self-Management

Glycemic regulation is managed via a few fundamental technologies. Early T2D glucose levels may be controlled by exercise and a healthy diet. Additional diabetic medications must be combined with diet and exercise therapy as the blood glucose level rises to control it. If these methods are unsuccessful, you can try combining two or more pills. If the first step is unsuccessful, blood glucose levels can be managed with insulin and diabetic medication.

4.2 Glucose Monitoring

Blood Glucose Monitoring

Clinicians are often concerned about poor compliance with routine blood glucose testing, especially in adolescents with T2D. The blood glucose test must be performed regularly and should be an essential part of a patient's routine approach to determining the proper management of T2D. Smart gadgets integrating with current validated portable devices through Bluetooth or remotely could deliver user-friendly visual interface data relating to blood glucose level (BGL) concentrations. By swiping with genuine transferrable devices over Bluetooth or wirelessly, available smart gadgets could provide user-friendly visual interface data regarding blood glucose level (BGL) concentrations. Growing awareness of a patient's BGL at a specific time of the day, before or after eating, or even when performing hard labour, exploits them.

Clinicians can quickly generate customized reminders using current technology to let T2D patients know that a BGL test is necessary at a particular time. After being taken using an encryption password, the BGL measurement is quickly transferred to the cloud network, saving the patients' time and effort. More than 5 years of insulin use among T2D patients increases the likelihood of severe hypoglycemia episodes (Noh et al., 2011). These affordable smart technologies support a patient's involvement in diabetes management care and prevent significant complications. According to Noh et al. (2011), severe hypoglycemia crises happen in 25% of T2D patients who have been on insulin for more than 5 years. Special logarithms might be used to predict critical conditions, such as a severe hypoglycemia crisis. In these cases, the smart device would automatically contact the necessary emergency services (Coda et al., 2018).

Self-Monitoring of Blood Glucose

The American Diabetes Association (ADA) stated that the following patients should use self-monitoring of blood glucose (SMBG): those receiving intensive insulin therapy (at least three times per day), those receiving non-intensive insulin therapy, those not receiving oral hypoglycemic agents, and those receiving diet therapy alone to reach the optimal glycemic target and optimal postprandial glycemic target. Despite the lack of sufficient evidence to authenticate that close monitoring in these patients is associated with a better outcome over the long term, SMBG can help patients with type 2 diabetes achieve better glycemic control, especially at the beginning of therapy or after reconciliation. It is yet unknown how SMBG affects people with type 2 diabetes who are just under dietary treatment (Montagnana et al., 2009).

Flash Glucose Monitoring

Since 2016, glucose monitoring has been performed using a device called the FreeStyle Libre (Hoss & Budiman, 2017) (Abbott Diabetes Care, Alameda, California), which is a factory-calibrated 14-day flash glucose monitor. A tiny circular sensor with a thin fiber is placed into the patient's upper arm, and then it records the patient's glucose level once every minute. The glucose monitoring system stores three daily scans at 8-hour intervals so that it can record the glucose results for a whole day. The ambulatory glucose profile (AGP), which is generated by scanning the sensor with a smartphone, is a quick and nearly inconspicuous method of monitoring glucose trends during the day and night. In insulin-treated type 2 diabetic patients, numerous randomized controlled trials and a meta-analysis indicated improvements in HbA1c and decreased time in hypoglycemia using flash glucose monitoring compared to SMBG (Yaron et al., 2019).

Increased resource usage with self-monitoring blood glucose (SMBG) compared to flash glucose monitoring estimates that total annual expenses of flash glucose monitoring in the UK for people with type 2 diabetes are 13 percent lower than SMBG. These estimates are based on the results of the REPLACE trial (Hellmund et al., 2018). Because of this, there will be fewer adverse effects, lesser need for finger-stick tests, and reduced hospital admissions, perhaps making this a more cost-effective option (Daly & Hovorka, 2021).

One of the shortcomings of the first-generation flash glucose monitoring device is that it does not provide any warnings in the event of hypo- or hyperglycemia. This problem has been resolved in the most recent version of the FreeStyle Libre 2, which also contains a reminder to check glucose levels and an alert for glucose numbers outside the normal range.

Continuous Monitoring of Glucose Levels

The key benefits of CGM over flash glucose monitoring are that it may provide alerts and alarms without physically scanning a sensor. HbA1c is sometimes substituted by the CGM system as the preferred metric for assessing glycemic management since it lacks reliability and information on glucose trends (Danne et al., 2017).

The ability of CGM technology to detect hypoglycemia makes it recommended by the International Consensus guidelines that it is taken into account in addition to HbA1c when assessing glycemia in patients with insulin-treated T2D (Danne et al., 2017).

The requirement for calibration with finger-stick glucose to counteract variations in sensor sensitivity over time had previously served as a barrier to implementing CGMs in clinical practice. By removing the need for frequent finger-stick testing, the Dexcom G648,49 (Dexcom, San Diego, California) CGM, which doesn't require calibration, has improved the quality of life for many users.

The ability to assess continual glucose readings may help promote healthy lifestyle choices and exercise motivation in addition to the glycemic benefits of CGM (Bailey et al., 2016).

Implantable Glucose Monitors

A little cylindrical-shaped sensor that is part of the 2016 invention of implantable CGM is inserted into the upper arm by a trained specialist. This method produces up to 180 days' worth of glucose data (Deiss et al., 2019). For patients who struggle with frequent sensor changes, the sensor's longer lifespan and implantability make it a superior choice. These individuals may benefit from "on-body" notifications provided by the transmitter's vibrating mechanism or those who have allergies to the common CGM adhesives. In persons with T2D who are not meeting glycemic objectives, CGM devices like flash glucose monitoring can assist lower HbA1c and/or reduce hypoglycemia. Masked CGMs are also highlighted to identify glycemic trends and allow early therapy modifications for those with T2D.

4.3   Insulin Delivery

Insulin Pens

Insulin pens are the most convenient approach for patients with type 2 diabetes to control their insulin (Klonoff & Kerr, 2018). In this method, subcutaneous insulin is injected from a cartridge using a needle that is thrown away after each usage. It is difficult for users and healthcare providers to understand glucose profiles or assess compliance with doses because of the requirement for manual blood glucose data recording and the lack of a link to an automated environment. Although this is a good strategy for managing insulin, it makes it difficult for both parties to understand glucose profiles.

The most up-to-date insulin pens come equipped with memories, caps, and attachments that can upload data to online platforms and keep track of dosages.

The "Memory" function of the insulin pen, which stores and displays information on prior bolus timing and amount, is especially helpful for people with a concomitant impairment or who are not actively managing their diabetes because of the difficulty of dosage administration. This is because the "Memory" function allows the user to more accurately calculate the timing and amount of subsequent boluses (Peyrot et al., 2012). Using an insulin pen that is Bluetooth-enabled and connected to smartphone apps, users can keep track of boluses, calculate how much insulin is left, monitor the temperature of their insulin, and receive dosage reminders (Sangave et al., 2019).

Insulin Pumps

As an alternative to several daily injections, insulin pump therapy, also known as continuous subcutaneous insulin infusion (CSII), was first presented in 1970 (Tamborlane et al., 1979; Pickup et al., 1978). However, the widespread availability of insulin pump therapy didn't happen for another 20 years. NICE advised against the use of CSII for individuals with T2D (NICE, TA151), and a joint statement from the American Diabetes Association and the European Association for the Study of Diabetes (EASD) in 2018 only explicitly mentioned a limited role for insulin pumps in a small proportion of individuals with T2D (Davies et al., 2018). In fact, the most recent update to the ADA Standards of Medical Care in Diabetes (2021) suggested that insulin pumps can be used as a treatment option for adults and children with T2D who are on multiple daily injections (MDI) and can manage the device. Recently, it has been established that CSII in T2D is more beneficial and secure than traditional insulin therapy (American Diabetes Association, 2021). Over the past 10 years, numerous studies have shown improvements in glycemic control, a reduction in the need for insulin, and higher quality of life for persons with T2D who take CSII as opposed to MDI or oral medicines (Frias et al., 2011). Given the higher cardiovascular risk in this group compared to individuals without diabetes, some studies comparing the use of CSII to MDI in T2D suggested a greater degree of weight gain with the former. Numerous studies have demonstrated the long-term cost-effectiveness of CSII therapy in T2D despite the greater initial expenses of insulin pump therapy compared to MDI. In the end, CSII is more affordable than MDI since it causes fewer diabetes-related complications and has lower daily insulin requirements (Roze et al., 2016).

4.4   Glucose-Responsive Insulin Delivery

A glucose-responsive "closed-loop" or "artificial pancreas" insulin delivery system that replicates the function of pancreatic cells may be able to significantly improve both the health and quality of life of diabetics. By administering glucose-responsive insulin, it is possible to circumvent the difficulty of customising insulin therapy in type 2 diabetes to keep glucose levels under control. Earlier glucose-responsive insulin delivery systems include the low-glucose-suspend feature, which prevents insulin supply from continuing below a predetermined glucose threshold, and the predictive low-glucose-suspend feature, which predicts impending hypoglycemia and stops insulin delivery in advance of its onset. Both of these features suspend insulin delivery (Rodbard, 2017). Since postprandial glycemic excursions are particularly challenging, the next stage of development was a "hybrid" closed-loop system. This system requires meal announcements and the user to initiate a pump-delivered meal bolus. The basal rate is controlled automatically for the remainder of the time between meals. Due to studies on the inpatient and outpatient use of this system by people with type 1 diabetes, the first hybrid closed-loop system for T1D was introduced in the USA in 2017 (670G pump; Medtronic Diabetes, Northridge, California). Medtronic Diabetes developed this system. Since then, many other hybrid closed-loop systems have been developed. NCT04025775, NCT04701424, and NCT04233229 are the three trials that are currently being carried out to determine whether or not the use of a fully closed-loop device for the treatment of type 2 diabetes in an outpatient setting is an option that is safe, practical, and affordable for this population outside of a hospital setting (Daly & Hovorka, 2021) (Tables 1 and 2).

5   Recent Status and Future Prospects

Although most diabetes technology research has been done in rich nations, it should be remembered that most T2D patients reside in low- and middle-income nations. Due to expense, low health literacy, and lack of access to education, many gadgets are out of reach for these individuals. Many people with T1D and T2D currently self-fund diabetic technologies because they are not frequently available from their healthcare providers in developed countries (Li & Hussain, 2020). Making sure that people from different countries and socioeconomic backgrounds have equal access to diabetes technologies is a major concern. In order to increase the user base for these technologies and enhance the quality of life for persons with weak technical competence or older people with T2D, less sophisticated technologies must include provisions for cognitive and sensory disabilities. Additional clinical studies, including this subgroup of T2D patients, are necessary.

On the other hand, current professional standards do not encourage routine usage in this particular demographic. To date, studies have demonstrated that using an insulin pump results in great outcomes for people with type 2 diabetes who have inadequate control when using MDI. The integration of insulin pumps with other diabetes technologies developed in the previous 10 years has prepared the way for closed-loop insulin delivery, which aims to provide optimal blood glucose control while reducing the amount of work that is required of the user. However, the widespread integration of these technologies into standard medical practice is hampered by limited access to insulin pumps. Further data from RCTs and health economics analyses are needed to promote their widespread use.

There are currently commercially available closed-loop systems for T1D patients that allow for glucose-responsive insulin delivery. More investigation is necessary to

Table 1 Evidence in favour of the therapeutic application of various treatments for adults with T1DM

Therapy	Outcomes (level of support)	Highest degree of support	References
Utilizing an insulin pump	Employing an insulin pump	Meta-analyses and systematic reviews of RCTs	Pickup and Sutton (2008), Monami et al. (2010)
	The risk of severe hypoglycemia is decreased similarly to how numerous daily injectable therapy reduces risk (A)		
	Reduced need for insulin (A)		
	Enhanced quality of life and happiness with treatment (A)		
	Decreased cardiovascular death rate (B)		
Checking glucose levels continuously:			
(a) Fast glucose checking	Non-severe hypoglycemia is reduced (A and C)	One RCT	Bolinder et al. (2016), Dover et al. (2017)
	Time to target glucose range is improved (A)		
	A reduction in the variability of glucose (A)		
	Lowering of HbA1c values (A and C)		
	Increased user satisfaction and quality of life (C)		
(b) Continuous real-time glucose measurement	Reduction in HbA1c levels (A)	Meta-analyses and systematic reviews of RCTs	Pickup et al. (2011), Hoeks et al. (2011), Juvenile Diabetes Research Foundation Continuous Glucose Monitoring Study Group (2010)
	Reduction in hypoglycemia, from mild to severe (A)		
	Shorter duration of hyperglycemia (A)		
	Enhanced standard of living (A)		
	Regardless of the insulin administration mode (pump or pen), benefits of continuous glucose monitoring are shown; nevertheless, they are dependent on the high frequency of sensor use (A)		
Insulin administration in response to blood sugar			
	Reduced risk of hypoglycemia, especially	RCTs	Weiss et al. (2015)

(continued)

Table 1 (continued)

Therapy	Outcomes (level of support)	Highest degree of support	References
Suspension based on a threshold	overnight, among those who are susceptible to it (A)		
	No discernible decline in overall glucose regulation (A)		
Low-glucose suspension with predictive	Additional decrease in the frequency and length of diurnal and nocturnal hypoglycemic episodes (A)	RCTs	Maahs et al. (2014)
Single-hormone hybrid closed-loop	In outpatient settings, safe use (A)	Comprehensive analysis and meta-analysis	Weisman et al. (2017)
	Lengthening of the target glucose range (A)		
	Shorter duration of hypoglycemia (A)		
	Shortened period of hyperglycemia (A)		
	A slight drop in HbA1c values (A)		

Randomized controlled trial abbreviated as "RCT"; diabetes mellitus type 1 is abbreviated as T1DM
Source: Tauschmann and Hovorka (2018)
[a]In accordance with the evidence grade method utilized by the American Diabetes Association (ADA) for clinical recommendations (ADA, 2017). The American Diabetes Association summarizes expert assessment and significant experimental evidence, regardless of blood glucose level, and clinically significant hypoglycemia is defined as that which is associated with considerable cognitive impairment. A blood glucose level is a threshold for further adjustment of insulin treatments in response to blood sugar and is regarded as an alert value

determine the effectiveness and safety of this treatment for T2DM (Daly & Hovorka, 2021).

6   Conclusion

The science for managing and caring for patients with T1DM is improving. Although most diabetes innovation research has been done in wealthy nations, it is essential to remember that most T2D patients reside in low- and middle-income nations. These people are unable to access most of these products because of cost, inadequate health awareness, and lack of access to basic. Improvements in outcomes have been made possible by glucose detection and insulin administration advancements. Making sure that people from different countries and socioeconomic backgrounds have equal access to diabetes technologies is a major concern. Additional data management features will enhance the ultimate level of service for

Table 2 Presently accessible smart pens for type 2 diabetes patients

Pen	Year of release	Manufacturer	Features	Bluetooth	Connection to online data repository/app	Compatible insulin
NovoPen 5	2015	Novo Nordisk	Shows the time and dosage of the most recent injection. Battery and memory condition (Novo Nordisk, 2020)	No	No	Penfill 3-mL cartridges
NovoPen 6	2019	Novo Nordisk	Monitors insulin dosages backed up treatment choices (Novo Nordisk, 2020)	No	On apple and android smartphones, near-field communication is available via compatible apps (diasend, mySugr)	Novo Nordisk Tresiba or Fiasp cartridges
ESYSTA BT	2016	Emperra	Shows the time and dose of the most recent injection. Time till next injection keeps track of the last 1000 injections' data and temperature alert (Emperra Digital Diabetes Care, 2017)	Yes	Connects to the mobile apps for apple and android smartphones and the cloud-based ESYSTA site	Eli Lilly, Novo Nordisk and Sanofi Aventis 3-mL cartridges
InPen	2016	Medtronic	Insulin tracking on board reminder for dose reminder for replacing the cartridge Reminder to check blood sugar temperature alert Battery life status records dose automatically (Gildon, 2018)	Yes	Connects to apple devices' InPen app integrates with Bluetooth-enabled CGMs and glucose metres	Insulin included in 3-mL lispro cartridges (Novolog, Humalog, Fiasp)
Pendiq 2.0	2017	Diamesco	It displays the amount of insulin left in the tank and sounds an alarm if you try to dose yourself twice.	Yes	Connects to the Dialife app on smartphones running apple and android. USB cable used to connect to the diasend platform	Eli Lilly, Novo Nordisk, Sanofi-Aventis, Berlin-Chemie

			(Pendiq Intelligent Diabetes Care, 2020).			
YpsoMate SmartPilot	2017	Ypsomed	Attach the YpsoMate pen and display the last injection's time and dose Provide direction during the injection process with app-based dose remembrance (YDS Delivery Systems, 2020)	Yes	Bluetooth injection event transfer to mobile application	Insulin pens that have been prefilled

Source: Daly and Hovorka (2021)

persons with T1DM and T2DM across all ages. It is necessary to do additional clinical studies with all age groups of T1DM and T2DM patients. Glucose management for patients with T2DM is still difficult, especially for those with a significant insulin deficiency and who need to use exogenous insulin. Clinically, safer and easier-to-use insulin administration techniques are required. Over the past 10 years, technological developments have made it possible to start insulin treatment and properly manage the disorder safely. Investigations on diabetic adult patients have indicated positive outcomes for highly individualized insulin therapy. In terms of increasing the number of users of these devices and enhancing the quality of life for those with poor technical knowledge or elderly persons, there is a necessity for new systems for the lower- and middle-class population. It is essential to make use of this in order to encourage the appropriate therapeutic intensification and achieve better results. It is necessary to conduct additional research to identify whether or not this can be applied to the diabetic outpatient population.

References

American Diabetes Asociation (ADA). (2021). Diabetes technology: Standards of medical care in diabetes—2021. *Diabetes Care, 44*(Suppl 1), S85–S99.

American Diabetes Association (ADA). (2015). Standards of medical Care in Diabetes—A bridged for primary Care providers. *Clin Diabetes., 33*(2), 97–111.

American Diabetes Association (ADA). (2017). Standards of medical care in diabetes-2017. *Diabetes Care, 40*(Suppl. 1), S1–S135.

Atkinson, M. A., Eisenbarth, G. S., & Michels, A. W. (2014). Type 1 diabetes. *Lancet, 383*, 69–82.

Bailey, K. J., Little, J. P., & Jung, M. E. (2016). Self-monitoring using continuous glucose monitors with real-time feedback improves exercise adherence in individuals with impaired blood glucose: A pilot study. *Diabetes Technology and Therapeutics, 18*(3), 185–193.

Battelino, T., Olsen, C. B., Schütz-Fuhrmann, I., Hommel, E., Hoogma, R., Schierloh, U., Sulli, N., & Bolinder, J. (2012). The use and efficacy of continuous glucose monitoring in type 1 diabetes treated with insulin pump therapy: A randomised controlled trial. *Diabetologia, 55*, 3155–3162.

Bergenstal, R. M., Klonoff, D. C., Garg, S. K., Bode, B. W., Meredith, M., Slover, R. H., Ahmann, A. J., Welsh, J. B., Lee, S. W., & Kaufman, F. R. (2013). Threshold- based insulin- pump interruption for reduction of hypoglycemia. *The New England Journal of Medicine, 369*, 224–232.

Blaslov, K., Naranda, F. S., Kruljac, I., & Renar, I. P. (2018). Treatment approach to type 2diabetes: Past, present and future. *WJD., 9*, 209–219.

Bogers, R. P., Bemelmans, W. J., & Hoogenveen, R. T. (2007). Association of overweight with increased risk of coronary heart disease partly independent of blood pressure and cholesterol levels: A meta-analysis of21 cohort studies including more than 300000 persons. *Arch Intern., 167*, 1720–1728.

Bolinder, J., Antuna, R., Geelhoed-Duijvestijn, P., Kröger, J., & Weitgasser, R. (2016). Novel glucose- sensing technology and hypoglycaemia in type 1 diabetes: A multicentre, non- masked, randomised controlled trial. *Lancet, 388*, 2254–2263.

Burckhardt, M. A., Smith, G. J., & Cooper, M. N. (2018). Real-world outcomes of insulin pumpcompared to injection therapy in a population-based sample of children with type 1 diabetes. *Pediatric Diabetes, 19*(8), 1459–1466.

Cafazzo, J. A., Casselman, M., Hamming, N., Katzman, D. K., & Palmert, M. R. (2012). Design of an mHealth app for the self- management of adolescent type 1diabetes: A pilot study. *Journal of Medical Internet Research, 14*, e70.

Cappon, G., Vettoretti, M., & Sparacino, G. (2019). Continuous glucose monitoring sensors for diabetes management: A review of technologies and applications. *Diabetes and Metabolism Journal, 43*(4), 383–397.

Christopher, F., Catherine, S. M., & Jennifer, K. Y. (2020). Type 1 diabetes in youth and technology-based advances in management. *Advances in Pediatrics, 67*, 73–91. https://doi.org/10.1016/j.yapd.2020.04.002

Coda, A., Sculley, D., Santos, D., Girones, X., & Acharya, S. (2018). Exploring the effectiveness of smart Technologies in the Management of type 2 diabetes mellitus. *Journal of Diabetes Science and Technology, 12*(1), 199–201.

Daly, A., & Hovorka, R. (2021). Technology in the management of type 2 diabetes: Present status and future prospects. *Diabetes, Obesity & Metabolism, 23*(8), 1722–1732.

Danne, T., Nimri, R., & Battelino, T. (2017). International consensus on use of continuous glucose monitoring. *Diabetes Care, 40*(12), 1631–1640.

Davies, M. J., D'Alessio, D. A., & Fradkin, J. (2018). Management of hyperglycaemia in type 2 diabetes, 2018. A consensus report by the American Diabetes Association (ADA) and the European Association for the Study of diabetes (EASD). *Diabetologia, 61*(12), 2461–2498.

Defronzo, R. A. (2009). Banting lecture. From the triumvirate to the ominous octet: A new paradigm for the treatment of type 2 diabetes mellitus. *Diabetes, 58*(4), 773–795.

Deiss, D., Szadkowska, A., & Gordon, D. (2019). Clinical practice recommendations on the routine use of ever sense, the first long-term implantable continuous glucose monitoring system. *Diabetes Technology & Therapeutics, 21*(5), 254–264.

DiMeglio, L. A., Molina, C. E., & Oram, R. A. (2018). Type 1 diabetes. *Lancet, 391*, 2449–2462.

Dominguez-Bendala, J., Lanzoni, G., & Klein, D. (2016). The human endocrine pancreas: New insights on replacement and regeneration. *Trends in Endocrinology and Metabolism, 27*(3), 153–162.

Dover, A. R., Stimson, R. H., Zammitt, N. N., & Gibb, F. W. (2017). Flash glucose monitoring improves outcomes in a type 1 diabetes clinic. *Journal of Diabetes Science and Technology, 11*, 442–443.

Duke, D. C., Barry, S., Wagner, D. V., Speight, J., Choudhary, P., & Harris, M. A. (2018). Distal technologies and type 1diabetes management. *The Lancet Diabetes and Endocrinology, 6*, 143–156.

El-Kaissi, S., & Sherbeeni, S. (2011). Pharmacological Management of Type 2 diabetes mellitus: An update. *Current Diabetes Reviews., 7*, 392–405.

Emperra Digital Diabetes Care. (2017). ESYSTA Personal and fully automatic. 2017. Accessed November 16, 2020. Available from https://www.emperra.com/en/esysta-product-system

Fagherazzi, G., & Ravaud, P. (2018). Digital diabetes: Perspectives for diabetes prevention, management and research. *Diabetes & Metabolism, 1042*, 1–8.

Frias, J. P., Bode, B. W., Bailey, T. S., Kipnes, M. S., Brunelle, R., & Edelman, S. V. (2011). A16-week open-label, multicenter pilot study assessing insulin pump therapy in patients with type 2 diabetes sub optimally controlled with multiple daily injections. *Journal of Diabetes Science and Technology, 5*(4), 887–893.

Garg, V. (2011). Noninsulin pharmacological management of type1 diabetes mellitus. *Indian Journal of Endocrinology and Metabolism, 15*, S5–S11.

Gildon, B. W. (2018). InPen smart insulin pen system: Product review and user experience. *Diabetes Spectrum: A Publication of the American Diabetes Association, 31*(4), 354–358.

Guh, D. P., Zhang, W., & Bansback, N. (2009). The incidence of co-morbidities related to obesity and overweight: A systematic review and meta-analysis. *BMC Public Health, 9*, 88.

Hellmund, R., Weitgasser, R., & Blissett, D. (2018). Cost calculation for a flash glucose monitoring system for adults with type 2 diabetes mellitus using intensive insulin-a UK perspective. *Eur. Endocrinol., 14*(2), 86–92.

Hoeks, L. B., Greven, W. L., & de Valk, H. W. (2011). Real- time continuous glucose monitoring system for treatment of diabetes: A systematic review. *Diabetic Medicine, 28*, 386–394.

Hoss, U., & Budiman, E. S. (2017). Factory-calibrated continuous glucose sensors: The science behind the technology. *Diabetes Technology & Therapeutics, 19*(S2), S44–S50.

Hunter, C. M. (2016). Understanding diabetes and the role of psychology in its prevention and treatment. *American Psychologist., 71*, 515–525.

International Diabetes Federation. (2020). IDF Diabetes Atlas, 9th ed., Brussels, Belgium. https://www.diabetesatlas.org. Accessed October 18, 2020.

Jonsson, S., Hedblad, B., & Engstrom, G. (2002). Influence of obesity on cardiovascular risk. Twenty-three-year follow-up of 22,025 men from an urban Swedish population. *International Journal of Obesity and Related Metabolic Disorders, 26*, 1046–1053.

Juvenile Diabetes Research Foundation Continuous Glucose Monitoring Study Group. (2008). Continuous glucose monitoring and intensive treatment of type 1 diabetes. *The New England Journal of Medicine, 359*, 1464–1476.

Juvenile Diabetes Research Foundation Continuous Glucose Monitoring Study Group. (2010). Effectiveness of continuous glucose monitoring in a clinical care environment: Evidence from the Juvenile Diabetes Research Foundation continuous glucose monitoring (JDRF-CGM) trial. *Diabetes Care, 33*, 17–22.

Kaissi, S. E., & Sherbeeni, S. (2011). Pharmacological Management of Type 2 diabetes mellitus: An update. *Current Diabetes Reviews., 7*, 392–405.

Katz, L. E. L., Bacha, F., Gidding, S. S., Weinstock, R. S., El-Ghormli, L., Libman, I., Nadeau, K. J., Porter, K., Marcovina, S., & McKay, S. (2018). Lipid profiles, inflammatory markers, and Insulin therapy in youth with type 2 diabetes. *The Journal of Pediatrics, 196*, 208–216.e2.

Khunti, K., & Seidu, S. (2019). Therapeutic inertia and the legacy of dysglycemia on the microvascular and macrovascular complications of diabetes. *Diabetes Care, 42*, 349–351.

Khunti, K., Wolden, M. L., Thorsted, B. L., Andersen, M., & Davies, M. J. (2013). Clinical inertia in people with type 2 diabetes: A retrospective cohort study of more than 80,000 people. *Diabetes Care, 36*(11), 3411–3417.

Klonoff, D. C. (2015). Cybersecurity for connected diabetes devices. *Journal of Diabetes Science and Technology, 9*(5), 1143–1147.

Klonoff, D. C., & Kerr, D. (2018). Smart pens will improve insulin therapy. *Journal of Diabetes Science and Technology, 12*(3), 551–553.

Lee, J. M., Newman, M. W., Gebremariam, A., Choi, P., Lewis, D., Nordgren, W., Costik, J., Wedding, J., West, B., Gilby, N. B., Hannemann, C., Pasek, J., Garrity, A., & Hirschfeld, E. (2017). Real-world use and self-reported health outcomes of a patient-designed do-it-yourself Mobile technology system for diabetes: Lessons for Mobile health. *Diabetes Technology & Therapeutics, 19*(4), 209–219.

Leslie, R. D. (2010). Predicting adult-onset autoimmune diabetes: Clarity from complexity. *Diabetes, 59*(2), 330–331.

Li, A., & Hussain, S. (2020). Diabetes technologies - what the general physician needs to know. *Clinical Medicine (London, England), 20*(5), 469–476.

Luna, B., & Feinglos, M. N. (2001). Oral agents in the Management of Type 2 diabetes mellitus. *American Family Physician, 63*(9), 1747–1756.

Maahs, D. M., Calhoun, P., Buckingham, B. A., Chase, H. P., Hramiak, I., Lum, J., Cameron, F., Bequette, B. W., Aye, T., Paul, T., Slover, R., Wadwa, R. P., Wilson, D. M., Kollman, C., & Beck, R. W. (2014). A randomized trial of a home system to reduce nocturnal hypoglycemia in type 1 diabetes. *Diabetes Care, 37*(7), 1885–1891.

Magliano, D. J., Sacre, J. W., Harding, J. L., Gregg, E. W., Zimmet, P. Z., & Shaw, J. E. (2020). Young onset type 2 diabetes mellitus—Implications for morbidity and mortality. *Nature Reviews. Endocrinology, 16*, 321–331.

Miller, C. K., Jean, J. L., Kristeller, L., Amy Headings, A., Nagaraja, H., & Miser, W. F. (2012). Comparative effectiveness of a mindful eating intervention to a diabetes self-management intervention among adults with type 2 diabetes: A pilot study. *Journal of the Academy of Nutrition and Dietetics, 112*(11), 1835–1842.

Monami, M., Lamanna, C., Marchionni, N., & Mannucci, E. (2010). Continuous subcutaneous insulin infusion versus multiple daily insulin injections in type 1 diabetes: A meta- analysis. *Acta Diabetologica, 47*(Suppl. 1), 77–81.

Montagnana, M., Caputo, M., Giavarina, D., & Lippi, G. (2009). Overview on self-monitoring of blood glucose. *Clinica Chimica Acta, 402*, 7–13.

Noh, R. M., Graveling, A. J., & Frier, B. M. (2011). Medically minimising the impact of hypoglycaemia in type 2 diabetes: A review. *Expert Opinion on Pharmacotherapy, 12*(14), 2161–2175.

Novo Nordisk. (2020). *Instructions for use: our pens and needles, 2019.* Accessed November 19, 2020. Available from https://www.novonordisk.com/our-products/pens-and-needles/instructions-foruse.html.

Novo Nordisk Blue sheet. (2010). *Quarterly perspective on diabetes and chronic diseases.* Available from: http://www.press.novonordisk-us.com/bluesheet-issue2/downloads/NovoNordisk_Bluesheet_Newsletter.pdf.

Pendiq Intelligent Diabetes Care. (2020). PENDIQ 2.0. 2017. Accessed November 20, 2020. Available from https://pendiq.com/en/home/

Penfornis, A., Personeni, E., & Borot, S. (2011). Evolution of devices in diabetes management. *Diabetes Technology & Therapeutics, 13*, S93–S102.

Peyrot, M., Barnett, A. H., Meneghini, L. F., & Schumm-Draeger, P. M. (2012). Insulin adherence behaviours and barriers in the multinational global attitudes of patients and physicians in insulin therapy study. *Diabetic Medicine, 29*(5), 682–689.

Pickup, J. C., Freeman, S. C., & Sutton, A. J. (2011). Glycaemic control in type 1 diabetes during real time continuous glucose monitoring compared with self monitoring of blood glucose: Meta-analysis of randomised controlled trials using individual patient data. *British Medical Journal, 343*, d3805.

Pickup, J. C., Keen, H., & Parsons, J. A. (1978). Continuous subcutaneous insulin infusion: An approach to achieving normoglycaemia. *British Medical Journal, 1*(6107), 204–207.

Pickup, J., Mattock, M., & Kerry, S. (2002). Glycaemic control with continuous subcutaneous insulin infusion compared with intensive insulin injections in patients with type 1 diabetes: Metaanalysis of randomised controlled trials. *British Medical Journal, 324*(7339), 705.

Pickup, J. C., & Sutton, A. J. (2008). Severe hypoglycaemia and glycaemic control in type 1 diabetes: Metaanalysis of multiple daily insulin injections compared with continuous subcutaneous insulin infusion. *Diabetic Medicine, 25*, 765–774.

Ramadasa, A., Queka, K. F., Chana, C. K. Y., & Oldenburgb, B. (2011). Web-based interventions for the management of type 2 diabetes mellitus: A systematic review of recent evidence. *International Journal of Medical Informatics, 80*, 389–405.

Reddy, M., Jugnee, N., Laboudi, A. E., Spanudakis, E., Anantharaja, S., & Oliver, N. (2018). A randomized controlled pilot study of continuous glucose monitoring and flash glucose monitoring in people with type 1 diabetes and impaired awareness of hypoglycaemia. *Diabetic Medicine, 35*, 483–490.

Rewers, M. J., Pillay, K., de Beaufort, C., Craig, M. E., Hanas, R., Acerini, C. L., & Maahs, D. M. (2014). ISPAD clinical practice consensus guidelines 2014. Assessment and monitoring of glycemic control in children and adolescents with diabetes. *Pediatric Diabetes, 15*(Suppl. 20), 102–114.

Rodbard, D. (2017). Continuous glucose monitoring: A review of recent studies demonstrating improved glycemic outcomes. *Diabetes Technology & Therapeutics, 19*(Suppl 3), S25–S37.

Roze, S., Duteil, E., & Smith-Palmer, J. (2016). Cost-effectiveness of continuous subcutaneous insulin infusion in people with type 2 diabetes in the Netherlands. *Journal of Medical Economics, 19*(8), 742–749.

Sangave, N. A., Aungst, T. D., & Patel, D. K. (2019). Smart connected insulin pens, caps, and attachments: A review of the future of diabetes technology. *Diabetes Spectrum: A Publication of the American Diabetes Association, 32*(4), 378–384.

Seidu, S., Cos, X., Brunton, S., Harris, S. B., Jansson, S. P. O., Cases, M. M., Neijens, A. M. J., Topsever, P., & Khunti, K. (2021). A disease state approach to the pharmacological management of Type2 diabetes in primary care: A position statement by primary Care diabetes Europe. *Primary Care Diabetes, 15*(1), 31–51.

Shah, V. N., Moser, E. G., Blau, A., Dhingra, M., & Garg, S. K. (2013). The future of basal insulin. *Diabetes Technology & Therapeutics, 15*, 727–732.

Shah, R. B., Patel, M., Maahs, D. M., & Shah, V. N. (2016). Insulin delivery methods: Past, present and future. *International Journal of Pharmaceutical Investigation, 6*(1), 1–9.

Sherr, J. L., Hermann, J. M., Campbell, F., Foster, N. C., Hofer, S. E., Allgrove, J., Maahs, D. M., Kapellen, T. M., Holman, N., Tamborlane, W. V., Holl, R. W., Beck, R. W., & Warner, J. T. (2016). Use of insulin pump therapy in children and adolescents with type 1 diabetes and its impact on metabolic control: Comparison of results from three large, transatlantic paediatric registries. *Diabetologia, 59*, 87–91.

Singhal, S., & Kumar, S. (2021). Current perspectives on Management of Type 2 diabetes in youth. *Children, 8*(1), 37.

Tamborlane, W. V., Sherwin, R. S., & Genel, M. (1979). Reduction to normal of plasma glucose in juvenile diabetes by subcutaneous administration of insulin with a portable infusion pump. *The New England Journal of Medicine, 300*(11), 573–578.

Tauschmann, M., & Hovorka, R. (2018). *Technology in the management of type 1 diabetes mellitus-current status and future prospects*. Endocrinology, Macmillan Publishers Limited, Part of Springer Nature.

Trawley, S., Browne, J. L., Hagger, V. L., Hendrieckx, C., Holmes-Truscott, E., Pouwer, F., Skinner, T. C., & Speight, J. (2016). The use of Mobile applications among adolescents with type 1 diabetes: Results from diabetes MILES youth—Australia. *Diabetes Technology & Therapeutics., 18*(12), 1–7.

Vallejo Mora, M. D. R., Carreira, M., Anarte, M. T., Linares, F., Olveira, G., & Romero, S. G. (2017). Bolus calculator reduces hypoglycemia in the short term and fear of hypoglycemia in the long term in subjects with type 1diabetes (CBMDI study). *Diabetes Technology & Therapeutics, 19*, 402–409.

Weisman, A., Bai, J. W., Cardinez, M., Kramer, C. K., & Perkins, B. A. (2017). Effect of artificial pancreas systems on glycaemic control in patients with type 1 diabetes: A systematic review and meta-analysis of outpatient randomised controlled trials. *The Lancet Diabetes and Endocrinology, 5*, 501–512.

Weiss, R., Garg, S. K., Bode, B. W., Bailey, T. S., Ahmann, A. J., Schultz, K. A., Welsh, J. B., & Shin, J. J. (2015). Hypoglycemia reduction and changes in hemoglobin A1c in the ASPIRE in-home study. *Diabetes Technology & Therapeutics, 17*, 542–554.

Wild, S., Roglic, G., Green, A., Sicree, R., & King, H. (2004). Global prevalence ofdiabetes: Estimates for the year 2000 and projections for 2030. *Diabetes Care, 27*(5), 1047–1053.

Wong, J. C., Neinstein, A. B., Spindler, M., & Adi, S. (2015). A minority of patients with type 1 diabetes routinely downloads and retrospectively reviews device data. *Diabetes Technology & Therapeutics, 17*, 555–562.

Yaron, M., Roitman, E., & Aharon-Hananel, G. (2019). Effect of flash glucose monitoring technology on glycemic control and treatment satisfaction in patients with type 2 diabetes. *Diabetes Care, 42*(7), 1178–1184.

YDS Delivery Systems. (2020). *Smart Pilot-Transforming YpsoMate into a smart product system, 2020*. Accessed November 19, 2020. Available from https://yds.ypsomed.com/en/injection-systems/smartdevices/smartpilot-for-ypsomate.html.

The Broader Aspects of Treating Diabetes with the Application of Nanobiotechnology

Rupak Roy, Aditi Chakraborty, Kartik Jana, Bhanumati Sarkar ⓘ, Paramita Biswas ⓘ, and Nithar Ranjan Madhu ⓘ

Abstract

When it comes to the topic of diabetes, the first thing that comes to mind is "diabetes mellitus." Another form of diabetes is diabetes insipidus, with different disease mechanisms that do not fall within this chapter's scope. Before we start discussing what diabetes is, let us first dig deep into history and find out how diabetes got its name and when people first knew about this disease. From various sources, we know that diabetes was known to ancient Egyptians, Greeks, and Indians. The word *mellitus* is a Latin word meaning sweet, and Greeks used the term *mellitus* to define something sweet. The Greeks knew about honey, which they called *mellita*; hence, the name mellitus comes from this term. Ancient Greeks used to taste urine as a method to test for diabetes. They also noticed that people with sweet urine tend to drink more fluids, and the fluid almost gets out of the body immediately like a siphon. They knew siphon by the term *diabetes*, and from there, we got the name "diabetes mellitus." The earliest record, written around 1500 BC, of diabetes we got is from *Ebers papyrus*, which was excavated from graves in Thebes, an ancient Egyptian city, around 1862 AD. Egyptologist Georg Ebers published the findings from the papyrus in 1874 AD. Diabetes was also known to Indians at around the same time by another

R. Roy · A. Chakraborty · K. Jana
SHRM Biotechnologies Pvt. Ltd., Kolkata, West Bengal, India

B. Sarkar
Department of Botany, Acharya Prafulla Chandra College, Kolkata, West Bengal, India

P. Biswas
Regional Research Sub Station (OAZ), Uttar Banga Krishi Viswavidyalaya, Malda, West Bengal, India

N. R. Madhu (✉)
Department of Zoology, Acharya Prafulla Chandra College, Kolkata, West Bengal, India

name—"madhumeha," which roughly translates to honey urine. Indian physicians by that time also noticed that urine from such patients attracted flies and ants, thereby devising the first clinical test for diabetes.

Keywords

Diabetes · Nanotechnology · Healthcare · Insulin

1   Introduction

Diabetes is presently a common disease. According to statistical analysis, around 60% of the population suffers from diabetes (Balakumar et al., 2016). In the long term, diabetes can affect different organs and may increase complications in patients. There are some chronic complications of diabetes, such as diabetic foot, peripheral neuropathy, segmental bone injury, diabetic retinopathy, diabetic nephropathy, and other complications. These complications will significantly impact patients having acute metabolic disorders, like diabetic ketoacidosis (DKA) and hyperosmolar hyperglycemia syndrome. Researchers are reviewing whether diabetes is caused by bad lifestyles or genetic inheritance. A few research papers explain that too much work pressure increases stress, which may cause hormone imbalance. Recent studies have shown that eating junk food may also be a reason for acquiring diabetes. Also, according to recent studies, it is expected that by 2030, diabetes will become the seventh largest cause of death (Yach et al., 2006). The complication of diabetes is irreversible, causing severe burden and injury to patients, both physically and economically. Diabetes people can be treated with oral medicine before each meal; they do this by pricking their fingers with a needle to monitor their blood glucose level, and this process is painful. Medicinal research, however, has developed new and innovative ideas for treating diabetes patients. Recent research has been done to enable innovative drugs to be easily absorbed into our bodies, and their size is in nanoscales.

Nanotechnology is a brand-new and cutting-edge technology of the twenty-first century. It is a recently developed discipline that has been creatively applied to fields, including physics, chemistry, biology, medicine, and environmental science. In recent years, nanotechnology has advanced the development of diabetic medication. Drugs between 1 and 100 nm in size can be created using nanotechnology (Nicolaou, 2018). Due to the use of nanoscales, modern drug development can now benefit from further research into how readily nanotechnology can be incorporated into small molecular sizes. Because of their tiny size, permeability, and ability to maintain strength, nanoparticles (NPs) can be absorbed by target organs, increasing treatment efficacy while reducing adverse effects and drug usage. Meanwhile, researchers have examined many nanoparticles with various consumption rates and particle sizes.

Gene therapy, nanosensors, and drug design are examples of biomedical applications. These are typically found in drug delivery systems. Characterizing nanomaterials has a number of one-of-a-kind impacts, including the quantum effect, which results in an increase in surface energy; small crystal size; and a rise in surface atomic ratio, which speeds up the interface effect (Sun, 2007). A recent study describes a glucose sensor as a component with a nanoscale that examines glucose using carbon nanostructures and nanoparticles. Additionally, nanoscale components frequently improve glucose sensors' sensitivity and temporal response and can result in sensors that enable continuous in vivo glucose monitoring (DiSanto et al., 2015). Insulin is a hormone that the body produces in the pancreas cell. This hormone plays an important role in the transport of glucose from the blood into the cells of the body, where it is converted into an energy source that is used by the body. When a person with diabetes is not properly treated, their body does not use insulin as it should. In this situation, blood glucose levels continue to rise. High blood sugar is the ailment being discussed here. Numerous health issues could result from this, some of which could be critical or severe.

2 Global Scenario of Diabetes

Compared to type 1 diabetes, type 2 diabetes is more prevalent. Type 1 affects more than one in ten people, while type 2 affects the remainder (Donnelly et al., 2005). With age, the prevalence of diabetes rises. Diabetes affects about 10.5% of the general population. The rate among people who are 65 years of age and older rises to 27%. Meanwhile, 8.5% of persons 18 years of age and older had diabetes in 2014 (Meo et al., 2017). A total of 1.5 million deaths were directly related to diabetes in 2019, and 48% of these deaths occurred in those under the age of 70. Premature diabetes mortality rates (mortality before the age of 70) increased by 5% between 2000 and 2016, as claimed by the World Health Organization.

The body's inefficient use of insulin leads to type 2 diabetes, also known as non-insulin-dependent or adult-onset diabetes. Type 2 diabetes affects more than 95% of those who have the disease. This particular type of diabetes is primarily brought on by increased body weight and inactivity. Although frequently less severe, type 2 diabetes symptoms might be comparable to type 1. As such, the condition may not be discovered until after it has developed problems. This type of diabetes was previously exclusively found in adults, but it is also increasingly common in kids today. In India, 77 million people were estimated to have diabetes in 2019, and by 2045, that number is projected to reach over 134 million. A little more than 57% of these people are still undiagnosed. The most common form of diabetes, type 2, can cause microvascular and macrovascular problems that can affect multiple organ systems (Pradeepa & Mohan, 2021). Diabetes affects both men and women equally. Over the next 5–10 years, the chance of developing type 2 diabetes is increased by a factor of 5 by the metabolic syndrome and is expected to increase by a factor of 2 (Meo et al., 2017).

The Global Diabetes Compact, introduced by WHO in April 2021, is a global project focused on assisting low- and middle-income countries. Its goal is to achieve lasting gains in diabetes prevention and care. National governments, United Nations (UN) agencies, nongovernmental organizations, businesses, academic institutions, philanthropic foundations, people with diabetes, and international donors are joining forces through the Compact to work toward a common goal of lowering the risk of diabetes and guaranteeing that everyone who is diagnosed with the disease has access to equitable, complete, affordable, and high-quality treatment and care.

The World Health Assembly adopted a Resolution in May 2021 that calls for improved diabetes prevention and management. It urges action in areas such as expanding access to insulin, fostering convergence and harmonization of regulatory requirements for insulin and other medications and healthcare products used to treat diabetes, and determining the viability and potential value of establishing a web-based tool to share information pertinent to the transparency of markets for diabetes medicines and healthcare products.

According to the first WHO Global report on diabetes, there are now 422 million persons worldwide who have the disease, nearly quadrupling since 1980. The rise in type 2 diabetes and its contributing variables, such as obesity and overweight, are substantially to blame for this sharp increase. Diabetes alone resulted in 1.5 million fatalities in 2012. Heart attack, stroke, blindness, kidney failure, and lower limb amputation are only a few of its complications. Governments are urged in the new report to make sure that people can make healthy decisions and that healthcare systems properly identify, treat, and care for diabetics. It motivates each of us to eat sensibly, exercise regularly, and keep our weight in check. According to the National Institutes of Health's (NIH's) National Institute of Diabetes and Digestive and Kidney Diseases, in the United States, 34.5% of persons aged 18 and above, or 88 million people, are prediabetic. This includes just (i) under 29 million adults between ages 18 and 44 (24.3% of US adults in this age group), (ii) more than 35 million adults between ages 45 and 64 (41.7% of US adults in this age group), and (iii) more than 24 million adults, or 46.6% of all US adults, who are 65 years of age or older. Men (37.4% of US adults) are more likely than women (29.2%) to have prediabetes, but regardless of race, ethnicity, or level of education, prediabetes is equally prevalent in men and women. In the United States, 18% of adolescents aged 12–18 have prediabetes, affecting more than one in six.

3 Pathophysiology of Diabetes and Its Types

Ancient Chinese and Indian physicians described diabetic patients who complained of poor breath and extreme thirst, which were likely caused by ketosis, a disorder unknown to them in those early days. Since then, issues with diabetes have gotten worse, to the point where it is now a multimetabolic condition that simultaneously affects several organs.

In general, diabetes is characterized by elevated blood sugar levels. In diabetology, researchers examine the root reasons for this elevated blood sugar level. Diabetes normally develops when cells cannot absorb glucose despite the presence of insulin, a hormone that regulates blood sugar and is produced by beta cells in the pancreatic islets of Langerhans. Along with clinical subtypes of diabetes, we will examine in the following sections several additional clinical problems that contribute to the causes of diabetes.

Only a few additional common kinds of diabetes, such as type 1 and type 2 diabetes and gestational diabetes, are known to the general public. There are many additional subtypes of diabetes; however, they affect only less than 5% of all the people with the disease.

3.1 Type 1 Diabetes

When the body's capacity to create enough insulin is somehow hampered, type 1 diabetes can occur. When those beta cells are damaged or subjected to an autoimmune attack, they are unable to produce insulin, which results in a drop in insulin production. Type 1 diabetes is nearly usually caused by an autoimmune attack on beta cells, but researchers are still trying to determine the precise reason why immune cells would attack beta cells and either fully destroy them or render them useless. Type 1 diabetes symptoms often appear within months and exhibit severe clinical signs. If untreated, blood sugar can rise quickly as a result of an abrupt insulin shortage and begin to harm other organs (Meo et al., 2017).

3.2 Type 2 Diabetes

More than 90% of people with diabetes have this form, making it the most prevalent type of diabetes. Type 2 diabetes is often referred to as middle-aged diabetes as it typically occurs in people after 30 years of age. However, there are cases of patients with type 2 diabetes appearing before the age of 30. This type of diabetes occurs when, despite sufficient insulin in the bloodstream, sugar molecules cannot properly make their way into body cells. Scientists termed this phenomenon insulin resistance (Pradeepa & Mohan, 2021). Type 2 diabetes is often associated with obesity and abnormally high body weight, lack of exercise, high carbohydrate diet, etc., suggesting this subtype is more of a lifestyle disorder.

3.3 Gestational Diabetes

Gestational diabetes occurs almost exclusively during pregnancy in those who never had diabetes. Typically, symptoms go away after the delivery of the baby but pose significant risks of developing again during successive pregnancies. Patients who suffer gestational diabetes once or more in their lifetime also risk getting type 2 diabetes.

3.4   MODY or Maturity-Onset Diabetes in Young

MODY is the first form of diabetes that is exclusively hereditary with defects in a single gene (HNF family) that result in the early onset of diabetes before adolescence. The common subtypes of MODY include HNF1-alpha, HNF1-beta, HNF4-alpha, and glucokinase. Other rare subtypes of diabetes include LADA (latent autoimmune diabetes in adults), steroid-induced diabetes, neonatal diabetes (different from type 1 diabetes, which is also known as juvenile diabetes), type 3c diabetes, Wolfram syndrome, and Alstrom syndrome.

3.5   Mechanism

The regulation of cellular processes such as protein transcription, translation, and posttranslational activity by insulin is crucial. Additionally, superoxide free radicals produced by hyperglycemia-induced protein glycation and oxidative stress are linked to the etiology of diabetes. Reactive products that are created as a result of the production of reactive oxygen species may lead to lipid peroxidation and serious harm to the molecules and structures of cells. The detection of glycosylated hemoglobin A1c (HbA1c) levels, which provide a snapshot of blood glucose levels over 3 months, is a component of the clinical diagnosis of diabetes. There is a greater need for efficient disease treatment as the prevalence of type 2 diabetes mellitus rises globally.

Diabetes has a significant negative impact on cardiovascular mortality. Numerous physio-pathological alterations in the cardiovascular system are linked to it. The largest risk factors for coronary artery disease (CAD) include hemostatic disorders and endothelial dysfunction, as shown in Fig. 1, while microangiopathy, myocardial

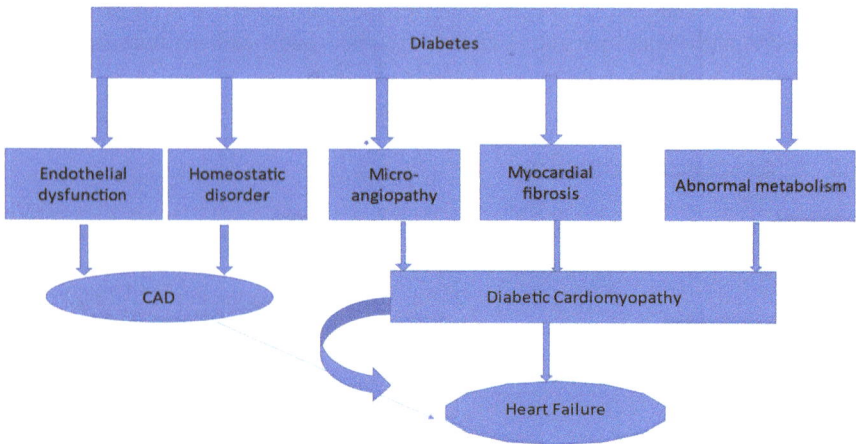

Fig. 1 Diabetes risk factors

fibrosis, and aberrant myocardial metabolism have been linked to the development of diabetic cardiomyopathy (Lotfy et al., 2017).

3.6 The Diabetic Signaling Pathway Process

Numerous signaling pathways, including the insulin signaling system, the AMP-activated protein kinase (AMPK) pathway, the peroxisome proliferator-activated receptor (PPAR) regulation pathway, and the chromatin modification pathway, have been linked to the pathophysiology of diabetes, according to increasing research. Thus, the primary source of prospective new therapeutic targets for treating metabolic disorders and diabetes has turned out to be these signaling pathways (Chu, 2022).

3.7 Insulin

Figures 2 and 3 illustrate that insulin resistance in diabetes is partially mediated by decreased insulin receptor (IR) expression levels. Following this, there is a reduction in the interaction of the regulatory subunit (P85) of phosphoinositide-3 kinase (PI3K) with IRS-1 and defective tyrosine phosphorylation of IR and subsequent tyrosine phosphorylation of IRS-1. Its catalytic component (P110) is subsequently deactivated as a result (Shaeena & Mani, 2021). Because of this, when the PI3K signaling pathway is diminished, the protein kinase AKT is activated, which causes a decrease in glucose transport. The glycogen synthase kinase (GSK) 3 pathway is subsequently activated by PI3K to control glycogen and lipid synthesis and promote

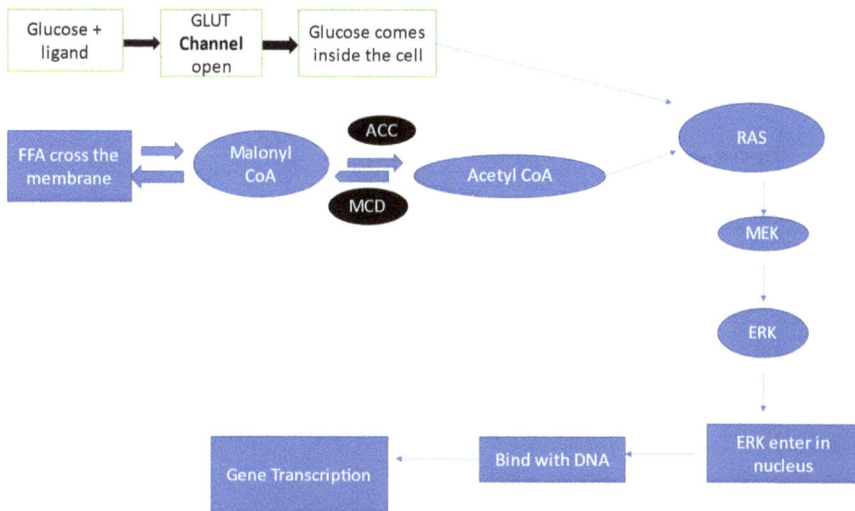

Fig. 2 Glucose pathway

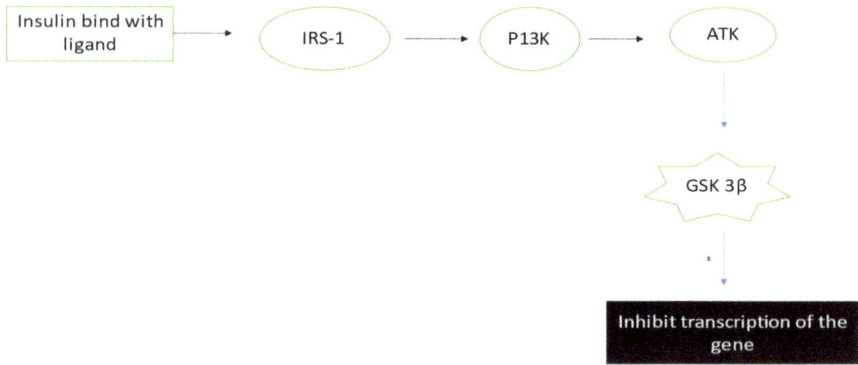

Fig. 3 Insulin pathway

glucose absorption. The Ras/MEK/ERK pathway is another mechanism through which PI3K controls cell proliferation (Shaeena & Mani, 2021).

3.8 AMPS

A key energy sensor is AMP-activated protein kinase (AMPK). The gluconeogenic enzymes phosphoenolpyruvate carboxykinase (PEPCK) and glucose 6-phosphatase (G6Pase) are deactivated by activated AMPK, reducing hepatic glucose synthesis. By activating glucose transporters, it increases glucose absorption (GLUT). By inhibiting acetyl-CoA carboxylase (ACC) and activating malonyl-CoA decarboxylase (MCD), AMPK also increases lipid metabolism by lowering malonyl-CoA levels (Landree et al., 2004). By activating AMPK in peripheral tissues, the natural substance berberine (BBR) has been shown to lower body weight, enhance glucose tolerance, and facilitate insulin action.

3.9 PPAR

The ligand-activated transcription factors known as peroxisome proliferator-activated receptors (PPARs) have been used as therapeutic targets for developing drugs to treat metabolic syndrome. PPARs come in three different isoforms. The liver, heart, muscle, and kidney all express PPAR, which controls the transport and metabolism of fatty acids. Adipose tissue, muscle, and macrophages express PPAR, which controls adipogenesis and lipid synthesis (Gupta et al., 2021). Involved in fat oxidation, energy expenditure, and lipid storage, PPAR is widely expressed (Gupta et al., 2021).

4   Introduction to Nanotechnology

Nanotechnology is nowadays the most popular and effective field in medicinal, atomic, nanoscale, and nanomaterial production. Nanotechnology is the kind of research and development that uses atoms at a molecular level. Nanotechnology is applicable in the range between 0.1 and 100 nm. A new particle arrangement can be made by arranging and combining different particles. Nanomaterials are highly active, and various interactions exist between various nano units. This property gives nanomaterials their distinctive effects, including the interface effect brought on by a rapid increase in the surface atomic ratio (Mazari et al., 2021), the quantum effect brought on by an increase in surface energy, the size effect resulting from a large specific surface area, and the side effect caused by the small crystal structure and small size.

Given the small molecular size and multiple functions, nanotechnology is used in the research of modern medicine and in trying to solve problems that earlier were not solved. A drug produced by nanotechnology is more target orientated. It reduces side effects and has the property of sustainable release, which can enhance the action time of the drug. Nanotechnology improves the stability of the drug action, facilitates drug storage, and helps invent new drug delivery routes. Drug development by nanotechnology has low doses and higher drug-loading facilities (Benn, 2009). This drug is easily diffused into the bloodstream without damaging the vascular endothelial tissue. The drug is protected from degradation by other enzymes, and due to this factor, the drugs have more efficacy than other drug administrations. They have many forms, like nanoparticles, nanoemulsions, nanomaterials, and nano-drug crystallization.

NP-related research has grown in popularity as nanotechnology has progressed. Nanocarrier medications must have the right particle size, low toxicity, biodegradability, and good biocompatibility in order to function. They can have a therapeutic impact, ensure concentration, precisely locate the lesion, and do not harm healthy cells (Chen et al., 2016; Dwivedi et al., 2012).

Nanoemulsion is a non-equilibrium system with an average particle size of 1–100 nm and the appropriate ratio of the oil phase, water phase, and surfactant. It is low viscous, highly effective, permeable, easy to prepare, and extremely safe. Enhancing bioavailability is possible. Due to its specific particle size, it also possesses targeting and gradual release effects (Pradeepa & Mohan, 2021). The medicine can be administered in various methods, including transdermally, orally, mucosally, intravenously, and more, depending on the features of the drug and the various emulsions (Lotfy et al., 2017).

Hydrogel, a three-dimensional network structure with particles smaller than 200 nm, makes up the nanomaterials (Chu, 2022). Nanomaterials can enclose active components to shield them from environmental deterioration because they have a three-dimensional network structure with the same hydrogel, high water content, and swelling characteristics (Shaeena & Mani, 2021).

In the area of oral protein and peptide absorption, nanotechnology has many potential uses. Certain ligands can alter the properties of nanocarriers. In order to diagnose the condition and track how far the disease has progressed, nanoparticles are utilized to transport ribosomal ribonucleic acid (RNA) and proteins from one location to another. In addition to nanocarriers, the respiratory system offers another effective pathway for the administration of drugs. In the recent past, numerous strategies have been presented for the noninvasive monitoring of blood glucose, which is something that nanotechnology has the potential to make possible. Insulin administration in the treatment of diabetes is now being researched using some different nanoparticle kinds, such as (1) polymeric biodegradable nanoparticles, (2) polymeric micelles, (3) ceramic nanoparticles, (4) liposomes, and (5) dendrimers (Ahmed, 2015).

5   Conventional Treatment Approach Today

Since diabetes was discovered, humans have tried to control it with varying degrees of success. The prognosis of diabetes had not changed much until the late 1980s–1990s when lots of treatments became available and understanding of the disease mechanism grew. Ancient physicians noticed that people who developed diabetes were typically a rich section of the population, who did not work much and had more food with variety. Back in those days, food for rich people included a high carbohydrate diet and sweets; hence, they developed diabetes more. Therefore, physicians predominantly tried to control diabetes with a lifestyle change, with a diet that has more proteins, fats, and carbohydrates. Even to this day, diet and lifestyle changes have been as much important as drugs. Although there is no cure for diabetes even today, diabetes is largely controlled by the use of various groups of drugs and, of course, artificial insulin.

Here, we will discuss the available class of drugs used to treat diabetes, along with insulin, which too comes with a varying degree of bioavailability and a tweaked mode of action to achieve better control of the disease.

5.1   Biguanides

Biguanides are a class of antihyperglycemic drugs that help lower blood sugar levels by decreasing the amount of sugar produced in the liver (decreases hepatic gluconeogenesis), thereby reducing sugar molecule absorption in the intestines. Metformin is the only antihyperglycemic drug in the biguanides class and also the primary choice for a drug against type 2 diabetes.

5.2   Meglitinides or Glinides (Prandial Glucose Regulators)

Meglitinides/glinides are orally administered drugs used exclusively for treating type 2 diabetes. Meglitinides work by acting on sulfonylurea receptors (SUR1) on the beta cells of the pancreas that are associated with adenosine triphosphate

(ATP)-sensitive potassium channel. Meglitinides binding to SUR1 receptors bring about a depolarization effect within the beta cells, opening a voltage-gated calcium channel, thereby causing insulin to release. Meglitinides can only work if beta cells are functional, and therefore, they cannot be used against type 1 diabetes, where beta cells do not produce insulin. A few drugs of the glinide class are repaglinide and nateglinide (Meglitinide—An Overview|ScienceDirect Topics, n.d.; Patlak, 2002).

5.3 Sulphonylureas

Sulphonylureas were first discovered in 1942 quite by accident. While studying sulfonamide antibiotics, scientist Marcel Janbon and his coworkers noticed that sulfonamides induced hypoglycemia in animals, leading to later research on sulfonamides as potential antidiabetic medicine. Sulfonylureas work similarly to meglitinides, targeting the same potassium on the beta cells of the pancreas, causing a similar depolarization effect. This depolarization leads to a rise in intracellular calcium ions, which in turn stimulate the production of more insulin from the beta cells (Seino, 2012; Derosa & Maffioli, 2012). Like meglitinides, sulfonylureas too cannot be used on patients with type 1 diabetes or with injury in and around the pancreas. Examples of sulfonylurea drugs are tolbutamide, tolazamide, glipizide, glibenclamide, and glimepiride.

5.4 Alpha-Glucosidase Inhibitors

Alpha-glucosidase inhibitors act as competitive inhibitors for the enzyme alpha-glucosidase, found in the linings of the small intestine, which is responsible for converting complex polysaccharides (present in carbohydrate diets) to simple glucose molecules and other monosaccharides for easy absorption by the intestine. Alpha-glucosidase inhibitors, by inhibiting alpha-glucosides, reduce the amount of glucose production from polysaccharides, thereby lowering blood sugar levels. Examples of alpha-glucosidase inhibitors are acarbose, miglitol, and voglibose (Bradley, 2002).

5.5 Glitazones or Thiazolidinedione

Glitazones are a new class of drugs used to treat diabetes mellitus type 2. Glitazones act on peroxisome proliferator-activated receptors, a group of nuclear receptors, resulting in the decrease of fatty acids in circulation, thereby forcing cells to depend on the oxidation of carbohydrates and reducing blood sugar levels. These family drugs are pioglitazone, rosiglitazone, and lobeglitazone (Gupta & Kalra, 2011).

5.6   DPP-4 Inhibitors/Gliptins

Gliptins, otherwise known as dipeptidyl-peptidase inhibitors, are a relatively new class of drugs with comparatively fewer side effects. Members of this class are sitagliptin, vildagliptin, saxagliptin, linagliptin, etc. (Hansen et al., 2010).

5.7   Incretin Mimetics/GLP-1 Analogs

Drugs called incretin mimetics mimic the effects of incretin hormones like glucagon-like peptide-1, which increases the amount of insulin produced by the pancreatic beta cells. The intestinal peptide hormones glucose-dependent insulinotropic polypeptide (GIP) and glucagon-like peptide-1 (GLP-1) are released after eating. Their major job is to increase the pancreas's synthesis of insulin when glucose is present. Lixisenatide, a GLP-1 analog, mimics the natural hormone GLP-1 and encourages insulin secretion (Adeghate & Kalász, 2011).

5.8   Amylin Analogs

Amylin, also known as islet amyloid polypeptide (IAPP), is produced alongside insulin from the beta cells of the pancreas. Amylin-analog-like pramlintide, when introduced with insulin in type 1 diabetes patients, reduces postprandial hyperglycemia compared to insulin alone. Pramlintide also slows the rate at which the stomach empties, giving a full stomach feeling, thereby decreasing food intake and indirectly regulating glucose (Seoudy et al., 2021).

5.9   SGLT2 Inhibitors/Gliflozin Sodium-Glucose Cotransporter 2

Sodium-glucose cotransporter 2 is responsible for the reabsorption of almost 90% of glucose in the renal system. In patients with diabetes, when sodium-glucose transport protein 2 (SGLT2) inhibitors block this enzyme, glucose reuptake is stopped, and glucose is released through urine—better known by the term glycosuria. SGLT2 inhibitors, by promoting glycosuria, can control high blood sugar levels but come with a side effect as they cannot be used for patients with renal problems. Some members of this class of drug are canagliflozin, dapagliflozin, and empagliflozin (Nie et al., 2020).

6 Nanotechnology as a Future Prospect in Diabetes Treatment

Nanotechnology is being thought of as a potential drug delivery system to counter existing problems with diabetic medications. Nanotechnology deals with particle sizes ranging from 10^{-9} m to 10^{-7} m. When it comes to treatment options with nanoparticles, especially in the case of diabetes, nanoparticles themselves cannot be used as drugs. Instead, they are thought of as drug molecule carriers. Further below, we will discuss in detail the nanoparticles that can be used as drug molecule carriers and what their advantages are over conventional drugs still in use.

6.1 Vesicular System

Vesicular systems are modern drug delivery systems mainly colloidal in nature and are made of an aqueous core, as well as one or more lipid bilayers. Vesicular systems have hydrophilic agents that comprise the inner core, and lipophilic drug molecules are enclosed in the lipid bilayer. Upon entry into the system, drugs cross this lipid bilayer at a specific site, leading to region-specific drug delivery (Akbarzadeh et al., 2013). A few types of vesicular systems potentially treating diabetes are discussed below.

6.2 Liposomes

Liposomes are spherical, phospholipid bilayered vesicles that are of great use in modern medicine, especially in the field of drug delivery. Liposomes are highly biocompatible and hence are thought to be more tolerant in vivo. Studies have found that drugs, when encapsulated within liposomes and administered in the body, exhibit a sustained/prolonged release effect, thereby giving better drug bioavailability (Yücel & Aktaş, 2018; Bhosale et al., 2013). A study by Yucel et al. suggests that when a drug named resveratrol was incorporated within nanoliposomes and delivered in streptozotocin-induced diabetic beta-TC3 cells, it exhibited remarkable antidiabetic properties, as well as prolonged antioxidant activity, compared with the drug alone (Meena & Smita, 2021). Although liposomes come with many advantages, their main disadvantages lie in high synthesis costs and poor stability in the gastrointestinal (GI) tract due to the effects of various enzymes and hydrochloric acid.

6.3 Nanocochleates

To overcome the shortcomings of liposome-based drug delivery systems, nanocochleates are considered alternatives since they can be stable in the GI tract due to their unique structure. Apart from the oral drug delivery route, nanocochleates can be delivered via almost every other possible route, which includes parenteral, rectal, topical, sublingual, mucosal, nasal, ophthalmic, subcutaneous, intramuscular, intravenous, transdermal, spinal, intra-articular, intra-arterial, bronchial, lymphatic, and intrauterine/intravaginal routes (Yücel & Atmar, 2018). Nanocochleates are made up of negatively charged liposomes, especially phospholipids, in fusion with cations like ca^{+2}, and Mg^{+2}, along with the desired drug molecule. Nanocochleates are cylindrically shaped, quite resembling the shape of a snail shell or a cigar. The cation fusees the phospholipid layers, which take the shape of cylindrical sheets. The unique spiral shape helps the drug to bind without chemical bonding, thereby keeping the drug molecule unaltered. Since the phospholipid bilayer contains both hydrophobic and hydrophilic ends, it can carry all three types of drugs, including hydrophobic, hydrophilic, and amphiphilic drugs, giving nanocochleates supreme versatility (Arunachalam et al., 2012). As with liposomes, Yucel et al. worked with nanocochleate too, developing resveratrol embedded in nanocochleates, which they tested on beta-TC3 cells. They found out that nanocochleate-embedded drugs worked better than the liposome system, even when a lower concentration of resveratrol is used, which again suggests that less amount of the parent drug can be used, keeping the desired results unchanged (Sharma et al., 2017).

6.4   Niosomes

Niosomes are another type of vesicular system made up of nonionic surfactants and cholesterol/lipids. Niosomes are quite similar to liposomes and can be used to carry amphibolic and lipophilic drugs. Since niosomes are nonionic, biodegradable, biocompatible, and nonimmunogenic, they promise drug delivery vehicles in the near future (Alam et al., 2018). A recent study by Sharma et al., which prepared lycopene-loaded noisome, shows sustained and prolonged release properties, suggesting that lycopene-loaded niosomes can reduce drug intake frequency. Lycopene-loaded niosomes also showed antidiabetic properties in vivo, quite similar to glibenclamide, and prolonged and sustained release properties. These can be good candidates for antidiabetic drugs (Lu et al., 2019). Embelin, a naturally occurring compound when embedded within niosomes using a thin-film hydration technique, displayed an array of antidiabetic properties in streptozotocin (STZ)-induced Wistar rats, which includes an increase in superoxide dismutase (antioxidant effect), a hypoglycemic effect comparable to repaglinide, decreased lipid peroxidation, and glutathiones (Yu et al., 2016). Thus, it can be concluded that niosomes can be used in drug formulation to combat diabetes.

6.5   Phytosomes

Phytosomes are phyto-phospholipid complexes that are made up of phyto-compounds and phospholipids. Although phytosomes and liposomes look quite similar, they mainly differ in the active compound's position. Unlike in liposomes, where the active compound is present inside the system, in phytosomes, the active compound is present in the membrane itself, which greatly increases bioavailability (Zielińska et al., 2020). Scientists Yu et al. prepared a berberine-phospholipid complex using rapid solvent evaporation and self-assembly, which showed excellent oral bioavailability, suppressed glucose fasting levels, and improved systemic hyperlipidemia metabolism in DB/DB diabetic mice model. Berberine as a parent molecule, although having antidiabetic properties, comes with low bioavailability and poor gastrointestinal absorption. Still, when incorporated into a phytosome system, its effects increased drastically, almost threefold at times, which clearly shows the efficacy of phytosomes as drug molecule carriers (Chauhan et al., 2018).

6.6   Polymeric Nanoparticles

Polymeric nanoparticles are colloidal, ranging from 1 to 1000 nm in diameter. Typically, drug molecules are embedded within the nanoparticles' core. Reports show that these drug-entrapped nanoparticles increase bioavailability, have better survival in harsh gastrointestinal tracts, and express control release properties. Polymeric nanoparticles are made of different materials, depending on the type of drug used and the administration routes. Based on that, they are of two types: inartificial-polymer-type nanoparticles and synthetic-polymer-based nanoparticles (Sonia & Sharma, 2012).

6.7   Inartificial-Polymer-Type Nanoparticles

Inartificial polymers, as the name suggests, are made from natural sources and are reported to be nontoxic and biocompatible. These are also cost-effective and hence widely used in drug research to entrap drugs and tweak their efficacy.

6.8   Chitosan-Based Nanoparticles

Chitosan is made alkaline deacetylation of chitin-found easily in nature from the fungal cell walls, insects, and exoskeletons of crustaceans. Chitosan is a type of polysaccharide that is naturally polycationic. Chitosan is a special choice when it comes to delivering antidiabetic drug molecules since it is easy to modify, has a low immunogenic effect, and is mostly nontoxic. Chauhan et al. reported that curcumin-chitosan nanoparticles had a superior effect on GLUT4 translocation compared to curcumin alone. Panwar et al. reported that ferulic acid-chitosan nanoparticles increase oral bioavailability in vivo and increased antidiabetic properties against streptozotocin-induced diabetic mice than ferulic acid alone. Chitosan can open tight junctions between epithelial cells; this remarkable property makes them suitable drug candidates to control a wide range of diseases. Despite having a lot of benefits, chitosan nanoparticles leak out entrapped drugs readily and dissolve quickly in the presence of gastric acids (Nie et al., 2020; Chauhan et al., 2018; Sonia & Sharma, 2012).

6.9   Alginate/Chitosan-Based Nanoparticles

Alginate is a hydrophilic anionic copolymer found in the cell walls of brown algae and used widely with chitosan due to its unique ability to form a gel in an aqueous medium. Alginate-chitosans form polyelectrolyte complexes through electrostatic interactions between oppositely charged groups. Alginate-chitosan complex shows a prolonged release profile as it slows down encapsulated drug release through many factors. Another study by Maity et al. made a novel alginate-coated chitosan core-shell nanocarrier system for the oral delivery of naringenin to streptozotocin-induced diabetic mice and reported effective control of embedded naringenin from fast-pass metabolism. Overall, the study indicates that the drugs, when embedded within nanoparticles, showed greater control in maintaining glucose homeostasis in streptozotocin-induced diabetic mice, thereby suggesting better hypoglycemic effects (Nie et al., 2020).

6.10   Gum-Based NPs and Gum-Chitosan-Based NPs

Natural gums like guar gum, acacia gum, locust bean gum, konjac gel, and xanthan gum can form a hydrogel upon exposure to water and are remarkably stable in a broad pH range. These properties made them suitable candidates for the drug delivery system. Thymoquinone-loaded gum rosins were developed by Rani et al. using nanoprecipitation methods and utilizing only half of the amount of thymoquinone used in natural form, but they showed better antihyperglycemic activity in type 2 diabetes mice. Another study by Rani et al., which used glycyrrhizin-loaded nanoparticles (gum arabica+chitosan), showed excellent antihyperglycemic and antilipidemic effects in type 2 diabetes mice, and the effects were quite similar to standard metformin (Nie et al., 2020; Sonia & Sharma, 2012; Rani et al., 2019).

6.11   Dextran-Based NPs

Dextran comes from sucrose-rich environments of lactobacillus, streptococcus, and leuconostoc. Dextran is a negatively charged polysaccharide with excellent water solubility. However, due to the low affinity between hydrophilic polymeric matrix and lipophilic drugs, Kapoor et al. modified dextran to make it amphiphilic by connecting the hexadecyl chain with ether bonds. Hence, O-hexadecyl-dextran is formed, which can enclose the drug (berberine) in its hydrophobic pocket in an aqueous medium. Even when berberine is used 20-fold less in concentration along with O-hexadecyl-dextran, it shows high efficiency in preventing glucose-induced oxidative stress, mitochondrial depolarization, and apoptosis (Hamid et al., 2015; Kapoor et al., 2014).

6.12 Synthetic-Polymer-Based NPS

Synthetic polymer nanoparticles are made up of various agents, including polyvinyl alcohol (PVA), poly-lactic acid (PLA), poly-lactic-co-glycolic acid (PLGA), and poly ε-caprolactone, along with a combination of these two. As a result, polymer-based nanoparticles come in a large variety with desirable properties and a wide range of control over synthetic processes (Yücel & Atmar, 2018).

6.13 PLGA-Based NPS

Of all synthetic nanoparticles, PLGA-based compounds are the most famous as they have great potential in drug delivery and are used to treat a wide range of diseases, including cancer, diabetes, and heart diseases, to name a few. Some important features of PLGAs are (1) biocompatibility and biodegradability and (2) easy surface modification for better interaction in vivo. Also, they (3) can protect the drug from degradation and (4) can carry both hydrophobic and hydrophilic drugs. Finally, (v) the sustained release profile can be changed according to drug requirements (Arunachalam et al., 2012). When encapsulated within PLGAs, a drug—pelargonidin—was found to protect alloxan-induced hyperglycemic L6 cells from oxidative stress, mitochondrial dysfunction, deoxyribonucleic acid (DNA) damage, and, more importantly, glucose homeostasis imbalance. It is also found that nano-pelargonidin is ten times more effective in terms of volume used than pelargonidin alone (Sharma et al., 2017). Another drug complex, quercetin-loaded PLGA nanoparticles, increased oral bioavailability by more than 500% compared with quercetin alone. Furthermore, nanoparticle-loaded drug administration was reduced to once in 5 days compared to daily administration for quercetin alone (Alam et al., 2018). Thus, it can be concluded that nanoparticle-based drugs not only reduce dosage but also increases bioavailability, allows better absorption in the system, and provides better control of the disease. A study suggests that PLGA-loaded drugs are readily taken up through a Peyer's patch via M-cells into the lymphatic system and then into circulation, thereby avoiding fast-pass metabolism. This explains why PLGA-based drugs show more efficacy than parent drug molecules alone (Lu et al., 2019).

6.14 PLA-, PCL-, and PVA-Based NPS

Polylactic acid (PLA) is made from lactic acid monomer in three steps: (i) lactic acid production either through fermentation or chemical processes, (ii) the purification of lactic acid and the subsequent synthesis of its cyclic dimer (lactide), and (iii) lactide

ring-opening polymerization (ROP), otherwise known as lactic acid polycondensa-
tion. One advantage of PLA is that it is biodegradable, thus finding its use in medical
and agricultural fields (Yu et al., 2016). Stevioside is a noncaloric sweetener also
known for its antidiabetic properties; it cannot be used therapeutically because of
poor bioavailability and gastrointestinal absorption. When stevioside is encapsulated
into pluronic-F-68-PLA nanoparticles, its bioavailability remarkably increases many
folds, along with increased sustained release properties (Zielińska et al., 2020).

Polycaprolactone is a biodegradable semi-crystalline aliphatic polyester that has
been approved for use as an implanted biomaterial in a variety of biomedical
applications by the Food and Drug Administration (FDA). It has also been employed
as a vehicle for the long-term delivery of a variety of medicinal compounds.
Polycaprolactone was first produced in the 1930s by ring-opening polymerization
of ε-caprolactone with various anionic or cationic catalysts or by ring-opening
polymerization of 2-methylene-1-3-dioxepane with a free radical catalyst.
Polycaprolactone can copolymerize with various polymers to form miscible blends;
this copolymerization changes the chemical properties of polycaprolactone and may
also affect other properties, such as solubility, degradation pattern, and crystallinity,
resulting in a tailored polymer with the desired drug delivery properties. For oral
delivery of insulin, Damage et al. blended nanoparticles of biodegradable
polycaprolactone and a polycationic nonbiodegradable acrylic polymer; their
antidiabetic impact was evaluated in vivo. The outcomes showed an expanded
serum insulin level with an improved glycemic reaction for a prolonged period
(Chauhan et al., 2018).

6.15   Polyethylene Glycol Surface Modification

Covering the outer layer of nanoparticles with polyethylene glycol (PEG), or
"PEGylation," is increasingly used and has quite gained popularity when it comes
to working on the effectiveness of medication and high-quality delivery to the
intended cells and tissues. PEGylating proteins as a starting point for further
developing fundamental course time and decreasing immunogenicity, the effect of
PEG coatings on the destiny of foundationally managed nanoparticle details has
been widely studied. Stake coatings on nanoparticles safeguard the surface from
accumulation, opsonization, and phagocytosis, delaying foundational flow time
within the system (Sonia & Sharma, 2012). In general, PEG is hydrophilic, with
very less toxicity and blood-compatible polymer, allowing us to modify it into a
large variety of surfaces, especially nanoparticles. El-Naggar et al. reportedly created
PLA-PEG nanoparticles, where PLA is the hydrophobic and PEG is the hydrophilic
part. They encapsulated curcumin into a hydrophobic polymer portion and stabilized
it using the cationic surfactant CTAB (cetyl-trimethylammonium bromide) in
deionized water. It is found that curcumin-based PLA-PEG nanoparticles give better
results as compared to curcumin alone or polymer nanoparticle without PEGylation.
PLA-PEG nanoparticles, as compared to curcumin alone/polymer NP were able to
control liver damage successfully, lower plasma glucose levels, and elevated insulin

levels in streptozotocin-induced diabetic rats, which demonstrates the efficiency of PEGylation. Although some report suggests that PEG causes immunogenic responses in some peoples, PEG is still the most sought-after polymeric coating of nanomedicines with FDA approval (Benn, 2009; Bassas-galia et al., 2017; Rani et al., 2019).

6.16 Micelles

Micelles range in size from 5 to 50 nm and are made of amphiphilic molecules in an aqueous medium. The hydrophobic part of the micelle holds the parent drug, and the hydrophilic part forms the micelle's outer layer, giving shape and structure. One good use of micelles is in the case of the transplantation of islet beta cells. If islet beta cells are transferred alone, they die quickly following hypoxia and apoptosis, which is when micelles come into play. Scientists Han et al. developed a peptide micelle curcumin delivery system using an oil-in-water emulsion/solvent evaporation method that effectively protects islet beta cells in vitro during transplantation. Zhang et al. used the dialysis method to create an amentoflavone-loaded micelle system made of N-vinyl pyrrolidone, as well as maleic acid guerbet alcohol mono-ester (amphiphilic copolymer), which showed increased oral bioavailability by 3.2 times than amentoflavone alone. It is also to be noted that the antidiabetic effects of amentoflavone are quite similar to those of metformin in insulin-resistant key mice.

6.17 Lipid-Based Nanoparticles

Numerous drug molecules are nonsolvent in artificially and naturally delicate fluid frameworks or present serious aftereffects. Lipid-based nanoparticle (LBNP) frameworks address one of the most encouraging colloidal transporters for bioactive natural atoms. LBNPs benefit in numerous ways as they involve low manufacturing costs, give excellent oral bioavailability to water-insoluble drugs, and are more efficiently absorbed in the GI tract (Hamid et al., 2015). Lipid-based nanoparticles are mainly of two types based on their internal structure: solid lipid nanoparticles (SLNs) and nanostructured lipid carriers (NLCs) (Benn, 2009).

6.18 Solid Lipid Nanoparticles (SLNs)

Solid lipid nanoparticles are at the bleeding edge of the quickly creating area of nanotechnology with a few expected applications in drug delivery, clinical medication, research, and other disciplines (Kapoor et al., 2014). SLNs began developing in the early 1990s in search of alternatives to existing drug delivery systems. SLNs are made up of solid lipid bodies compared to liposomes, with their liquid lipid system. This solid lipid system gives the upper hand in the form of controlled drug release and better protects the parent drug molecule encapsulated (Washington et al., 2016).

When myricitrin, a plant-derived drug with antioxidant properties, was incorporated into SLNs and tested in vivo in streptozotocin-nicotinamide diabetic mice, the new drug molecule showed better antioxidant and antidiabetic properties when compared with standard medicines, like metformin (Mir et al., 2017). Another study with resveratrol-SLN in diabetic rats showed better antidiabetic effects, as well as reduced SNAP receptor (SNARE) protein production, which is linked with insulin resistance via GLUT4 immobilization (Samadder et al., 2016). Xu et al., who worked with berberine-SLN, reported a maximum drug concentration of 20 times higher in the liver than when berberine is used alone, suggesting that SLNs can improve bioavailability to new heights (Chitkara et al., 2012).

6.19   Nanostructured Lipid Carriers (NLCs)

NLCs address the second era of lipid-based nanocarriers, created from SLNs, which include a blend of solid and liquid lipids. This system was created to beat the impediments of SLNs; subsequently, NLCs have higher medication stacking limits and could likewise keep away from drug removal during capacity by keeping away from lipid crystallization because of the presence of fluid lipids in the NLC system. While SLNs are made of solid lipids, NLCs are a combination of solid and liquid lipids, for example, glyceryl tricaprylate, ethyl oleate, isopropyl myristate, and glyceryl oleate. The mean molecule sizes are exceptionally like SLNs, by and large in the scope of 10–1000 nm, and are impacted by the manufacturing process and raw materials used. The fundamental benefits of these nanoparticles are that they can be stacked with hydrophilic and hydrophobic medications, offer better control of parent drug delivery, and display low in vivo toxicity (Hamid et al., 2015). Baicalin, an antihyperglycemic drug with low hydrophilic properties and a poorly absorbed profile, was stacked with NLCs using Precirol as the solid lipid and Miglyol as the liquid lipid and was tested for antidiabetic effects. In vivo, baicalin-NLC showed better hypoglycemic and hypolipidemic effects when compared with baicalin alone (Torché et al., 2000).

6.20   Nanoemulsions (NEs) and Self-Nano-Emulsifying Drug-Delivery Systems (SNEDDSs): NEs

Nanoemulsions (NEs) are a colloidal system made up of oil, emulsifying agents, and an aqueous phase, and their size range from 10 to 1000 nm. Three types of nanoemulsions are known: (1) oil in water nanoemulsion, (2) water in oil nanoemulsion, and (3) bicontinuous nanoemulsion. NEs can be excellent drug carrier systems because of their ability to carry both hydrophilic and hydrophobic drug molecules. Nanoemulsions and drugs are reported to increase absorption, and nanoemulsions' small particulate size can be credited for this remarkable property. (Benn, 2009; Mohseni et al., 2019) A study found that bitter gourd oil (prepared using emulsion phase inversion, followed by high-pressure homogenization), exhibited excellent antidiabetic properties, as well as antioxidative stress management properties, in alloxan-induced diabetic mice.(Jaiswal et al., 2015) A report from Xu et al. suggests that berberine-NE significantly increases permeation in GI by a factor of 5.5 times, unlike berberine alone, which has a poor absorption profile. Moreover, postprandial and fasting glucose levels were seen to have improved threefold when berberine-NEs were used (Xu et al., 2019).

6.21   SNEDDS

Self-nano-emulsifying drug-delivery systems (SNEDDSs) are one of the emerging drug delivery systems developed to facilitate drug delivery when the drug has a low solubility profile and the oral delivery route is the preferred choice of drug administration. SNEDDSs are a mixture of the oil phase (medium to long-chain triglycerides), water-soluble surface-active agents/surfactants, and cosurfactants. Karamanidou et al. prepared SNEDDS using Lauroglycol® FCC as the oily phase, Cremophor® EL as the surfactant, and Transcutol® P or Labrafil® M 1944 CS as the cosurfactant to successfully deliver insulin via the oral route. They reported that the insulin-SNEDDS formula improved enzymatic stability and showed a sustained insulin release profile (Buya et al., 2020). Another report found that trans-cinnamic acid SNEDDSs improved antidiabetic effects in alloxan-induced diabetic rats. However, the current available SNEDDSs have some drawbacks in the form of low encapsulation, high surfactant requirement, oxidation of lipid components, etc. To address these problems, different SNEDDSs are being developed. Solid SNEDDSs, super saturated SNEDDSs, and control release SNEDDSs are some SNEDDSs currently in development (Benn, 2009).

6.22   Metallic Nanoparticles

Metallic nanoparticles like gold NPs and silver NPs are becoming increasingly popular daily due to their multiple uses in treating various diseases. They are prepared through various methods, such as physical (plasma arcing, spray pyrolysis, pulsed laser desorption, lithographic techniques, etc.), chemical (electrodeposition, sol-gel process, Langmuir Blodgett method, etc.), and biological methods (Kuppusamy et al., 2016). It has been reported that by using metallic NPs, plant-based compounds with antidiabetic effects can be enhanced to a great length than by using plant-based compounds alone (Benn, 2009). A study from Chengdu University reported that Catathelasmaventricosum polysaccharide-seleniumNP shows higher antidiabetic effects than other selenium preparations (Nie et al., 2020). Further study results showed that *cassia fistula extract* formulated with gold nanoparticles significantly reduced liver enzymes, such as alanine transaminase, alkaline phosphatase, serum creatinine, and uric acid, in streptozotocin-induced diabetic mice. The gold-nanoparticle-treated diabetic model also showed a decrease in HbA1c (glycosylated hemoglobin) levels, suggesting excellent diabetic control properties of gold nanoparticles. Another study by Swarnalatha et al. found that *Sphaeranthus amaranthoides* biosynthesized silver nanoparticles inhibited alpha-amylase and acarbose sugar in a diabetic mouse model (Daisy & Saipriya, 2012; Pednekar et al., 2017). Hence, metallic nanoparticles can be considered potent drug carriers when treating diabetes.

6.23   Mesoporous-Silica-Based Nanoparticles

Mesoporous silica nanoparticles (MSNs) are prepared from reactions of tetraethyl orthosilicate with the micellar rod as a template. They are solid materials composed of a honeycomb-like porous structure consisting of hundreds of empty channels between them that have an enormous capacity to incorporate various drug molecules (Pednekar et al., 2017). Huang et al. reported that *Polyalthia longifolia* extract contained in MSNs reduced DPP4 activity and at the same time downregulated hyperglycemia without any adverse effects in diet-induced diabetic mice (Huang et al., 2017).

7   Conclusion

Although we are at the dawn of nanobiotechnology, it can be assumed that shortly, nanobiotechnology will enter every medical science discipline, especially the drug delivery discipline. From the above discussion, we can get an idea of how nanoparticles can drastically change how a drug molecule behaves inside our body. Many drug molecules are incompatible inside our body, either readily destroyed by our body or excreted, leaving much of the drug useless. Nanobiotechnology comes to the rescue in this particular aspect, with special abilities to either protect those drug molecules from harsh inside-body environments or to protect them from rapid excretion by body systems and increase their overall bioavailability and efficacy to a great extent. For instance, in the case of type 1 diabetes, the standard treatment of choice is insulin supplements, which patients have to take via injections, which is uncomfortable for most patients. If insulin can be administered orally, those patients can largely be relieved from the discomfort. But here is the catch—insulin cannot be administered via the oral route since it will be readily destroyed in the GI tract. This is exactly where nanoparticles come into play and could potentially change the game. Although drugs are there to treat diabetes, we can still not completely cure the disease, largely due to drug inefficiency and poorly understood pathophysiology of the entire disease, which includes every metabolic aspect of the disease. As we have seen earlier in this chapter, nanoparticle-encapsulated drugs can largely change this picture. However, much research needs to be done in this field since nanoparticles also have disadvantages. The disadvantages to be addressed in the near future through extensive research are as follows: (i) nanoparticles involve high manufacturing cost; (ii) some metallic nanoparticles may come with excessive metal depositions, which can cause adverse patient effects; (iii) most nanoparticles come from inorganic/synthetic sources, which are thought to cause toxicity or are incompatible with the body; (iv) and most research on nanoparticles is largely limited to in vitro studies, and animal trials and human trials need to be done to assess toxicity profile and safety. Today, some nanoparticle drug delivery systems have found their place on the market against a few diseases, but most of them are still under research, and in the near future, we can expect them to hit the market and tackle every possible disease with safety and efficiency. With nanobiotechnology in the field of diabetes treatment, we can expect to change the picture in the near future.

Conflict of Interest   The study's authors have not reported any apparent conflicts of interest.

References

Adeghate, E., & Kalász, H. (2011). Amylin analogues in the treatment of diabetes mellitus: Medicinal chemistry and structural basis of its function. *The Open Medicinal Chemistry Journal, 5*(Suppl 2), 78–81. https://doi.org/10.2174/1874104501105010078

Ahmed, E. M. (2015). Hydrogel: Preparation, characterization, and applications—A review. *Journal of Advanced Research, 6*(2), 105–121.

Akbarzadeh, A., et al. (2013). Liposome: Classification, preparation, and applications. *Nanoscale Research Letters, 8*(1), 102. https://doi.org/10.1186/1556-276X-8-102

Alam, M. S., Ahad, A., Abidin, L., Aqil, M., Mir, S. R., & Mujeeb, M. (2018). Embelin-loaded oral niosomes ameliorate streptozotocin-induced diabetes in Wistar rats. *Biomedicine & Pharmacotherapy, 97*, 1514–1520. https://doi.org/10.1016/j.biopha.2017.11.073. Epub 2017 Nov 20. PMID: 29793314.

Arunachalam, A., Jeganath, S., Yamini, K., & Tharangini, K. (2012). Niosomes: A novel drug delivery system. *International Journal of Novel Trends in Pharmaceutical Sciences, 2*(1), 25–31.

Balakumar, P., Maung-U, K., & Jagadeesh, G. (2016). Prevalence and prevention of cardiovascular disease and diabetes mellitus. *Pharmacological Research, 113*, 600–609.

Bassas-galia, M., Follonier, S., Pusnik, M., & Zinn, M. (2017). 2-natural polymers: A source of inspiration. In G. Perale & J. Hilborn (Eds.), *Bioresorbable polymers for biomedical applications* (pp. 31–64). Woodhead Publishing.

Benn, T. M. (2009). *The release of engineered nanomaterials from commercial products.* Arizona State University.

Bhosale, R., Ghodake, P., Mane, A., & Ghadge, A. (2013). Nanocochleates: A novel carrier for drug transfer. *Journal of Scientific and Innovative Research, 2*, 964–969.

Bradley, C. (2002). The glitazones: A new treatment for type 2 diabetes mellitus. *Intensive & Critical Care Nursing, 18*(3), 189–191. https://doi.org/10.1016/s0964-3397(02)00010-1. PMID: 12405274.

Buya, A. B., Beloqui, A., Memvanga, P. B., & Préat, V. (2020). Self-nano-emulsifying drug-delivery systems: From the development to the current applications and challenges in oral drug delivery. *Pharmaceutics, 12*(12), 1194. https://doi.org/10.3390/pharmaceutics12121194

Chauhan, P., Tamrakar, A. K., Mahajan, S., & GBKS, P. (2018). Chitosan encapsulated nanocurcumin induces GLUT-4 translocation and exhibits enhanced anti-hyperglycemic function. *Life Sciences, 213*, 226–235. https://doi.org/10.1016/j.lfs.2018.10.027. Epub 2018 Oct 18. PMID: 30343126.

Chen, W., Guo, M., & Wang, S. (2016). Anti prostate cancer using PEGylated bombesin containing, cabazitaxel loading nanosized drug delivery system. *Drug Development and Industrial Pharmacy, 42*(12), 1968–1976.

Chitkara, D., Nikalaje, S. K., Mittal, A., Chand, M., & Kumar, N. (2012). Development of quercetin nanoformulation and in vivo evaluation using streptozotocstreptozotocin-inducedt model. *Drug Delivery and Translational Research, 2*(2), 112–123. https://doi.org/10.1007/s13346-012-0063-5. PMID: 25786720.

Chu, A. J. (2022). Quarter-century explorations of bioactive polyphenols: Diverse health benefits. *Frontiers in Bioscience-Landmark, 27*(4), 134.

Daisy, P., & Saipriya, K. (2012). Biochemical analysis of Cassia fistula aqueous extract and phytochemically synthesized gold nanoparticles as hypoglycemic treatment for diabetes mellitus. *International Journal of Nanomedicine, 7*, 1189–1202. https://doi.org/10.2147/IJN.S26650

Derosa, G., & Maffioli, P. (2012). α-Glucosidase inhibitors and their use in clinical practice. *Archives of Medical Science, 8*(5), 899–906.

DiSanto, R. M., Subramanian, V., & Gu, Z. (2015). Recent advances in nanotechnology for diabetes treatment. *Wiley Interdisciplinary Reviews: Nanomedicine and Nanobiotechnology, 7*(4), 548–564.

Donnelly, L. A., Morris, A. D., Frier, B. M., Ellis, J. D., Donnan, P. T., Durrant, R., et al. (2005). Frequency and predictors of hypoglycaemia in type 1 and insulin-treated type 2 diabetes: A population-based study. *Diabetic Medicine, 22*(6), 749–755.

Dwivedi, P., Kansal, S., Sharma, M., et al. (2012). Exploiting 4-sulphate N-acetyl galactosamine decorated gelatine nanoparticles for effective targeting to professional phagocytes in vitro and in vivo. *Journal of Drug Targeting, 20*(10), 883–896.

Gupta, V., & Kalra, S. (2011). Choosing a gliptin. *Indian Journal of Endocrinology and Metabolism, 15*(4), 298–308. https://doi.org/10.4103/2230-8210.85583

Gupta, P., Taiyab, A., & Hassan, M. I. (2021). Emerging role of protein kinases in diabetes mellitus: From mechanism to therapy. *Advances in Protein Chemistry and Structural Biology, 124*, 47–85.

Hamid Akash, M. S., Rehman, K., & Chen, S. (2015). Natural and synthetic polymers as drug carriers for delivery of therapeutic proteins. *Polymer Reviews, 55*(3), 371–406. https://doi.org/10.1080/15583724.2014.995806

Hansen, K. B., Vilsbøll, T., & Knop, F. K. (2010). Incretin mimetics: A novel therapeutic option for patients with type 2 diabetes—A review. *Diabetes, Metabolic Syndrome and Obesity: Targets and Therapy, 3*, 155–163. PMID: 21437085; PMCID: PMC3047973.

Huang, P.-K., Lin, S.-X., Tsai, M. J., Leong, M., Lin, S. R., Kankala, R., et al. (2017). Encapsulation of 16-Hydroxycleroda-3,13-Dine-16,15-Olide in mesoporous silica nanoparticles as a natural dipeptidyl peptidase-4 inhibitor potentiated hypoglycemia in diabetic mice. *Nanomaterials, 7*(5), 112. https://doi.org/10.3390/nano7050112

Jaiswal, M., Dudhe, R., & Sharma, P. K. (2015). Nanoemulsion: An advanced mode of drug delivery system. *3 Biotechnology, 5*(2), 123–127. https://doi.org/10.1007/s13205-014-0214-0

Kapoor, R., Singh, S., Tripathi, M., Bhatnagar, P., Kakkar, P., & Gupta, K. C. (2014). O-hexadecyl-dextran entrapped berberine nanoparticles abrogate high glucose stress-induced apoptosis in primary rat hepatocytes. *PLoS One, 9*(2), e89124. https://doi.org/10.1371/journal.pone.0089124. PMID: 24586539; PMCID: PMC3930636.

Kuppusamy, P., Yusoff, M. M., Maniam, G. P., & Govindan, N. (2016). Biosynthesis of metallic nanoparticles using plant derivatives and their new avenues in pharmacological applications—An updated report. *Saudi Pharmaceutical Journal, 24*(4), 473–484. https://doi.org/10.1016/j.jsps.2014.11.013. Epub 2014 Dec 8. PMID: 27330378; PMCID: PMC4908060.

Landree, L. E., Hanlon, A. L., Strong, D. W., Rumbaugh, G., Miller, I. M., Thupari, J. N., et al. (2004). C75, a fatty acid synthase inhibitor, modulates AMP-activated protein kinase to alter neuronal energy metabolism. *Journal of Biological Chemistry, 279*(5), 3817–3827.

Lotfy, M., Adeghate, J., Kalasz, H., Singh, J., & Adeghate, E. (2017). Chronic complications of diabetes mellitus: A mini review. *Current Diabetes Reviews, 13*(1), 3–10.

Lu, M., Qiu, Q., Luo, X., Liu, X., Sun, J., Wang, C., et al. (2019). Phyto-phospholipid complexes (phytosomes): A novel strategy to improve the bioavailability of active constituents. *Asian*

Journal of Pharmaceutical Sciences, 14(3), 265–274. https://doi.org/10.1016/j.ajps.2018. 05.011

Mazari, S. A., Ali, E., Abro, R., Khan, F. S. A., Ahmed, I., Ahmed, M., et al. (2021). Nanomaterials: Applications, waste-handling, environmental toxicities, and future challenges–a review. *Journal of Environmental Chemical Engineering, 9*(2), 105028.

Meena, T., & Smita, B. (2021). Nanocochleates: A potential drug delivery system. *Journal of Molecular Liquids, 334*, 116115.,ISSN 0167-7322,. https://doi.org/10.1016/j.molliq.2021. 116115

Meglitinide—An overview|ScienceDirect Topics. (n.d.).

Meo, S. A., Usmani, A. M., & Qalbani, E. (2017). Prevalence of type 2 diabetes in the Arab world: Impact of GDP and energy consumption. *European Review for Medical and Pharmacological Sciences, 21*(6), 1303–1312.

Mir, M., Ahmed, N., & Rehman, A. U. (2017). Recent applications of PLGA bPLGA-basedtructures in drug delivery. *Colloids and Surfaces. B, Biointerfaces, 159*, 217–231. https://doi.org/10.1016/j.colsurfb.2017.07.038. Epub 2017 Jul 28. PMID: 28797972.

Mohseni, R., ArabSadeghabadi, Z., Ziamajidi, N., et al. (2019). Oral administration of resveratrol-loaded solid lipid nanoparticle improves insulin resistance through targeting expression of snare proteins in adipose and muscle tissue in rats with type 2 diabetes. *Nanoscale Research Letters, 14*, 227. https://doi.org/10.1186/s11671-019-3042-7

Nicolaou, K. C. (2018). The emergence and evolution of organic synthesis and why it is important to sustain it as an advancing art and science for its own sake. *Israel Journal of Chemistry, 58*(1–2), 104–113.

Nie, X., Chen, Z., Pang, L., Wang, L., Jiang, H., Chen, Y., Zhang, Z., Fu, C., Ren, B., & Zhang, J. (2020). Oral nano drug delivery systems for the treatment of type 2 diabetes mellitus: An available administration strategy for antidiabetic phytocompounds. *International Journal of Nanomedicine, 15*, 10215–10240. https://doi.org/10.2147/IJN.S285134

Patlak, M. (2002). New weapons to combat an ancient disease: Treating diabetes. *The FASEB Journal, 16*(14), 1853. https://doi.org/10.1096/fj.020974bkt. PMID 12468446. S2CID 3541224 9

Pednekar P. P., Godiyal S. C., Jadhav, K. R., & Kadam, V. J. (2017). Mesoporous silica nanoparticles: A promising multifunctional drug delivery system. In *Nanostructures for cancer therapy* (pp. 593–621). Elesvier https://doi.org/10.1016/b978-0-323-46144-3.00023-4

Pradeepa, R., & Mohan, V. (2021). Epidemiology of type 2 diabetes in India. *Indian Journal of Ophthalmology, 69*(11), 2932.

Rani, R., Dahiya, S., Dhingra, D., Dilbaghi, N., Kaushik, A., Kim, K. H., & Kumar, S. (2019). Antidiabetic activity enhancement in streptozotocin + nicotinamide-induced diabetic rats through combinational polymeric nanoformulation. *International Journal of Nanomedicine, 14*, 4383–4395. https://doi.org/10.2147/IJN.S205319. PMID: 31354267; PMCID: PMC6580421.

Samadder, A., Abraham, S. K., & Khuda-Bukhsh, A. R. (2016). NanopharmNano pharmaceutical using pelargonidin towards enhancement of efficacy for prevention of alloxan-induced DNA damage in L6 cells via activation of PARP and p53. *Environmental Toxicology and Pharmacology, 43*, 27–37. https://doi.org/10.1016/j.etap.2016.02.010. Epub 2016 Feb 12. PMID: 26943895.

Seino, S. (2012). Cell signaling in insulin secretion: The molecular targets of ATP, cAMP, and sulfonylurea. *Diabetologia, 55*(8), 2096–2108. https://doi.org/10.1007/s00125-012-25629. PMID 22555472. S2CID 7146975

Seoudy, A. K., Schulte, D. M., Hollstein, T., Böhm, R., Cascorbi, I., & Laudes, M. (2021). Gliflozins for the treatment of congestive heart failure and renal failure in type 2 diabetes. *Deutsches Ärzteblatt International, 118*, 122–129. https://doi.org/10.3238/arztebl.m2021.0016. Epub ahead of print. PMID: 33531116; PMCID: PMC8204375.

Shaeena, M. H., & Mani, R. K. (2021). Diabetes and nanotechnology–A recent advance in treatment of Diabetes. *Journal of University of Shanghai for Science and Technology, 23*(11), 445–453.

Sharma, P. K., Saxena, P., Jaswanth, A., Chalamaiah, M., & Balasubramaniam, A. (2017). Antidiabetic activity of lycopene niosomes: Experimental observation. *Journal of Pharmaceutical Sciences and Drug Development, 4*, 103.

Sonia, T. A., & Sharma, C. P. (2012). An overview of natural polymers for oral insulin delivery. *Drug Discovery Today, 17*(13–14), 784–792. https://doi.org/10.1016/j.drudis.2012.03.019. Epub 2012 Apr 9. PMID: 22521664.

Sun, C. Q. (2007). Size dependence of nanostructures: Impact of bond order deficiency. *Progress in Solid State Chemistry, 35*(1), 1–159.

Torché, A. M., Jouan, H., Le Corre, P., Albina, E., Primault, R., Jestin, A., & Le Verge, R. (2000). Ex vivo and in situ PLGA microspheres uptake by pig ileal Peyer's patch segment. *International Journal of Pharmaceutics, 201*(1), 15–27. https://doi.org/10.1016/s0378-5173(00) 00364-1. PMID: 10867261.

Washington, K., Kularatne, R., Karmegam, V., Biewer, M., & Stefan, M. (2016). Recent advances in aliphatic polyesters for drug delivery applications. *Wiley Interdisciplinary Reviews: Nanomedicine and Nanobiotechnology, 9*(4), e1446. https://doi.org/10.1002/wnan.1446

Xu, H. Y., Liu, C. S., Huang, C. L., Chen, L., Zheng, Y. R., Huang, S. H., & Long, X. Y. (2019). Nanoemulsion improves hypoglycemic efficacy of berberine by overcoming its gastrointestinal challenge. *Colloids and Surfaces. B, Biointerfaces, 181*, 927–934. https://doi.org/10.1016/j. colsurfb.2019.06.006. Epub 2019 Jun 4. PMID: 31382342.

Yach, D., Stuckler, D., & Brownell, K. D. (2006). Epidemiologic and economic consequences of the global epidemics of obesity and diabetes. *Nature Medicine, 12*(1), 62–66.

Yu, F., Li, Y., Chen, Q., He, Y., Wang, H., Yang, L., Guo, S., Meng, Z., Cui, J., Xue, M., & Chen, X. D. (2016). Monodisperse microparticles loaded with the self-assembled berberine-phospholipid complex-based phytosomes for improving oral bioavailability and enhancing hypoglycemic efficiency. *European Journal of Pharmaceutics and Biopharmaceutics, 103*, 136–148. https://doi.org/10.1016/j.ejpb.2016.03.019. Epub 2016 Mar 25. PMID: 27020531.

Yücel, Ç., Karatoprak, G. S., & Aktaş, Y. (2018). Nanoliposomal resveratrol as a novel approach to treatment of diabetes mellitus. *Journal of Nanoscience and Nanotechnology, 18*(6), 3856–3864. https://doi.org/10.1166/jnn.2018.15247. PMID: 29442719.

Yücel, Ç., Karatoprak, G. Ş., & Atmar, A. (2018). Novel resveratrol-loaded nanocochleates and effectiveness in the treatment of diabetes. *Fabad Journal of Pharmaceutical Sciences, 43*(2), 35–44.

Zielińska, A., Carreiró, F., Oliveira, A. M., Neves, A., Pires, B., Venkatesh, D. N., et al. (2020). Polymeric nanoparticles: Production, characterization, toxicology and ecotoxicology. *Molecules, 25*(16), 3731. https://doi.org/10.3390/molecules25163731

A Comprehensive Pharmacological Appraisal of Indian Traditional Medicinal Plants with Anti-diabetic Potential

Chandan Kumar Acharya, Balaram Das, Nithar Ranjan Madhu ⓘ, Somnath Sau, Manna De, and Bhanumati Sarkar

Abstract

Diabetes mellitus, a chronic metabolic dysfunction found in people of different age groups worldwide, is now seriously threatening mankind's health. Despite the application of insulin and other synthetic oral anti-diabetic drugs, there is a great need for the discovery and development of novel anti-diabetic drugs of plant origin as the synthetic drugs have more side effects in long-term use. Therefore, researchers engaged in discovering novel bioactive compounds from plants bearing anti-diabetic potential also have fewer unwanted side effects than conventional drugs. In this chapter, an attempt has been made to discuss the prospective medicinal plants comprising either plant extracts or isolated bioactive phytoconstituents bearing anti-diabetic potential, which has been reported in several in vitro, in vivo, or clinical studies. Because of this, the mechanism of action and

C. K. Acharya
Department of Botany, Dr CV Raman University, Bilsapur, Chhattisgarh, India

Department of Botany, Bajkul Milani Mahavidyalaya, Kolkata, West Bengal, India

B. Das
Department of Physiology, Belda College, Medinipur, West Bengal, India

N. R. Madhu (✉)
Department of Zoology, Acharya Prafulla Chandra College, Kolkata, West Bengal, India

S. Sau
Department of Nutrition, Egra S.S.B. College, Medinipur, West Bengal, India

Department of Physiology, Belda College, Medinipur, West Bengal, India

M. De
Department of Botany, Dr CV Raman University, Bilsapur, Chhattisgarh, India

B. Sarkar
Department of Botany, Acharya Prafulla Chandra College, Kolkata, West Bengal, India

163

the management of diabetes will be valuable to scientists, chemists, and pharmaceutical corporations for the discovery of novel anti-diabetic drugs in the future.

Keywords

Diabetes mellitus · Medicinal plants · Phyto-constituents · Pharmacology · Management of diabetes

1   Introduction

Plant-human interaction has long been a part of the history of civilization (Mukherjee et al., 2007). About 60% of medicine is naturally derived, i.e. directly plant or plant-based secondary metabolites. The recent trend shows more experimental work based on plants, around triple the amount from the past. Diabetes is a steadily rising trend globally, with around 422 million adults inflicted in 2014 compared to 108 million adults in 1980, which from 4.7% doubled to 8.5%. Early life was lost in around 43% of all deaths due to high blood glucose (World Health Organization (WHO) Global report on diabetes, 2020). In 2021, the International Diabetes Federation (IDF) Atlas 10 edition highlighted 537 million adults with diabetes between 20 and 79 years, and this number of cases will rise to around 643 million and 783 million by 2030 and 2045 years, respectively, and more will be affected in low- and middle-income countries, with loss of life accounting for 6.7 million.

It has traditionally been the most difficult process to find drugs from natural plants through product identification and proper application of that to a "lead drug," but recent multidisciplinary research approaches have placed a high priority on developing natural drugs for a variety of molecular targets and the treatment of various illnesses and disease conditions. In ancient times, people used rawer extraction, a purification technique, for plant drug discovery. At that time, an empirical technique was preferred over a rational method for discovering new medicines. Evidence from recorded history suggests that the Mesopotamian culture, which dates back to 2600 BC, utilized medicine derived from 1000 different plant species. Egyptian medicine dates back to 2900 BC, and some medical documents called the 'Ebers Papyrus', which were discovered around 1550 BC, describe 700 medications derived from plants. Ayurveda from India dates back to the first millennium BC, whereas the Chinese medical system is over a thousand years old. Greek and Roman cultures of medicine contributed significantly to the western world's extensive knowledge of herbal medicine. These cultures contribute their self-knowledge to the Arabian medicinal system, part of which comes from India's and China's old medical practices (Atanas et al., 2015). Though individuals still use them as unrecognized drug states, ethnographic data on a plant with medical benefits is a big interest to new research perspectives.

Ayurveda is a notable example of this source and is best suited for anti-diabetic drugs—metformin and more biguanide drug were made by copying like *galegine* from *Galega officinalis* L. plant. Plant phytochemical for drug testing is related to traditional in vitro and in vivo approaches. This pharmacological method is divided into forwarding pharmacological and reverse pharmacological approaches. The primary step in identifying and evaluating a pre-drug plant is to acquire animal data based on signs and symptoms or in vitro experiments. If the available biomolecules are active and interact with the target protein in vivo, or if searching phytochemical data for specific functional molecules and their application in animals or other species, then this is an example of a 'reverse approach' to the process of discovering new drugs from plants. The preliminary step in screening plant bioactives is the extraction of either aqueous or organic solvents. After that, fractionalization for the isolation of interactive compounds is performed using methods such as column chromatography, which is a traditional method. However, the most sensitive characterization is performed using nuclear magnetic resonance (NMR), gas chromatography-mass spectroscopy (GCMS), or high-performance liquid chromatography (HPLC), which can distinguish secondary plant metabolites more accurately (Atanas et al., 2015).

2   History of Indian Medicinal Plants for Diabetes Control

Traditional herbal medicine has existed worldwide since the prehistoric era. It was recorded in Ayurveda, Chinese, Greek, and Egyptian for various therapeutic uses; meanwhile, African and American native people use herbs in their cultural rituals. Ayurveda has become one of the ancient alternative Indian medicine systems in the period between 4000 BC and 1500 BC, where the history of medical knowledge about plants or herbal therapy was well recognized in two early textbooks, i.e. *Charaka Samhita* and *Sushruta Samhita.* Around 1000 BC, the Indian medical practice of Ayurveda referred to diabetes mellitus (DM) as madhumeha, which is a type of Prameha. At that time diabetes was a mythological view mentioned eating Havisha, a special type of food i.e., offered during the Yagna occasion by Daksha Prajapati. In the Vedic period, around 600 BC, this illness was referred to as 'Asrava', and its detailed description was recorded in Charaka Samhita, Sushruta Samhita, and Vagbhatta (Singh, 2011). Since then, indigenous herbal medicines as raw forms have been strong reflections. For example, *Gymnema sylvestre* (Asclepiadaceae) is native to the tropical forest of South West India. This plant leaf extract was used as anti-diabetic medicine in the contemporary period of Sushruta (600 BC) (Laha & Paul, 2019). But in the early nineteenth century, scientific advancement with time beings industrialists, scientists, and researchers focused on synthetic pharmaceuticals, which show the declinature of herbal value. However, recent initiatives are being utilized to revive and widely promote traditional plant medicine in the mainstream medical system while considering quality and safety issues (Parasuraman et al., 2014). A WHO-estimated report says that about 80% of the world population from third-world countries trust traditional plant medicine for health care (Mathew & Babu, 2011; Amalraj & Gopi, 2016). About 60% of rural people in India use herbal medicine (Amalraj & Gopi, 2016). The

Indian subcontinent is a rich diversity of 45,000 plant species, and 15,000 medicinal flora have been recognized. Knowingly, out of the medicinal plants, 7000–7500 species have been used by community people to cure diseases. Ayurveda recorded about 700 types of medicinal plants used in the ancient era (Parasuraman et al., 2014). Importantly, literature data shows more than 400 hypoglycaemic plants have been discovered, but still surprising to discover new diabetic plants serve as an alternative in the most attractive field. Information from an ethnobotanical survey suggests that 800 plants may pose potent anti-diabetic properties. Out of these, three have been evaluated as beneficial for treating type-2 diabetes mellitus—*Trigonella foenum-graecum*, *Pterocarpus marsupium*, and *Momordica charantia* (Patel et al., 2012; Ponnusamy et al., 2011). Because even the most advanced allopathic system cannot entirely cure diabetes and oral hypoglycaemic medicines have caused a number of unpleasant side effects, the entire globe is concentrating its efforts now on herbal treatments for the disease (Srivastava et al., 2012). That is why in recent years, considerable importance has been given to plant herbal medicine for diabetes management. Herbal preparation makes more polyherbal formulations due to extra therapeutic benefits than a single herbal formulation. The combination of herbs improves the pharmacological activity of polyherbal formulation (Parasuraman et al., 2014). Examples include the Indian company Diabet (Herbal Galenicals), which manufactures anti-diabetic herbal medicines using a variety of plant extracts, including *Strychnos potatorum*, *Tamarindus indica*, *Tribulus terrestris*, *Curcuma longa*, *Coscinium fenestratum*, and *Phyllanthus reticulates*, among others (Umamaheswari et al., 2010). Other anti-diabetic treatments include Diabrid, Diakyur, Dihar, Dianex, Diashis, Diasulin, and Diasol, all of which have been shown to have effects that are compatible with those of regular allopathic drugs (Srivastava et al., 2012).

3   Pharmacological Management of Diabetes Mellitus

According to the World Health Organization (WHO), diabetes mellitus is a chronic and metabolic disorder characterized by a rise in blood glucose (or blood sugar) level, which leads over time to serious damage to the eyes, blood vessels, kidneys, heart, nerves, etc. This disease is divided into type-1 diabetes and type-2 diabetes. In patients with type-1 diabetes, once known as insulin-dependent diabetes or juvenile diabetes chronic conditions, the body does not produce enough (or any) insulin; that is why these patients are required to inject insulin directly. Type-2 diabetes is the most common type of diabetes, which is usually found in adults. Diabetes develops when the human body either develops insulin resistance or fails to produce sufficient amounts of insulin. After a while, diabetes mellitus is accompanied by specific vascular and neuropathic complications (Ngugi et al., 2012). The frequency of type-2 diabetes in the population has dramatically risen in the past three decades in countries of all income levels. It is critical to the survival of people with diabetes, although they access affordable treatment, including insulin. There is a global consensus that the rate of increase in diabetes and obesity should be slowed down by the year 2025 (WHO, 2022a, 2022b). Thus, nowadays, there is a great need to manage diabetes to avoid or reduce chronic complications arising from it and to prevent acute complications like hyper- and hypoglycemia, blindness, limb amputation, and cardiovascular diseases (CVDs).

4   Pharmacological Management

Over 61 million people in India have diabetes, so it can be said that India is the 'capital of diabetes'. It is challenging to treat diabetes and manage all the complications simultaneously because of inadequate health care and a lack of sufficient facilities (Jacob & Narendhirakannan, 2019). In the past, medical practitioners used insulin and sulfonylureas to treat patients with diabetes mellitus type 2. But nowadays, several other medications are available on the market to treat type-2 diabetes. These are acarbose, migital, repaglinide, metformin hydrochloride, troglitazone, and rosiglitazone. In the near future, hopefully several other agents will be available to treat patients with diabetes. To control diabetic patients' blood sugar levels without requiring insulin injections, physicians will need to use these agents. Here are some common uses of these drugs.

4.1   Insulin Therapy

Insulin is a peptide hormone secreted by beta cells located in the islets of the pancreas. Insulin is released into the bloodstream as a consequence of the ingestion of food particles. This step assists in the movement of glucose from the food particles we have consumed into cells so that it can be used as a source of energy.

4.2   For Patients with Type-1 Diabetes

Due to the autoimmune reaction of the body of patients with type-1 diabetes mellitus, insulin-producing cells have been destroyed, so the body produces little or no insulin. In this scenario, it is necessary to replace the insulin in the body on a daily basis by injecting it.

4.3   For Patients with Type-2 Diabetes

In the case of type-2 diabetes, the body produces insulin but does not work correctly. In order to maintain stable blood sugar levels and postpone the requirement for tablets and insulin, lifestyle changes are necessary. When the patient requires insulin, it is important to understand that this is just the condition's natural progression (Diabetes Australia, 2022).

4.4   Sulfonylureas

Patients who still have some function in their pancreatic beta cells may experience a reduction in their blood glucose level as a result of the ability of sulfonylureas to stimulate increased insulin secretion from their pancreatic beta cells. The stimulation of pancreatic beta cells by sulfonylureas can result in an increase in the amount of insulin that is secreted. In the earlier stages of type-2 diabetes, sulfonylureas are a more effective medication for patients who have an increased pancreatic beta-cell function (Costello & Shivkumar, 2021).

4.5   Metformin Hydrochloride

Metformin is prescribed to patients who have type-2 diabetes. Metformin is an anti-diabetic medication that is available in both immediate-release and extended-release forms. Additionally, it is frequently combined with other anti-diabetic medications in a variety of combination products. The treatment and prevention of polycystic ovary syndrome (PCOS), the management of antipsychotic-induced weight gain, gestational diabetes, the prevention of type-2 diabetes, etc. are some of the off-label indications that have been approved for the use of this medication. Metformin is currently the only anti-diabetic medication that is recommended for pre-diabetic patients by the American Diabetes Association. Anti-aging effects, cancer prevention, and neuroprotection are just some of the possible applications for this compound.

4.6   Acarbose

Acarbose is a medication prescribed to patients with type-2 diabetes. This medication should be taken orally before each meal to reduce the amount of sugar that remains in the blood. Diabetes mellitus can lead to many complications over time, but taking this medication can help prevent those complications.

5   Potential Plant Used for Diabetes Control

Diabetes is a chronic lifestyle disease that is not completely curable but controllable and manifests macronutrients–carbohydrate, protein, and fat metabolic impairments. It will become the world's most-wanted killer disease in the coming 25 years. Literature source describes over 400 medicinal plant species that have hypoglycaemic effects. Plants are diverse sources of important bioactive components; many of these are potentially used in diabetes management (Malviya et al., 2010). The following figure describes the potential plants with their family bearing anti-diabetic activity (Fig. 1).

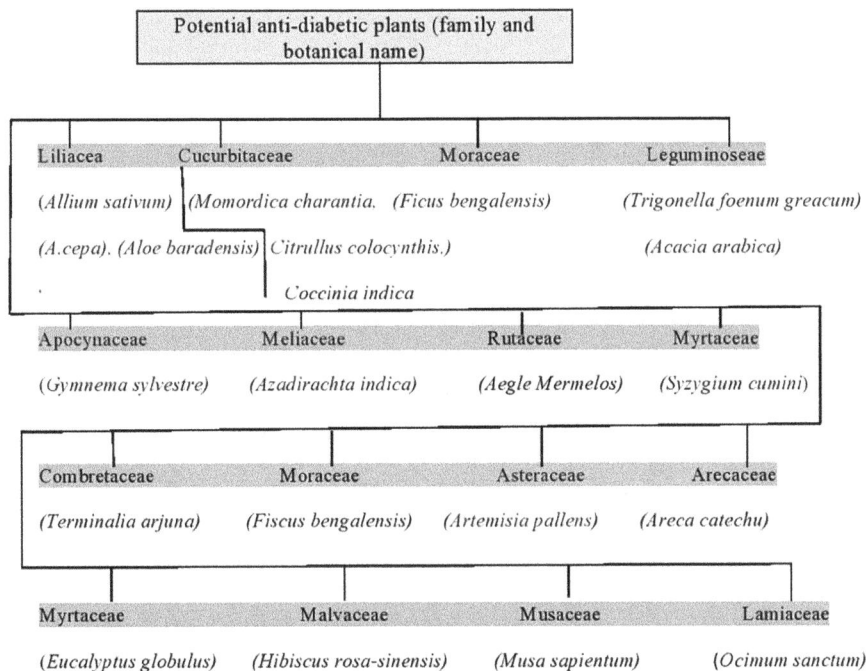

Fig. 1 The potential plants with their family bearing anti-diabetic activity

6   Medicinal Plants and Their Active Parts in Diabetes Mellitus Management

Due to the presence of therapeutically important active photo components, medicinal plants have a significant potential to treat a wide range of diseases and conditions. In recent years, various developing and underdeveloped countries have relied on medicinal plants to cure several diseases since herbal medicines are accessible, cost-effective, and without any or fewer side effects. Diabetes mellitus is a serious metabolic disorder in human beings. There are several medicines available on the market to alleviate the disease. However, these drugs are more expensive and are associated with several complications to the human body. So contemporary researchers have investigated various medicinal plants with anti-diabetic activities. Here, we will discuss various active medicinal plant parts used to treat diabetes mellitus (Tables 1, 2 and 3).

Table 1 Leaves of medicinal plants bearing anti-diabetic potential

Sl. no.	Scientific name	Common name	Family	Activities	References
1	*Acacia nilotica* (L.) Delile	Gum arabic tree	Leguminosae	Antioxidant, anti-diabetic, anti-hyperlipidaemic	Jacob and Narendhirakannan (2019), Hegazy et al. (2013)
2	*Aloe vera* (L.) Burm.f.	Ghritkumari	Asphodelaceae	Anti-hyperglycaemic	Jacob and Narendhirakannan (2019), Huseini et al. (2012)
3	*Annona squamosa* L.	Sugar apple	Annonaceae	Hypoglycaemic, anti-diabetic, anti-lipidaemic	Jacob and Narendhirakannan (2019), Gupta et al. (2008)
4	*Averrhoa bilimbi* L.	Bilimbi	Oxalidaceae	Anti-hyperglycaemic, anti-hyperlipidaemic, hypoglycaemic, anti-diabetic	Jacob and Narendhirakannan (2019), Kurup and Mini (2017)
5	*Azadirachta indica* A. Juss.	Neem tree	Meliaceae	Hypoglycaemic, β-cell regeneration	Jacob and Narendhirakannan (2019), McCalla et al. (2015)
6	*Beta vulgaris* L.	Beet	Amaranthaceae	Anti-hyperglycaemic, hypoglycaemic	Jacob and Narendhirakannan (2019), Kabir et al. (2015)
7	*Biophytum sensitivum* (L.) DC.	Little tree plant	Oxalidaceae	Hypoglycaemic	Jacob and Narendhirakannan (2019), Ananda et al. (2012)
8	*Boerhavia diffusa* L.	Spreading hogweed	Nyctaginaceae	Anti-diabetic	Jacob and Narendhirakannan (2019), Singh et al. (2011)
9	*Bombax ceiba* L.	Cotton tree	Bombacaceae	Anti-hyperglycaemic, anti-hyperlipidaemic	Jacob and Narendhirakannan (2019), Guang-Kai et al. (2017)
10	*Cajanus cajan* (L.) Millsp.	Pigeon pea	Fabaceae	Hypoglycaemic	Jacob and Narendhirakannan (2019), Uchegbu and Ishiwu (2016)
11	*Camellia sinensis* (L.) Kuntze	Tea plant	Theaceae	Anti-hyperglycaemic, anti-hyperglycaemic, and anti-diabetic	Jacob and Narendhirakannan (2019), Satoh et al. (2015)
12	*Centaurium erythraea* Rafn	Common centaury	Gentianaceae	Anti-hyperglycaemic	Jacob and Narendhirakannan (2019), Stefkov et al. (2014)
13	*Catharanthus roseus* (L.) G.Don.	Madagascar periwinkle	Apocynaceae	Anti-diabetic, hyperlipidaemic, anti-hyperglycaemic	Jacob and Narendhirakannan (2019), Al-Shaqha et al. (2015)

14	*Dillenia indica* L.	Elephant apple	Dilleniaceae	Anti-diabetic, hypolipidaemic	Jacob and Narendhirakannan (2019), Kumar et al. (2011)
16	*Eucalyptus globulus* Labill.	Blue gum	Myrtaceae	Anti-hyperglycaemic	Jacob and Narendhirakannan (2019), Ahlem et al. (2009)
17	*Ipomoea batatas* (L.) Lam.	Sweet potato	Convolvulaceae	Hypoglycaemic	Jacob and Narendhirakannan (2019), Mohanraj and Sivasankar (2014)
18	*Lithocarpus polystachyus* (Wall. ex A.DC.) Rehder	Sweet tea	Fagaceae	Hypoglycaemic	Jacob and Narendhirakannan (2019), Hou et al. (2011)
19	*Mangifera indica* L.	Mango	Anacardiaceae	Anti-diabetic, hypoglycaemic	Jacob and Narendhirakannan (2019), Ganogpichayagrai et al. (2017)
20	*Memecylon umbellatum* Burm.f.	Ironwood	Melastomataceae	Anti-diabetic	Jacob and Narendhirakannan (2019), Sunil et al. (2017)
21	*Morus alba* L.	White mulberry	Moraceae	Anti-hyperglycaemic, anti-hyperlipidaemic, anti-glycaemic	Jacob and Narendhirakannan (2019), Jiao et al. (2017)
22	*Nelumbo nucifera* Gaertn.	Indian lotus	Nelumbonacea	Hypoglycaemic, anti-hyperlipidaemic	Jacob and Narendhirakannan (2019), Liu et al. (2013)
23	*Olea europaea* L.	Wild olive	Oleaceae	Anti-hyperglycaemic, hypolipidaemic, hypoglycaemic	Jacob and Narendhirakannan (2019), Wainstein et al. (2012)
24	*Solanum virginianum* L.	Yellow fruit nightshade	Solanaceae	Anti-hyperglycaemic	Jacob and Narendhirakannan (2019), Poongothai et al. (2011)
25	*Vitex negundo* L.	Chinese chaste tree, horseshoe vitex	Lamiaceae	Anti-hyperglycaemic	Jacob and Narendhirakannan (2019), Sundaram et al. (2012)
26	*Viscum album* L.	Mistletoe	Viscaceae	Anti-hyperglycaemic, insulinotropic	Jacob and Narendhirakannan (2019), Eno et al. (2008)

Table 2 Fruits of Medicinal plants bearing anti-diabetic potential

Sl. no.	Scientific name	Common name	Family	Activities	References
1	*Abelmoschus esculentus* (L.) Moench	Lady's finger	Malvaceae	Anti-diabetic, antioxidant, and anti-hyperlipidaemic	Jacob and Narendhirakannan (2019), Mishra et al. (2016)
2	*Annona squamosa* L.	Sugar apple	Annonaceae	Hypoglycaemic, anti-diabetic, anti-lipidaemic	Jacob and Narendhirakannan (2019), Gupta et al. (2008)
3	*Averrhoa bilimbi* L.	Bilimbi	Oxalidaceae	Anti-hyperglycaemic, anti-hyperlipidaemic, hypoglycaemic, anti-diabetic	Jacob and Narendhirakannan (2019), Kurup and Mini (2017)
4	*Berberis vulgaris* L.	Barberry	Berberidaceae	Hypoglycaemic, anti-diabetic	Jacob and Narendhirakannan (2019), Meliani et al. (2011)
5	*Beta vulgaris* L.	Beet	Amaranthaceae	Anti-hyperglycaemic, hypoglycaemic	Jacob and Narendhirakannan (2019), Kabir et al. (2015)
6	*Capparis decidua* (Forssk.) Edgew.	Karira	Capparidaceae	Anti-diabetic, hypolipidaemic	Jacob and Narendhirakannan (2019), Sharma et al. (2010)
7	*Capsicum annuum* L.	Capsicum	Solanaceae	Insulinotropic	Jacob and Narendhirakannan (2019), Islam and Choi (2008)
8	*Citrullus colocynthis* (L.) Schrad	Bitter apple, wild gourd	Cucurbitaceae	Hypoglycaemic, anti-hyperglycaemic	Jacob and Narendhirakannan (2019), Barghamdi et al. (2016)
9	*Lantana camara* L.	Big sage, wild sage	Verbenaceae	Anti-hyperglycaemic, anti-diabetic	Jacob and Narendhirakannan (2019), Venkatachalam et al. (2011)
10	*Mangifera indica* L.	Mango	Anacardiaceae	Anti-diabetic, hypoglycaemic	Jacob and Narendhirakannan (2019), Ganogpichayagrai et al. (2017)
11	*Morus alba* L.	White mulberry	Moraceae	Anti-hyperglycaemic, anti-hyperlipidaemic, anti-glycaemic	Jacob and Narendhirakannan (2019), Jiao et al. (2017)
12	*Phyllanthus emblica* L. (Gaertn.)	Amla	Phyllanthaceae	Anti-diabetic	Acharya et al. (2021)

13	*Piper nigrum* L.	Black pepper	Piperaceae	Anti-hyperglycaemic	Jacob and Narendhirakannan (2019), Atal et al. (2012)
14	*Psidium guajava* L.	Guava	Myrtaceae	Anti-hyperglycaemic, anti-hyperlipidaemic	Jacob and Narendhirakannan (2019), Huang et al. (2011)
15	*Punica granatum* L.	Pomegranate	Punicaceae	Anti-diabetic, anti-hyperlipidaemic, anti-glycation	Jacob and Narendhirakannan (2019), Kumagai et al. (2015)

Table 3 Root of medicinal plants bearing anti-diabetic potential

Sl. no.	Scientific name	Common name	Family	Activities	References
1	*Averrhoa bilimbi* L.	Bilimbi	Oxalidaceae	Anti-hyperglycaemic, anti-hyperlipidaemic, hypoglycaemic, anti-diabetic	Jacob and Narendhirakannan (2019), Kurup and Mini (2017)
2	*Berberis vulgaris* L.	Barberry	Berberidaceae	Hypoglycaemic, anti-diabetic	Jacob and Narendhirakannan (2019), Meliani et al. (2011)
3	*Caesalpinia bonduc* (L.) Roxb.	Fever nut	Caesalpiniaceae	Hypoglycaemic, anti-hyperglycaemic, anti-hypercholesterolemic, anti-hypertriglyceridaemic, hypolipidaemic	Jacob and Narendhirakannan (2019), Sayyed and Wadkar (2018)
4	*Clausena anisata* (Willd.) Hook.f. ex Benth.	Horsewood	Rutaceae	Hypoglycaemic	Jacob and Narendhirakannan (2019), Ojewole (2002)
5	*Cheilocostus speciosus* (J.Koenig) C.D.Specht	Cane reed	Costaceae	Anti-hyperglycaemic, hypoglycaemic, hypolipidaemic	Jacob and Narendhirakannan (2019), Eliza et al. (2009)
6	*Helicteres isora* L.	Indian screw tree	Sterculiaceae	Anti-diabetic, hypolipidaemic, anti-hyperglycaemic	Jacob and Narendhirakannan (2019), Venkatesh et al. (2004)
7	*Ipomoea batatas* (L.) Lam.	Sweet potato	Convolvulaceae	Hypoglycaemic	Jacob and Narendhirakannan (2019), Mohanraj and Sivasankar (2014)
8	*Ophiopogon japonicus* (Thunb.) Ker Gawl.	Dwarf lilyturf	Asparagaceae	Anti-diabetic, hypoglycaemic	Jacob and Narendhirakannan (2019), Li et al. (2012)
9	*Salacia reticulata* Wight	Marking nut tree	Celastraceae	Anti-diabetic, anti-hyperlipidaemic	Jacob and Narendhirakannan (2019), Stohs and Ray (2015)
10	*Tinospora sinensis* (Lour.) Merr.	Malabar gulbel	Menispermaceae	Hypoglycaemic and hypolipidaemic	Jacob and Narendhirakannan (2019), Prince and Menon (2003)

7   Phyto-Chemistry of Medicinal Plants

Plants used for medicinal purposes are an abundant source of various vital bioactive photo components as well as bio-nutrients. These bioactive phytochemicals are generally called active principles or plant secondary metabolites. Plant secondary metabolites generally group according to their biosynthetic pathways and can be grouped into three large groups: alkaloids, phenolic compounds, and terpenes (Francesca et al., 2019).

7.1   Alkaloids

The word 'alkaloid' comes from the Arabic word 'al-qali', which refers to the plant from which soda was first isolated. The term 'alkaloid' has been in use since the nineteenth century (Kaur & Arora, 2015). The term 'alkaloid' was first used in 1819 by the German chemist Carl F. W. Meissner to refer to a substance derived from plants that possessed alkaline characteristics. According to, alkaloids are base-type compounds that contain an N-atom at any position in the molecule and do not include N in an amide or peptide bond. Furthermore, alkaloids are distinguished from amines and peptides by the absence of an amide bond. According to Gutiérrez et al.'s research from 2020, alkaloids are abundant in bacteria, fungi, plants, and animals, and they are formed when acids combine with alkaloids to make salts. One of the defining characteristics of alkaloids is that they do not contain nitrogen in the amide or peptide bonds that they form. Since ancient times, purgatives, anti-tussives, and sedatives that are derived from plants have been used as part of folklore medicine to treat a wide variety of diseases and ailments. Alkaloids have played a significant role in the treatment of these conditions. The low MW (molecular weight) structures that makeup alkaloids account for approximately 20% of the secondary metabolites that are plant-derived. It is estimated that approximately 12,000 different alkaloids have been isolated to this point from various plant families across the kingdom of plants (Kaur & Arora, 2015). Alkaloids are typically produced by plants in order to facilitate their existence in the ecosystem, the formation of seeds, and their escape from a variety of predators. Alkaloids are the most abundant and diverse group of secondary metabolites. They also have anti-diabetic activity because they inhibit enzymes such as beta-amylase, beta-glucosidase, dipeptidyl peptidase-IV, aldose reductase, and protein tyrosine phosphatase-1B, among others.

The anti-hyperglycaemic activity of the roots of *Aerva lanata* Linn. (Amaranthaceae) was investigated in an experiment conducted by Agrawal et al. (2013). The results of the experiment showed that the plant exhibits this kind of activity due to the presence of alkaloids known as canthin-6-one derivatives. Sangeetha, Priya, and Vasanthi conducted an investigation in vitro for the purpose of determining the anti-hyperglycaemic potential of the stem of *Tinospora cordifolia* in 2013. Because it contains palmatine and alkaloid, the stem of *Tinospora cordifolia* has been shown to have anti-hyperglycaemic properties when it has been extracted with petroleum ether.

The anti-diabetic potential of alkaloids such as vindoline, vindolidine, vindolicine, and vindolinine was demonstrated by Tiong et al. (2013) when they were extracted from the leaves of *Catharanthus roseus*. In another experiment, discovered that the seed of *Brassica oleracea* var. *capitata* contains 2,3-Dicyano-5,6-diphenyl pyrazine, which is an alkaloid that has anti-diabetic activities.

7.2   Phenolic Compounds

Phenolic compounds, phenolics, and polyphenolics are all umbrella terms that refer to the same thing: chemical compounds that have at least one aromatic ring and one or more hydroxyl substituents. Phenolic compounds, phenolics, and polyphenolics are all examples of these types of compounds. This encompasses functional derivatives like esters, methyl ethers, glycosides, and other compounds that are very similar to these. Over 8000 distinct types of phenolic compounds can be found in the plant kingdom (Bhuyan & Basu, 2017; Ho, 1992; Cartea et al., 2011). Whereas the term 'plant phenolics' refers to the natural secondary metabolites that arise through biogenesis from either the shikimic acid pathway or the phenylpropanoid pathway, which directly provides phenylpropanoids, or the 'poly-ketide' acetate/malonate pathway, which also accomplishes a very wide range of physiological roles in medicinal plants, the phenylpropanoids that are directly provided by the phenylpropa (Quideau et al., 2011; Harborne, 1989).

Ever since the course of evolutionary consequences of various plant lineages, higher plants have synthesized several thousand known phenolic compounds. This has allowed plants to tolerate adverse environmental challenges over the course of evolutionary time. When plants are subjected to a variety of environmental stresses, such as pathogen infections, herbivores, low temperatures, high light, and nutrient deficiency, the phenolic compounds found in the plants play an important role in the defence mechanisms that the plants employ. This can increase the number of free radicals and other oxidative species produced by the plants (Lattanzio, 2013).

7.3   Terpenes

Terpenes are the most numerous and diverse group of naturally occurring chemical compounds that can be found almost exclusively in plants. They are also known as terpenoids and isoprenoids. Terpenes, such as sterols and squalenes, have also been discovered in animal tissues. The majority of terpenes have structures that are multicyclic and contain oxygen-containing functional groups. Terpenoids account for approximately 60% of all known natural products that play a role in the secondary metabolism of plants. Despite the fact that the nomenclature of this compound is sometimes used interchangeably with 'terpenes', the compounds in question are hydrocarbons, whereas terpenoids have additional functional groups, most of which contain the element oxygen. Plants have a wide variety of pigments, aromas, and flavours, all of which are caused by the presence of terpenes

Table 4 Some medicinal plants with responsible phytochemicals bearing anti-diabetic activities are presented here (Alam et al., 2022)

Name	Family	Bioactive compounds
Allium sativum L.	Amaryllidaceae	Allicin, alliin, diallyl trisulfide, S-allyl cysteine, allyl mercaptan, ajoene
Aegle marmelos Correa	Rutaceae	Aegeline
Artocarpus heterophyllus Lam.	Moraceae	Gallic acid, catechin, caffeic acid, rutin, and quercetin
Bauhinia forficata Link	Fabaceae	Kaempferitrin
Beta vulgaris L.	Chenopodiaceae	Betavulgarosides (II, III, IV), apigenin 8-C-b-D-glucopyranoside (vitexin), acacetin 8-C-b-D-glucopyranoside, acacetin 8-C-a-L-rhamnoside
Bougainvillea spectabilis Willd.	Nyctaginaceae	D-pinitol
Cecropia obtusifolia Bertol.	Urticaceae	Isoorientin, chlorogenic acid
Centella asiatica (L.) Urb.	Apiaceae	Asiaticoside (triterpene saponin compound), madecassic acid, asiatic acid, brahmoside, and brahminoside (glycosides)
Lagerstroemia speciosa (L.) Pers.	Lythraceae	Corosolic acid (2a-hydroxyursoloic acid)
Laminaria japonica Aresch	Laminariaceae	Butyl-isobutyl-phthalate, polysaccharides
Mangifera indica L.	Anacardiaceae	Mangiferin, kaempferol
Salacia chinensis L.	Celastraceae	Salasones A, B, and C; salaquinone A; salasol A; 22-dihydroxyolean12-en-29-oic acid; tingenone; tingenine B; regeol A; triptocalline A
Salacia reticulate Wight	Celastraceae	Salacinol, kotalanol
Scoparia dulcis L.	Scrophulariaceae	Scutellarein, apigenin, luteolin, scopadulcic acid B, betulinic acid, scoparic acid A
Stevia rebaudiana Bertoni	Asteraceae	Stevioside

(Cox-Georgian et al., 2019). The presence of terpenoids in plants is one factor that contributes to their aromatic qualities, and as a result, they play an important role in traditional herbal medicine. In addition, terpenes serve a number of other important functions in plants, including signalling, thermoprotection, flavouring, pigmentation, and solvent production. Terpenes also have a number of applications in the medical field (Cox-Georgian et al., 2019) (Table 4).

8 Mechanism of Action in Diabetes Management

1. *Role in the Normalization of Insulin Production*: The pharmacokinetic activity of Indian traditional plants, especially insulin production from pancreatic beta cells or similar insulin-related functions, is sourced from various literature surveys (Patel et al., 2012) (Table 5).
2. *Role in Insulin Resistance and Associated Metabolic Disorders*: Noninsulin-dependent diabetes mellitus (NIDDM) or insulin resistance can lead to carbohydrate, protein, and fat metabolic derangement or syndrome, which is at the forefront of today's research (Wilcox, 2005). Therapeutic drugs alone challenge to manage this syndrome. There has been a long history of traditional plants used in diabetic medicine, evident in Ayurveda, Unani, and Siddha medical systems (Ozturk, 2018). However, the mechanism of action in various parts of the medicinal plant will interest to know the importance of these plants, which mentioning bellows.

8.1 *Terminalia arjuna* (Roxb.)

The common name of *Terminalia arjuna* (Roxb.) is arjuna. It is used traditionally in various medical applications, such as for anti-microbial purposes, human immuno-deficiency virus (HIV) treatment, bone fracture treatment, hypolipidaemic drugs, anti-inflammatory, cardio protection, antioxidant, and diabetes treatment. This plant is 60–80 ft in height and is distributed in India, Myanmar, Sri Lanka, and Mauritius. Tannins, polyphenols, flavonoids, saponins, triterpenoids, sterols, and mineral constituents are just some phytochemicals found in arjuna in high concentrations. The active parts of this plant are stem bark and root bark (Amalraj & Gopi, 2016). In normal conditions, insulin decreases carbohydrate catabolism and promotes an anabolic pathway. Arjuna bark extract has similar functions in the condition of insulin limitation. These are enhanced hexokinase, glucokinase, and phosphofructo-kinase enzyme activity. Hexokinase presents all cells and involves the phosphoryla-tion of glucose to glucose 6-phosphates. On the other hand, it lowers the activity of gluconeogenic enzymes, viz. glucose-6-phosphatase and fructose-1, 6-diphosphatase, in kidney and liver cells. This may indicate insulin secretion by secretagogue in bark extract; otherwise, enzyme controlling function cannot proceed (Ragavan & Krishnakumari, 2006).

The medicinal use of arjuna has evidence of documentation from Charaka to his decedent Chakradatta, Bhavamishra, and present-day Ayurveda practitioners. Medi-cation with *T. arjuna* preparation may be effective in CVD. It has been suggested that the soluble fibre, sitostanol content, and flavonoid antioxidant found in this plant are responsible for its hypocholesterolemic effects. It reduces platelet aggregation by inhibiting ca^{2+} release and CD62P gene expression in platelets (Maulik & Talwar, 2012). Applying bark extract on streptozotocin (STZ) induced diabetic rats in 8 weeks positively impacted left ventricular pressure, reducing inflammatory cytokines, oxidative stress, blood lipid profile, and myocardial injuries. In

Table 5 List of traditional plant materials having hypoglycaemic or insulin secretory activity related to their phytoconstituents with in vivo and in vitro studies

Sl. no.	Botanical name	Mechanisms of action
1	*Aloe barbadensis*	1. Hypoglycaemic effects by stimulating insulin synthesis or release at the beta cells of the pancreas in a diabetic rat due to the bitterness of *Aloe vera* (Singh, 2011; Modak et al., 2007).
2	*Agrimonia eupatoria*	1. Stimulation of insulin synthesis and release is glucose dependent (Bhushan et al., 2010; Bnouham et al., 2006).
3	*Allium sativum* (Alliaceae)	1. S-allyl cysteine (SAC) is active in insulin stimulation and hyperglycaemia prevention. 2. SAC increases pancreatic insulin secretion in normal rats. 3. Garlic might be an inhibitor of DPP-4 (dipeptidyl peptidase-4), which activates insulin secretion and deactivates glucagon secretion, resulting in reduced blood glucose level (Modak et al., 2007; Bnouham et al., 2006, Noor et al., 2013).
4	*Aegle marmelos* (Rutaceae)	1. Beal fruit extract maintains blood glucose levels through decreasing glycogenolysis and increasing neo-glucogenesis in the liver (Sharma et al., 2011).
5	*Azadirachta indica* (neem)	1. Glucagon-like peptide-1 hormones can enhance insulin synthesis when the extract inactivates dipeptidyl peptidase IV (DPP-IV). 2. Aqueous leaf extracts improve serum insulin through the normalization of beta cells because oxidative stress destroys the beta cells of the pancreas and causes a shortfall in insulin levels (Yarmohammadi et al., 2021).
6	*Gymnema sylvestre*	1. The hyperactivation of beta cells increases in size and degenerates morphologically, so leaf extract treatment helps regenerate or repair beta cells and insulin release. This will restore pancreas β cells' normal activity and normalize insulin production, so hyperinsulinaemia associated with type-1 DM can be alleviated (Baskaran et al., 1990).
7	*Momordica charantia* (Cucurbitaceae)	1. Unripe karela fruit juice extract can partially stimulate insulin release through β-cell sensitization in the pancreas, which is not even interfered with by L-epinephrine (Raman & Lau, 1996; Grover et al., 2002). 2. P-insulin is an insulin-like substance that works similarly in place of human insulin and reduces blood glucose. 3. This fruit extract down-regulates the activation of mitogen-activated protein kinases (MAPKs) and stress induces protein kinase/c-Jun N-terminal kinase (SAPK/JNK), p38, and p44/42, the activity of NF-κB, etc., thereby fortifying pancreatic β cells from oxidative stress and normalizing insulin production (Joseph & Jini, 2013).
8	*Syzygium cumini* (*Eugenia jambolana*)	1. Jamun seed extracts proven in vivo and in vitro increase insulin production (insulin secretagogue) (Zulcafli et al., 2020).

(continued)

Table 5 (continued)

Sl. no.	Botanical name	Mechanisms of action
9	*Terminalia arjuna* (Roxb.)	1. *Terminalia arjuna* has insulinotropic effects. The bark extract raises intracellular ca^{2+} ions, indicating β-cell depolarization and insulin release. 2. They keep blood glucose lower, inhibiting starch digestion by α-amylase and α-glucosidase enzymes. 3. Arjuna bark extract maintains insulin levels by inhibiting glucose binding to insulin, which leads to weak insulin receptor interaction at tissue sites (Thompson et al., 2014).
10	*Ficus benghalensis*	1. The bark extract of *Ficus benghalensis* contains leucopelargonidin, which increases insulin production for prolonged use and reduces blood sugar. A change in metabolic fate infers this effect. 2. Due to insulin, the inadequacy of the cytochrome P-450 enzyme system in the liver alters and affects lipid peroxidation. This bark extract normalizes the condition (Gayathri & Kannabiran, 2008).
11	*Trigonella foenum-graecum*	1. Fenugreek contains diosgenin saponin increases insulin production secretion through regeneration or repair of β cells. 2. 4-hydroxy isoleucine derived from fenugreek seeds enhances insulin secretion (Ramji & Foka, 2002; Baset et al., 2020).
12	*Coccinia indica*	1. Like insulin, leaf extracts regulate high blood glucose through secretion or by controlling glucose metabolism (Grover et al., 2002; Balaraman et al., 2010).

conclusion, it reduces diabetic-associated cardiopathy, vascular thrombosis, and inflammation (Amalraj & Gopi, 2016).

8.2   *Aloe vera*

Bioactive AIII and prototinosaponin AIII compounds increase glucose utilization through liver gluconeogenesis (Bnouham et al., 2006).

8.3   *Allium cepa*

Allium cepa is commonly known as onion, and it benefits patients with type-2 diabetes mellitus. Insulin insensitivity is a consequence of type-2 diabetes mellitus and may also cause hypoglycemia as a result of the exhaustion of pancreatic beta cells. During the last 10 years, research focuses on the pathogenic role of type-2 DM and insulin resistance due to lipid toxicity and low-grade inflammation. Because a higher number of free fatty acids may directly activate macrophage cells to release pro-inflammatory cytokines like TNF-α, IL-1β and IL6. Excessive stimulation of inflammatory cytokines leads to direct insulin resistance in muscle cells' insulin receptors. In addition, an increase in free fatty acid may deposit in the liver cell, causing fatty liver, free radicals induce oxidative stress, and activation of stress induce signalling pathway making storage glycogen breakdown and hyperglycemia. Onion peel extract may ameliorate hyperglycaemia and insulin resistance by increasing glucose uptake in the peripheral tissue by stimulating GLUT-4 and INSR gene expression in muscle tissues. Onion peel extract is highly evidenced in quercetin in the dry outer part, which increases insulin sensitization capabilities as well as lowered plasma free fatty, suppression of inflammatory cytokines and oxidative injury of liver cells (Jung et al., 2011; Noor et al., 2013; Ozougwu, 2011).

8.4   Aegle marmelos

Aegle marmelos, whose common name is bael, is an Indian traditional medicinal plant. Its leaf and fruit have been used for treating diabetes since ancient times. In recent times, advanced research established that the ingestion of bael fruit extracts heals the beta cells of the pancreatic gland and causes insulin sensitivity, gives antioxidant effects and lowers the risk of diabetic-associated hyperlipidemia. In type-2 DM, the insulin resistance triggers are higher free fatty acid (as a source of fatty meal diet), inflammatory cytokines like IL-6, TNF-β, oxidative stress induce beta-cell destruction, down-regulation insulin receptors, etc. For compensatory mechanisms, insulin over-secretion and reduced insulin extraction by the liver lead to hyperinsulinemia, which indicates beta-cell abnormality and hyperglycemia. Recent research suggests that *Aegle marmelos* fruit extracts increase insulin sensitivity through the up-regulation of the peroxisome proliferator-activated receptor-γ expression (PPARγ) and phosphoinositide 3-kinase (PI3 kinase) and increase the tyrosine phosphorylation of insulin receptors. As a result, the amount of glucose taken up by muscle, liver, and peripheral tissues increases. Dyslipidemia is a complication of insulin resistance and is evidence of increased free fatty acid flow, total cholesterol, total triglyceride, LDL-c, and lower HDL-c. The risk factor analysis suggests a high-fat diet and insulin resistance. Because insulin resistance increases 3-hydroxy-3-methylglutaryl (HMG), coenzyme A (CoA) re-educates enzyme synthesis and inhibits lipoprotein lipase (LPL) enzyme activity so that a higher amount of free fatty acid mobilized from peripheral tissue deport. Normally, insulin promotes lipogenesis and inhibits lipolysis. So the possible mechanisms of *Aegle marmelos* fruit extracts are for increasing liver triglyceride (TG) synthesis, activating LPL enzymes for the clearance of TG and free fatty acids from the blood vessels to the peripheral organs and tissues, and decreasing dietary cholesterol absorption and synthesis. These are the anti-dyslipidemia effects of bael fruit extracts. The antioxidant enzyme activity of superoxide dismutase (SOD) is also increased by bael fruit extracts, which helps prevent lipid peroxidation by interfering with the formation of superoxides and hydroperoxides. And finally, bael fruit extracts protect pancreas integrity and prevent beta-cell destruction due to their antioxidant nature (Sharma et al., 2011; Sabu & Kuttan, 2004).

8.5   *Azadirachta indica* (Neem)

Insulin resistance is synonymous with metabolic syndrome and includes the collective effects of obesity, insulin resistance, hypertension, high blood cholesterol, triglyceride, etc. Neem in parts of leaf, bark, stem, and flower, are very effective in high blood pressure, insulin resistance, antioxidants, dyslipidemia, etc. For blood pressure lowering, several possible mechanisms demonstrate that:

1. Leave extract of neem up to regulate gene expression of the extracellular signal-regulated kinase (ERK 1 and 2) in the smooth endothelial muscle of blood vessels and blocking of ca^{2+} channel for smooth muscle contraction (Shah et al., 2014; Omóbòwálé et al., 2018),
2. Gene expression of nuclear factor erythroid 2–related factor 2 (Nrf2) as a transcription factor for synthesis of anti-oxidant enzymes that prevent high blood pressure (Howden, 2013).
3. Elevation of nitric oxide (vasodilator) (Omóbòwálé et al., 2019).

The fact of neem leaf extract in lipid-lowering describes that prevent free radical induce (ROS) oxidation of cholesterol, TG, fatty acid, and phospholipid otherwise oxidation leads to raise bad cholesterol level and reduce HDL-c. This decreases LDL-c, TG, and phospholipids and increases high-density lipoprotein (HDL) levels (Zuraini et al., 2006; Peer et al., 2008). The stem bark and root extract of neem applied for the control of obesity inhibit pancreatic lipase and α-glycosidase, which are responsible for weight gain (Mukherjee & Sengupta, 2013). Neem powder is used as a dietary supplement in the treatment of type-2 diabetes mellitus. It better controls blood glucose and HbA1c in patients ages 18–70 years. This leaf extract reduces glucose production during starch digestion by inactivating lingual α-amylase and intestinal glycosidase and induces the glycolysis pathway by activating hexokinase (Tadera et al., 2006). This is related to controlling high blood glucose during insulin resistance. Neem is used to prepare poly-herbal formulas (PHFs) and treat DM. This extract also up-regulates the GLUT-4 gene in muscle cells, increasing glucose uptake from blood vessels and maintaining blood glucose homeostasis (Yarmohammadi et al., 2021).

8.6   *Gymnema sylvestre*

The leaf of *Gymnema sylvestre* is an Indian traditional medicinal plant mentioned in the Sushruta (600 BC) and is now also recommended by many Ayurveda practitioners. At that time, this plant belonged to the 'sala saradi' group and was preferred for type-2 DM treatment with dietary restriction. This plant is also known as a sugar destroyer. The anti-diabetic role of the leaf extract is active compound gymnemic acid which has anti-hyperinsulinemia, insulin secretion, regeneration or repair of pancreatic β cells, and liver and kidney cell protection. The suggested mechanisms are as follows:

1. Managing high blood sugar during hyperinsulinemia through liver glycogen synthesis by activating the glycogen synthetase enzyme and inhibiting glucose 6-phosphatase, which converts glucose 6-phosphate to glucose; it decreases the activity of other insulin-independent enzymes, viz. glycogen phosphorylase, gluconeogenic enzymes, fructose 1,6- diphosphatase, and sorbitol dehydrogenase, by which glucose release in to blood vessels is lower (Baskaran et al., 1990; Khan et al., 2019).
2. Reducing glycosylated haemoglobin (HbA1c).
3. Preventing glucose absorption by the small intestine, thus lowering fasting sugar in the blood; in lowering glucose absorption, the gymnemic acid matches structurally with a glucose molecule, which is why it blocks glucose-absorbing receptors in the small intestine (Khan et al., 2019).
4. Reducing high lipid profile, such as LDL-c, VLDL-c, TG, and HDL-c, and the secondary development of atherosclerosis risks; this is due to the presence of acidic polyphenolic substances such as flavonoids, saponins, tannins, etc. (Khan et al., 2019).

8.7   *Momordica charantia*

Momordica charantia is commonly known as bitter gourd, karela, etc. It plays an important role in anti-hyperglycaemic and hypoglycaemic conditions and diabetic complications. It has bioactive compounds in fruits derived P insulin, charantin, and seed-derived vicine. Charantin is a mixture of two steroidal sitosteryl glucoside and stigmasteryl glucoside which have significant hypoglycaemic effects. The oral dose of 50 mg/kg charantin has 42% lower blood glucose. Bitter gourd fruits extract vital anti-diabetic functions in a number of ways:

1. It reduces Na^+- and K^+-dependent glucose absorption in the intestine.
2. Karela juice has clinical insulin-mimetic functions, such as enhancing glucose uptake in muscle cells and in the liver by increasing glycogenesis.
3. Karela seed protein extracts stimulate lipogenesis or inhibit lipolysis, neoglucogenesis, etc.
4. Hepatic and red blood cell glucose-6-phosphate dehydrogenase (G6PD-6-PDH) activity utilized excess glucose in hexose mono phosphate shunt pathway (HMP) that is the role of hypoglycaemic. It inhibits the activities of liver fructose 1, 6-diphosphatase and glucose-6-phosphatase that prevent free glucose release into blood stream i.e., anti-hypoglycaemic role.
5. Insulin sensitization through AMP-activated protein kinase phosphorylation in the target receptors may affect karela juice consumption. (Joseph & Jini, 2013; Raman & Lau, 1996).

8.8 *Syzygium cumini*

Both the seeds and bark of the Jamun stem show promise as an anti-diabetic medicine in humans. The various parts (leaves, bark, fruits, and seeds) of *Syzygium cumini* are traditionally used for diabetes treatment. It may reduce blood glucose by inhibiting α-glycosidase for di- and oligosaccharide digestion and inhibitory effects on pancreatic amylase. The activities of plant seeds control blood sugar through anti-hyperglycaemic and hypoglycaemic, where increased glycogen synthesis in liver and muscle cells and inactivation of hepatic gluconeogenesis (glucose-6-phospha-tase) is called anti-hyperglycaemic and increased activity of key glycolytic enzymes (hexokinase) called hypoglycemia (Zulcafli et al., 2020). Mycaminose component shows the anti-hyperglycaemic effect of seeds. Oral administration of seed powder continues for 3 months, effectively managing various diabetic symptoms such as polyurea, polyphagia, weakness and weight loss. This seed extract increases cellular glucose uptake by up-regulating gene expression of GLUT4 and protecting β-cell dysfunction. The pathogenesis of β-cell dysfunction is due to insulin resistance inducing inflammatory cytokines such TNF-α, IL-6, etc. Apart from glucose homeo-stasis, this seed extract possibly reduces dyslipidemia through the gene expression of peroxisome proliferator-activated receptor alpha (PPARα) and peroxisome proliferator-activated receptor gamma (PPAR γ) activities in the liver. That will affect insulin sensitization through increased mobilization of free fatty acid and storage. Endogenous cholesterol synthesis is inhibited by inactive HMG CoA reductase and decreased intestinal fat absorption by the function of seed extract (Baliga et al., 2011; Grover et al., 2002).

8.9 *Ficus benghalensis*

During diabetes, insulin resistance causes changes in the metabolic enzymes found in the liver. These changes can include glycogen synthase, glucokinase, lactate dehydrogenase (LDH), succinate dehydrogenase, and malate dehydrogenase. That leads to impaired glycogenesis, tricarboxylic acid (TCA) cycle, etc., which may be alleviated by bark extract. This suggests that gene expression of enzyme protein in response to the released insulin and utilizes blood glucose (Gayathri & Kannabiran, 2008). Because of its amylase inhibitory action during the digestion of starch, the pelargonidin derivative glycoside that was isolated from the bark extract has the potential to lower fasting blood glucose levels. Leucocyanidin derivative, which was isolated from the bark of *F. benghalensis*, has hypoglycaemic properties with a dose of 100 mg/kg in normal, which is why it is better to manage diabetes as combination therapy with insulin, and even lower the requirement of insulin therapy (Grover et al., 2002).

8.10   *Trigonella foenum-graecum*

Fenugreek seeds are similarly medicinal plants good for diabetic management and associated chronic illnesses. With regard to blood lipid control, methi seed extracts could reduce LDLL-c, VLDL-c, and TG and increase HDL-c. This is possible due to the crude fibres and saponin in methi seeds as these components may increase faecal bile and cholesterol excretion (Geberemeskel et al., 2019). Fenugreek seeds manage blood glucose and lipid profile in several ways:

1. They increase glucose uptake through the gene expression of the GLUT-2 transporter.
2. They enhance the messenger RNA (mRNA) transcription of CCAAT/enhancer-binding protein (C/EBPδ) and peroxisome proliferator-activated receptor-γ (PPAR-γ) protein. These protein receptors are located in the liver, adipocyte, intestines, lungs, kidneys, and myeloid cells. Due to the activation of C/EBP and PPAR-γ family receptors in adipose tissue and liver, increased lipogenesis and excess FFA, TG and cholesterol will mobilize.
3. They activate C/EBPα-mediated glycogen synthetase enzymes, which promote glycogenesis (Ramji & Foka, 2002; Baset et al., 2020).
4. Galactomannan, a carbohydrate-containing bioactive substances (45–60%), reduces postprandial blood glucose by blocking α-amylase and pancreatic lipase.
5. An oral dose of fenugreek seeds may reduce renal complications by protecting the glomerular basement membrane and lowering the creatinine level. It also reduces TG and increases HDL-c levels. Fenugreek seeds have an antioxidant potential; the legumes raise the level of glutathione peroxidase and catalase, preventing lipid peroxidation (Baset et al., 2020).

8.11   *Coccinia indica*

It has been used in Indian traditional medicine from Ayurveda and the Unani systems. Beta-sitosterol, an active compound in leaves and pectins in fruits, has hypoglycaemic effects. Pectins hypoglycaemic mechanism in normal conditions is activated through glycogen synthetase and inactivation of glycogen phosphorylase. Leaf-containing beta-sitosterol deactivates liver fructose 1, 6 bisphosphatase, glucose 6-phosphatase, and LDH and increases lipoprotein lipase activity which is anti-hyperglycemic anti-cholesterolemic effect. Oral administration of leaf extract may reduce fasting sugar due to halting starch-breaking enzymes, insulin secretion, etc. (Grover et al., 2002; Balaraman et al., 2010) (Fig. 2).

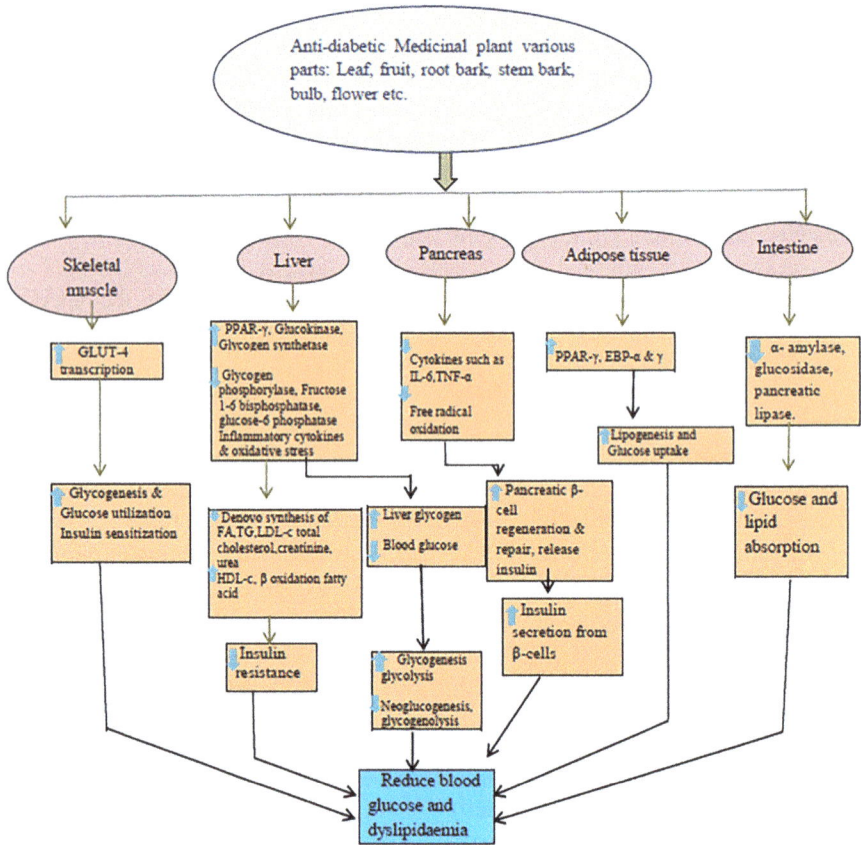

Fig. 2 Mechanism of action of the anti-diabetic activity of different Indian medicinal plants

9   Future Prospects of Indian Medicinal Plants in Diabetes

It is estimated that approximately 33 million adults have been suffering from diabetes mellitus in recent years in India, and it is likely to increase by 57.2 million by 2025 (Manukumar et al., 2016). Plant-derived active ingredients are applied to patients, along with some medications, to treat diabetes mellitus. But these active principles are practically extractable with different solvent phases, so it is essential to isolate individual phyto-components to confirm their activity on diabetes mellitus. Therefore, more research is needed. Several medicinal plants play an important role in treating diabetes mellitus and its associated complications, but herbal formulations' major drawback is that these active ingredients are not well defined. Their molecular interactions and mode of action should be more carefully investigated for future novel drug discoveries. Recently, more initiatives have been undertaken to fight diabetes mellitus. Abdulazeez, in the year 2013, remarked that the cells that produce insulin might be derived from stem cells, and he is confident that shortly scientists may come to light with solutions for the use of stem cells as therapeutic agents to cure diabetes (Abdulazeez, 2013). Pancreas and cell of islet transplantation is currently an outstanding achievement at the University of Illinois Chicago (UIC).

10　Conclusion

In this chapter, we discussed the Indian traditional medicinal plants for the treatment of diabetes. There are so many Indian traditional plants with anti-diabetic roles that have been reported. The effect is attributed to the plant phytochemicals, single compounds as well as extracts. The main plant phytochemicals responsible for anti-diabetic activity are alkaloids, phenolic acids, flavonoids, glycosides, etc. Identification of anti-diabetic properties on Indian traditional ethnographic plants is remarkable progress for human welfare and evidence-based short-term and long-term implication has also been integrated in past, present and future also. To see the observable effects, multiple in vitro and in vivo experiments have been conducted. Various mechanisms are described for the anti-diabetes role of medicinal plant parts, explaining the beneficial effects of phytochemicals, such as their role in the normalization of Insulin production, Role in insulin resistance and associated metabolic disorders, regulation of glucose and lipid metabolism, stimulating β cells and ROS protective action. However, still many aspects such as a mark popularization over pharmaceutical drugs, study at the genomic level to examine the role of plant secondary metabolites in diabetes, limited availability of toxicological safety data, etc. remain to be aspirated areas. Only a few medicinal plants have been studied for efficacy in humans. Thus, more efficient clinical studies are warranted for further validation.

References

Abdulazeez, S. S. (2013). Diabetes treatment: A rapid review of the current and future scope of stem cell research. *Saudi Pharmaceutical Journal, 613*(4), 1–8. https://doi.org/10.1016/j.jsps.2013. 12.012

Acharya, C. K., Madhu, N. R., Khan, N. S., & Guha, P. (2021). Improved reproductive efficacy of *Phyllanthus emblica* L. (Gaertn.) on testis of male Swiss mice and a pilot study of its potential values. *International Journal of Food Sciences and Nutrition, 10*(4), 7–14.

Agrawal, R., Sethiya, N. K., & Mishra, S. H. (2013). Antidiabetic activity of alkaloids of Aerva lanata roots on streptozotocin-nicotinamide induced type-II diabetes in rats. *Pharmaceutical Biology, 51*(5), 635–642. https://doi.org/10.3109/13880209.2012.761244

Ahlem, S., Khaled, H., Wafa, M., Sofiane, B., Mohamed, D., Jean-Claude, M., & Abdelfattah el, F. (2009). Oral administration of *Eucalyptus globulus* extract reduces the alloxan-induced oxidative stress in rats. *Chemico-Biological Interactions, 181*(1), 71–76. https://doi.org/10. 1016/j.cbi.2009.06.006

Alam, S., Sarker, M. M. R., Sultana, T. N., Chowdhury, M. N. R., Rashid, M. A., Chaity, N. I., Zhao, C., Xiao, J., Hafez, E. E., Khan, S. A., & Mohamed, I. N. (2022). Antidiabetic phytochemicals from medicinal plants: Prospective candidates for new drug discovery and development. *Frontiers in Endocrinology, 13*, 800714. https://doi.org/10.3389/fendo.2022. 800714

Al-Shaqha, W. M., Khan, M., Salam, N., Azzi, A., & Chaudhary, A. A. (2015). Anti-diabetic potential of Catharanthus roseus Linn. and its effect on the glucose transport gene (GLUT-2 and GLUT-4) in streptozotocin induced diabetic Wistar rats. *BMC Complementary and Alternative Medicine, 15*(1), 379.

Amalraj, A., & Gopi, S. (2016). Medicinal properties of Terminalia arjuna (Roxb.) wight & amp; Arn.: A review. *Journal of Traditional and Complementary Medicine, 7*(1), 65–78.

Ananda, P. K., Kumarappan, C. T., Christudas, S., & Kalaichelvan, V. K. (2012). Effect of biophytum sensitivum on streptozotocin and nicotinamide-induced diabetic rats. *Asian Pacific Journal of Tropical Biomedicine, 2*(1), 31–35.

Atal, S., Agrawal, R. P., Vyas, S., Phadnis, P., & Rai, N. (2012). Evaluation of the effect of piperine on blood glucose level in alloxan-induced diabetic mice. *Acta Poloniae Pharmaceutica, 69*(5), 965–969.

Atanas, G., Waltenberger, A. B., Pferschy-Wenzig, E. M., Linder, T., Wawrosch, C., Uhrin, P., Temml, V., Wang, L., Schwaiger, S., Heiss, E. H., Rollinger, J. M., Schuster, D., Breuss, J. M., Bochkov, V., Mihovilovic, M. D., Kopp, B., Bauer, R., Dirsch, V. M., & Stuppner, H. (2015). Discovery and resupply of pharmacologically active plant-derived natural products: A review. *Biotechnology Advances, 33*, 1582–1614.

Balaraman, A. K., Singh, J., Dash, S., & Maity, T. K. (2010). Antihyperglycemic and hypolipidemic effects of Melothria maderaspatana and Coccinia indica in Streptozotocin induced diabetes in rats. *Saudi Pharmaceutical Journal, 18*(3), 173–178. https://doi.org/10. 1016/j.jsps.2010.05.009

Baliga, M. S., Bhat, H. P., Baliga, B. R. V., Wilson, R., & Palatty, P. L. (2011). Phytochemistry, traditional uses and pharmacology of *Eugenia jambolana* lam. (black plum): A review. *Food Research International, 44*, 1776–1789.

Barghamdi, B., Ghorat, F., Asadollahi, K., Sayehmiri, K., Peyghambari, R., & Abangah, G. (2016). Therapeutic effects of Citrullus colocynthis fruit in patients with type II diabetes: A clinical trial study. *Journal of Pharmacy & Bioallied Sciences, 8*(2), 130–134.

Baset, M. E., Ali, T. I., Elshamy, H., El-Sadek, A. M., Sami, D. G., Badawy, M. T., & Abdellatif. (2020). Anti-diabetic effects of fenugreek (Trigonella foenum-graecum): A comparison between oral and intraperitoneal administration—an animal study. *International Journal of Functional Nutrition, 1*, 2. https://doi.org/10.3892/ijfn.2020.2

Baskaran, K., Ahamath, B. K., Shanmugasundaram, K. R., & Shanmugasundaram, E. R. B. (1990). Antidiabetic effect of a leaf extract from *Gymnema sylvestre* in non-insulin-dependent diabetes mellitus patients. *Journal of Ethnopharmacology, 30*(3), 295–305.

Bhushan, M. S., Rao, C. H. V., Ojha, S. K., Vijayakumar, M., & Verma, A. (2010). An analytical review of plants for anti-diabetic activity with their phytoconstituent & mechanism of action. *International Journal of Pharmaceutical Sciences and Research, 1*(1), 29–46.

Bhuyan, D. J., & Basu, A. (2017). Phenolic compounds, potential health benefits and toxicity. In Q. V. Vuong (Ed.), *Utilisation of bioactive compounds from agricultural and food waste* (pp. 27–59). CRC Press.

Bnouham, M., Ziyyat, A., Mekhfi, H., Tahri, A., & Legssyer, A. (2006). Medicinal plants with potential antidiabetic activity-a review of ten years of herbal medicine research (1990–2000*). International Dubai Diabetes and Endocrinology Journal, 14*(1), 1–25.

Cartea, M. E., Francisco, M., Soengas, P., & Velasco, P. (2011). Phenolic compounds in brassica vegetables. *Molecules, 16*(1), 251–280.

Costello, R. A., & Shivkumar, A. S. (2021). *Stat Pearls* [Internet]. Accessed June 6, 2022, from https://www.ncbi.nlm.nih.gov/books/NBK513225/

Cox-Georgian, D., Ramadoss, N., Dona, C., & Basu, C. (2019). *Therapeutic and medicinal uses of terpenes.*

Diabetes-Australia Insulin. (2022). Accessed June 6, 2022, from https://www.diabetesaustralia. com.au/living-with-diabetes/medicine/insulin/

Eliza, J., Daisy, P., Ignacimuthu, S., & Duraipandiyan, V. (2009). Antidiabetic and antilipidemic effect of eremanthin from *Costus speciosus* (Koen.) Sm., in STZ-induced diabetic rats. *Chemico-Biological Interactions, 182*(1), 67–72.

Eno, A. E., Ofem, O. E., Nku, C. O., Ani, E. J., & Itam, E. H. (2008). Stimulation of insulin secretion by *Viscum album* (mistletoe) leaf extract in streptozotocin-induced diabetic rats. *African Journal of Medicine and Medical Sciences, 37*(2), 141–147.

Francesca, G. M. R., Daniela, E. G. F., & Vivian, M. C. (2019). Secondary metabolites in plants: Main classes, phytochemical analysis and pharmacological activities. *Bionatura, 4*(4), 1000–1009. https://doi.org/10.21931/RB/2019.04.04.11

Ganogpichayagrai, A., Palanuvej, C., & Ruangrungsi, N. (2017). Antidiabetic and anticancer activities of Mangifera indica cv Okrong leaves. *Journal of Advanced Pharmaceutical Technology & Research, 8*(1), 19–24.

Gayathri, M., & Kannabiran, K. (2008). Antidiabetic and ameliorative potential of Ficus bengalensis bark extract in streptozotocin induced diabetic rats. *Indian Journal of Clinical Biochemistry, 23*(4), 394–400. https://doi.org/10.1007/s12291-008-0087-2

Geberemeskel, G. A., Debebe, Y. G., & Nguse, N. A. (2019). Antidiabetic effect of fenugreek seed powder solution (*Trigonella foenum graecum* L.) on hyperlipidemia in diabetic patients. *Journal of Diabetes Research, 2019*, 1–8. https://doi.org/10.1155/2019/8507453

Grover, J. K., Yadav, S., & Vats, V. (2002). Medicinal plants of India with anti-diabetic potential. *Journal of Ethnopharmacology, 81*(1), 81–100.

Guang-Kai, X. U., Xiao-Ying, Q. I., Guo-Kai, W. A., Guo-Yong, X. I., Xu-Sen, L. I., Chen-Yu, S. U., Bao-Lin, L. I., & Min-Jian, Q. I. (2017). Antihyperglycemic, antihyperlipidemic and antioxidant effects of standard ethanol extract of Bombax ceiba leaves in high-fat-diet-and streptozotocin-induced type 2 diabetic rats. *Chinese Journal of Natural Medicines, 15*(3), 168–177.

Gupta, R. K., Kesari, A. N., Diwakar, S., Tyagi, A., Tandon, V., Chandra, R., & Wata, G. (2008). In vivo evaluation of anti-oxidant and anti-lipidemic potential of Annona squamosa aqueous extract in type 2 diabetic models. *Journal of Ethnopharmacology, 118*(1), 21–25.

Harborne, J. B. (1989). General procedures and measurement of total phenolics. In J. B. Harborne (Ed.), *Methods in plant biochemistry* (Plant phenolics) (Vol. 1, pp. 1–28). Academic Press.

Hegazy, G. A., Alnoury, A. M., & Gad, H. G. (2013). The role of Acacia arabica extract as an antidiabetic, antihyperlipidemic, and antioxidant in streptozotocin-induced diabetic rats. *Saudi Medical Journal, 34*(7), 727–733.

Ho, C. T. (1992). Phenolic compounds in food. In C. T. Ho, C. Y. Lee, & M. T. Hung (Eds.), *Phenolic compounds in food and their effects on health* (pp. 2–7).

Hou, S. Z., Chen, S. X., Huang, S., Jiang, D. X., Zhou, C. J., Chen, C. Q., Liang, Y. M., & Lai, X. P. (2011). The hypoglycemic activity of *Lithocarpus polystachyus* Rehd. Leaves in the experimental hyperglycemic rats. *Journal of Ethnopharmacology, 138*(1), 142–149.

Howden, R. (2013). Nrf2 and cardiovascular defense. *Oxidative Medicine and Cellular Longevity, 2013*, 104308.

Huang, C. S., Yin, M. C., & Chiu, L. C. (2011). Antihyperglycemic and antioxidative potential of Psidium guajava fruit in streptozotocin-induced diabetic rats. *Food and Chemical Toxicology, 49*(9), 2189–2195.

Huseini, H. F., Kianbakht, S., Hajiaghaee, R., & Dabaghian, F. H. (2012). Anti-hyperglycemic and anti-hypercholesterolemic effects of Aloe vera leaf gel in hyperlipidemic type 2 diabetic patients: A randomized double-blind placebo-controlled clinical trial. *Planta Medica, 78*(04), 311–316.

Islam, M. S., & Choi, H. (2008). Dietary red chilli (*Capsicum frutescens* L.) is insulinotropic rather than hypoglycemic in type 2 diabetes model of rats. *Phytotherapy Research, 22*(8), 1025–1029.

Jacob, B., & Narendhirakannan, R. T. (2019). Role of medicinal plants in the management of diabetes mellitus: A review. *Biotech, 9*(1), 4. https://doi.org/10.1007/s13205-018-1528-0

Jiao, Y., Wang, X., Jiang, X., Kong, F., Wang, S., & Yan, C. (2017). Antidiabetic effects of Morus alba fruit polysaccharides on high-fat diet- and streptozotocin-induced type 2 diabetes in rats. *Journal of Ethnopharmacology, 199*, 119–127.

Joseph, B., & Jini, D. (2013). Antidiabetic effects of Momordica charantia (bitter melon) and its medicinal potency. *Asian Pacific Journal of Tropical Disease, 3*(2), 93–102.

Jung, J. Y., Lim, Y., & Moon, M. S. (2011). Onion peel extracts ameliorate hyperglycemia and insulin resistance in high fat diet/streptozotocin-induced diabetic rats. *Nutrition & Metabolism, 8*(1), 18. https://doi.org/10.1186/1743-7075-8-18

Kabir, A. U., Samad, M. B., Ahmed, A., Jahan, M. R., Akhter, F., Tasnim, J., Hasan, S. N., Sayfe, S. S., & Hannan, J. M. (2015). Aqueous fraction of Beta vulgaris ameliorates hyperglycemia in diabetic mice due to enhanced glucose stimulated insulin secretion, mediated by acetylcholine and GLP-1, and elevated glucose uptake via increased membrane bound GLUT4 transporters. *PLoS One, 10*(2), e0116546.

Kaur, R., & Arora, S. (2015). Alkaloids-important therapeutic secondary metabolites of plant origin. *Journal of Critical Reviews, 2*(3), 1–8.

Khan, F., Sarker, M. M. R., Ming, L. C., Mohamed, I. N., Zhao, C., Sheikh, B. Y., Tsong, H. F., & Rashid, M. A. (2019). Comprehensive review on phytochemicals, pharmacological and clinical potentials of Gymnema sylvestre. *Frontiers in Pharmacology, 10*, 1223. https://doi.org/10.3389/fphar.2019.01223

Kumagai, Y., Nakatani, S., Onodera, H., Nagatomo, A., Nishida, N., Matsuura, Y., Kobata, K., & Wada, M. (2015). Anti-glycation effects of pomegranate (Punica granatum L.) fruit extract and its components in vivo and in vitro. *Journal of Agricultural and Food Chemistry, 63*(35), 7760–7764.

Kumar, S., Kumar, V., & Prakash, O. (2011). Antidiabetic, hypolipidemic and histopathological analysis of *Dillenia indica* (L.) leaves extract on alloxan induced diabetic rats. *Asian Pacific Journal of Tropical Medicine, 4*(5), 347–352. https://doi.org/10.1016/S1995-7645(11)60101-6

Kurup, S. B., & Mini, S. (2017). Averrhoa bilimbi fruits attenuate hyperglycemia-mediated oxidative stress in streptozotocin-induced diabetic rats. *Journal of Food and Drug Analysis, 25*(2), 360–368.

Laha, S., & Paul, S. (2019). Gymnema sylvestre (Gurmar): A potent herb with anti-diabetic and antioxidant potential. *The Pharmacogenomics Journal, 11*(2), 201–206.

Lattanzio, V. (2013). Phenolic compounds: Introduction. In K. G. Ramawat & J. M. Merillon (Eds.), *Natural products*. https://doi.org/10.1007/978-3-642-22144-6_57

Li, P. B., Lin, W. L., Wang, Y. G., Peng, W., Cai, X. Y., & Su, W. W. (2012). Antidiabetic activities of oligosaccharides of Ophiopogonis japonicas in experimental type 2 diabetic rats. *International Journal of Biological Macromolecules, 51*(5), 749–755.

Liu, S., Li, D., Huang, B., Chen, Y., Lu, X., & Wang, Y. (2013). Inhibition of pancreatic lipase, α-glucosidase, α-amylase, and hypolipidemic effects of the total flavonoids from Nelumbo nucifera leaves. *Journal of Ethnopharmacology, 149*(1), 263–269.

Malviya, N., Jain, S., & Malviya, S. (2010). Antidiabetic potential of medicinal plants. *Acta Poloniae Pharmaceutica—Drug Research, 67*(2), 113–118.

Manukumar, H. M., Shiva Kumar, J., Chandrashekar, B., Raghava, S., & Umesha, S. (2016). Evidences for diabetes and insulin mimetic activity of medicinal plants: Present status and future prospects. *Critical Reviews in Food Science and Nutrition, 57*, 2712. https://doi.org/10.1080/10408398.2016.1143446

Mathew, L., & Babu, S. (2011). Phytotherapy in India: Transition of tradition to technology. *Current Botany, 2*(5), 1722.

Maulik, S. K., & Talwar, K. K. (2012). Therapeutic potential of Terminalia Arjuna in cardiovascular disorders. *American Journal of Cardiovascular Drugs, 12*, 157–163.

McCalla, G., Parshad, O., Brown, P. D., & Gardner, M. T. (2015). Beta cell regenerating potential of Azadirachta indica (neem) extract in diabetic rats. *The West Indian Medical Journal, 65*(1), 13–17. https://doi.org/10.7727/wimj.2014.224

Meliani, N., Dib, M. E. A., Allali, H., & Tabti, B. (2011). Hypoglycaemic effect of Berberis vulgaris L. in normal and streptozotocin-induced diabetic rats. *Asian Pacific Journal of Tropical Biomedicine, 1*(6), 468–471.

Mishra, N., Kumar, D., & Rizvi, S. I. (2016). Protective effect of Abelmoschus esculentus against alloxan-induced diabetes in Wistar strain rats. *Journal of Dietary Supplements, 13*(6), 634–646.

Modak, M., Dixit, P., Londhe, J., Ghaskadbi, S., Paul, A., & Devasagayam, T. (2007). Indian herbs and herbal drugs used for the treatment of diabetes. *Journal of Clinical Biochemistry and Nutrition, 40*(3), 163–173.

Mohanraj, R., & Sivasankar, S. (2014). Sweet potato (*Ipomoea batatas* [L.] lam)—A valuable medicinal food: A review. *Journal of Medicinal Food, 17*(7), 733–741.

Mukherjee, P. K., Rai, S., Kumar, V., Mukherjee, K., Hylands, P. J., & Hider, R. C. (2007). Plants of Indian origin in drug discovery. *Expert Opinion on Drug Discovery, 2*(5), 633–657.

Mukherjee, A., & Sengupta, S. (2013). Indian medicinal plants known to contain intestinal glucosidase inhibitors also inhibit pancreatic lipase activity-an ideal situation for obesity control by herbal drugs. *Indian Journal of Biotechnology, 12*(1), 32–39.

Ngugi, M. P., Njagi, J. M., Kibiti, C. M., & Miriti, P. M. (2012). Pharmacological Management of Diabetes Mellitus. *Asian Journal of Biochemical and Pharmaceutical Research, 1*(2), 375–381.

Noor, A., Bansal, V. S., & Vijayalakshmi, M. A. (2013). Current update on anti-diabetic biomolecules from key traditional Indian medicinal plants. *Current Science, 104*(6), 721–727.

Ojewole, J. A. (2002). Hypoglycaemic effect of Clausena anisata (Willd) hook methanolic root extract in rats. *Journal of Ethnopharmacology, 81*(2), 231–237.

Omóbòwálé, T. O., Oyagbemi, A. A., Ogunpolu, B. S., Ola-Davies, O. E., Olukunle, J. O., & Asenuga, E. R. (2019). Antihypertensive effect of polyphenol-rich fraction of Azadirachtaindica on Nω-Nitro-L-arginine methyl ester-induced hypertension and cardiorenal dysfunction. *Drug Research, 69*, 12–22.

Omóbòwálé TO, Oyagbemi, A. A., Alaba, B. A., Ola-Davies, O. E., Adejumobi, O. A., & Asenuga, E. R. (2018). Ameliorative effect of Azadirachtaindica on sodium fluoride-induced hypertension through improvement of antioxidant defence system and upregulation of extracellular signal regulated kinase 1/2 signaling. *Journal of Basic and Clinical Physiology and Pharmacology, 29*, 155–164.

Ozougwu, J. C. (2011). Anti-diabetic effects of Allium cepa (onions) aqueous extracts on alloxan-induced diabetic Rattus novergicus. *Journal of Medicinal Plants Research, 5*(7), 1134–1139. http://www.academicjournals.org/JMPR

Ozturk, M. (2018). A comparative analysis of the medicinal plants used for diabetes mellitus in the traditional medicine in Turkey, Pakistan, and Malaysia. In M. Ozturk & K. Hakeem (Eds.), *Plant and human health* (Vol. 1). Springer. https://doi.org/10.1007/978-3-319-93997-1_11

Parasuraman, S., Thing, G. S., & Dhanaraj, S. A. (2014). Polyherbal formulation: Concept of ayurveda. *Pharmacognosy Reviews, 8*(16), 73–80.

Patel, D. K., Prasad, S. K., Kumar, R., & Hemalatha, S. (2012). An overview on antidiabetic medicinal plants having insulin mimetic property. *Asian Pacific Journal of Tropical Biomedicine, 2*(4), 320–330.

Peer, P. A., Trivedi, P. C., Nigade, P. B., Ghaisas, M. M., & Deshpande, A. D. (2008). Cardioprotective effect of Azadirachtaindica A Juss on isoprenaline induced myocardial infarction in rats. *International Journal of Cardiology, 126*, 123–126.

Ponnusamy, S., Ravindran, R., Zinjarde, S., Bhargava, S., & Kumar, A. R. (2011). Evaluation of traditional Indian antidiabetic medicinal plants for human pancreatic amylase inhibitory effect in vitro. *Evidence-based Complementary and Alternative Medicine, 2011*, 515647.

Poongothai, K., Ponmurugan, P., Ahmed, K. S., Kumar, B. S., & Sheriff, S. A. (2011). Antihyperglycemic and antioxidant effects of solanum xanthocarpum leaves (field grown & in vitro raised) extracts on alloxan induced diabetic rats. *Asian Pacific Journal of Tropical Medicine, 4*(10), 778–785.

Prince, P. S. M., & Menon, V. P. (2003). Hypoglycaemic and hypolipidaemic action of alcohol extract of *Tinospora cordifolia* roots in chemical induced diabetes in rats. *Phytotherapy Research, 17*(4), 410–413.

Quideau, S., Deffieux, D., Douat-Casassus, C., & Pouyse'gu L. (2011). Plant polyphenols: Chemical properties, biological activities, and synthesis. *Angewandte Chemie, International Edition, 50*(3), 586–621.

Ragavan, B., & Krishnakumari, S. (2006). Antidiabetic effect of T. arjuna bark extract in alloxan induced diabetic rats. *Indian Journal of Clinical Biochemistry, 21*(2), 123–128.

Raman, A., & Lau, C. (1996). Anti-diabetic properties and Phytochemistry *Momordica charantia* L. (Cucurbitaceae). *Phytomedicine, 2*(4), 349–362.

Ramji, D. P., & Foka, P. (2002). CCAAT/enhancer-binding proteins: Structure, function and regulation. *The Biochemical Journal, 365*(Pt-3), 561–575.

Sabu, M. C., & Kuttan, R. (2004). Antidiabetic activity of aegle marmelos and its relationship with its antioxidant properties. *Indian Journal of Physiology and Pharmacology, 48*(1), 81–88.

Satoh, T., Igarashi, M., Yamada, S., Takahashi, N., & Watanabe, K. (2015). Inhibitory effect of black tea and its combination with acarbose on small intestinal α-glucosidase activity. *Journal of Ethnopharmacology, 161*, 147–155.

Sayyed, F. J., & Wadkar, G. H. (2018). Studies on in-vitro hypoglycemic effects of root bark of *Caesalpinia bonducella*. *Annales Pharmaceutiques Françaises, 76*(1), 44–49.

Shah, A. J., Gilani, A. H., & Hanif, H. M. (2014). Neem (Azadirachta indica) lowers blood pressure through a combination of Ca^+ channel blocking and endothelium-dependent muscarinic receptors activation. *International Journal of Pharmacology, 10*, 418–428.

Sharma, A. K., Bharti, S., Goyal, S., Arora, S., Nepal, S., Kishore, K., Joshi, S., Kumari, S., & Arya, D. S. (2011). Upregulation of PPARγ by Aegle marmelos ameliorates insulin resistance and β-cell dysfunction in high fat diet fed-Streptozotocin induced type 2 diabetic rats. *Phytotherapy Research, 25*, 1457–1465.

Sharma, B., Salunke, R., Balomajumder, C., Daniel, S., & Roy, P. (2010). Antidiabetic potential of alkaloid rich fraction from Capparis decidua on diabetic mice. *Journal of Ethnopharmacology, 127*(2), 457–462. https://doi.org/10.1016/j.jep.2009.10.013

Singh, L. W. (2011). Traditional medicinal plants of Manipur as anti-diabetics. *Journal of Medicinal Plant Research: Planta Medica, 5*(5), 677–687.

Singh, P. K., Baxi, D., & Doshi, A. V. R. (2011). Antihyperglycaemic and renoprotective effect of *Boerhaavia diffusa* L. in experimental diabetic rats. *Journal of Complementary and Integrative Medicine, 8*(1), 1–20. https://doi.org/10.2202/1553-3840.1533

Srivastava, S., Lal, V. K., & Pant, K. K. (2012). Polyherbal formulations based on Indian medicinal plants as antidiabetic phytotherapeutics. *Phytopharmacology, 2*, 115.

Stefkov, G., Miova, B., Dinevska-Kjovkarovska, S., Stanoeva, J. P., Stefova, M., Petrusevska, G., & Kulevanova, S. (2014). Chemical characterization of *Centaurium erythrea* L. and its effects on carbohydrate and lipid metabolism in experimental diabetes. *Journal of Ethnopharmacology, 152*(1), 71–77.

Stohs, S. J., & Ray, S. (2015). Anti-diabetic and anti-hyperlipidemic effects and safety of Salacia reticulata and related species. *Phytotherapy Research, 29*(7), 986–995. https://doi.org/10.1002/ptr.5382

Sundaram, R., Naresh, R., Shanthi, P., & Sachdanandam, P. (2012). Antihyperglycemic effect of iridoid glucoside, isolated from the leaves of Vitex negundo in streptozotocin-induced diabetic rats with special reference to glycoprotein components. *Phytomedicine, 19*(3–4), 211–216.

Sunil, V., Shree, N., Venkataranganna, M. V., Bhonde, R. R., & Majumdar, M. (2017). The antidiabetic and antiobesity effect of *Memecylon umbellatum* extract in high fat diet induced obese mice. *Biomedicine & Pharmacotherapy, 89*, 880–886.

Tadera, K., Minami, Y., Takamatsu, K., & Matsuoka, T. (2006). Inhibition of α-glucosidase and α-amylase by flavonoids. *Journal of Nutritional Science and Vitaminology, 52*(2), 149–153.

Thompson, A. T., Opeolu, O. O., Peter, R. F., & Yasser, H. A. A. W. (2014). Aqueous bark extracts of Terminalia arjuna stimulates insulin release, enhances insulin action and inhibits starch digestion and protein glycation in vitro. *Austin Journal of Endocrinology and Diabetes, 1*(1), 1001.

Tiong, S. H., Looi, C. Y., Hazni, H., Arya, A., Paydar, M., & Wong, W. F. (2013). Antidiabetic and antioxidant properties of alkaloids from *Catharanthus roseus* (L.) G.Don. *Molecules, 18*(8), 9770–9784.

Uchegbu, N. N., & Ishiwu, C. N. (2016). Germinated pigeon pea (Cajanus cajan): A novel diet for lowering oxidative stress and hyperglycemia. *Food Science & Nutrition, 4*(5), 772–777.

Umamaheswari, S., Joseph, L. D., Srikanth, J., Lavanya, R., Chamundeeswari, D., & Reddy, C. U. (2010). Antidiabetic activity of a polyherbal formulation (DIABET). *International Journal of Pharmacy and Pharmaceutical Sciences, 2*, 1822.

Venkatachalam, T., Kumar, V. K., Selvi, P. K., Maske, A. O., Anbarasan, V., & Kumar, P. S. (2011). Antidiabetic activity of *Lantana camara* Linn fruits in normal and streptozotocin-induced diabetic rats. *Journal of Pharmacy Research, 4*(5), 1550–1552.

Venkatesh, S., Reddy, G. D., Reddy, Y. S. R., Sathyavathy, D., & Reddy, B. M. (2004). Effect of Helicteres isora root extracts on glucose tolerance in glucose-induced hyperglycemic rats. *Fitoterapia, 75*(3), 364–367.

Wainstein, J., Ganz, T., Boaz, M., Bar Dayan, Y., Dolev, E., Kerem, Z., & Madar, Z. (2012). Olive leaf extract as a hypoglycemic agent in both human diabetic subjects and in rats. *Journal of Medicinal Food, 15*(7), 605–610.

WHO. (2022a). *Global report on diabetes*, Accessed July 15, 2022, from https://www.who.int/publications/i/item/9789241565257

WHO. (2022b). *Health topics on diabetics*, Accessed July, 15, 2022, from https://www.who.int/health-topics/diabetes#tab=tab_1

Wilcox, G. (2005). Insulin and insulin resistance. *Clinical biochemist Reviews, 26*(2), 19–39.

Worldwide toll of diabetes. (2020). Diabetesatlas.org. Accessed August 7, 2020, from https://www.diabetesatlas.org/en/sections/worldwide-toll-of-diabetes.html

Yarmohammadi, F., Mehri, S., Najafi, N., Salar, A. S., & Hosseinzadeh, H. (2021). The protective effect of Azadirachta indica (neem) against metabolic syndrome: A review. *Iranian Journal of Basic Medical Sciences, 24*(3), 280–292.

Zulcafli, A. S., Lim, C., Ling, A. P., Chye, S., & Koh, R. (2020). Antidiabetic potential of Syzygium sp.: An overview. *The Yale Journal of Biology and Medicine, 93*(2), 307–325.

Zuraini, A., Vadiveloo, T., Hidayat, M. T., Arifah, A., Sulaiman, M., & Somchit, M. (2006). Effects of neem (*Azadirachta indica*) leaf extracts on lipid and C-reactive protein concentrations in cholesterol-fed rats. *Journal of Natural Remedies, 6*, 109–114.

Diabetes Management: From "Painful" Pricks to "Pain-Free" Bliss

Bhuvaneswari Ponnusamy, Ponnulakshmi Rajagopal, Raktim Mukherjee, Swetha Panneerselvam, and Selvaraj Jayaraman

Abstract

"Diabetes management" is a collective word given to various plans and measures taken in order to maintain glucose homeostasis. Hyperglycemia is one of the socioeconomic problems that has become prevalent globally. Lifestyle modifications and the lack of physical activities lead to the development of a hyperglycemic condition termed diabetes mellitus. This disorder is associated with comorbidities such as retinopathy, nephropathy, neuropathy, and cardiovascular diseases. Thus, the management of this disorder has become a hectic event. The era of insulin discovery lighted up the path for diabetes management. The pre-insulin discovery period recorded a number of mortality cases due to diabetes. However, the discovery of insulin reduced the mortality cases and extended patient lifetimes. Yet strategies in diabetes management and insulin administration were painful. In the last few decades, a number of treatment options and strategies were made to implement pain-free diabetes management. Many innovative user-friendly methods for insulin administration are being made to comfort users. The innovations made in the mode of administration and technological

B. Ponnusamy · S. Panneerselvam · S. Jayaraman (✉)
Centre of Molecular Medicine and Diagnostics (COMManD), Department of Biochemistry, Saveetha Dental College & Hospital, Saveetha Institute of Medical & Technical Sciences, Saveetha University, Chennai, India
e-mail: selvarajj.sdc@saveetha.com

P. Rajagopal
Department of Central Research Laboratory, Meenakhsi Ammal Dental College and Hospitals, Meenakhsi Academy of Higher Education and Research, Chennai, India

R. Mukherjee
Shree PM Patel Institute of PG Studies and Research in Science, Sardar Patel University, Anand, India

development paved the way for the introduction of pain-free methods for insulin administration, like microneedle, transdermal patches, chemical permeation enhancers, sonophoresis, iontophoresis, electroporation, and vesicle formation. This chapter focuses on the transformation of painful strategies in diabetes management for the pain-free, blissful management of diabetes.

Keywords

Insulin · Insulin discovery · Innovations in insulin delivery methods

1   Introduction

Diabetes mellitus is a metabolic disorder that is characterized by uncontrolled hyperglycemic conditions due to the defect in insulin release, uptake, and resistance. This defect or imbalance in insulin regulatory role can be root from disturbances in biomolecule metabolism (carbohydrates, lipids, and proteins). Diabetes mellitus is categorized into several subtypes, namely, type 1 diabetes mellitus, type 2 diabetes mellitus, gestational diabetes, maturity-onset diabetes of the young, neonatal diabetes, and secondary causes, such as endocrinopathies, drugs, etc. (Sapra & Bhandari, 2022). Of these, the three main types of diabetes noted are type 1 diabetes mellitus, type 2 diabetes mellitus, and gestational diabetes. The progression of the disease can lead to various complications, such as cardiovascular diseases (Deshpande et al., 2008), (Fox, 2004), neuropathy (Candrilli et al., 2007), nephropathy (Andersen et al., 1983), retinopathy (Harris & Leininger, 1993), and diabetic foot ulcers (Reiber et al., 1998). It is estimated that globally, one in 11 adults has diabetes, of which 90% are type 2 diabetes mellitus patients (Sapra & Bhandari, 2022). Diabetes mellitus is a deep-rooted condition impacting 463 million people worldwide (International Diabetes Federation, 2019), and it is estimated that 150–200 million people globally are dependent on insulin therapy for their health (Garg et al., 2018). Type 2 diabetes is the commonest form of diabetes among the world population, characterized by defects in insulin secretion and insulin resistance. In the western world, type 2 diabetes prevails in up to 7% of the total population, and in the global range, it is estimated to be 5–7% (Chaluvaraju et al., 2012). As type 2 diabetes has become a global epidemic, the World Health Organization (WHO) has launched the Global Diabetes Compact in order to manage diabetes worldwide.

The most important discovery among various biomedical events in the last century is the discovery of "insulin" (Mayer et al., 2007). Diabetes mellitus is characterized by insufficiency in insulin secretion and resistance of the cells toward insulin, which occurs due to diminished beta cell function, autoimmunity, inflammation, glucolipotoxicity, oxidative stress, genes epigenetic factors, insulin resistance, aging, environmental factors, etc. (Cernea & Raz, 2020). Thus, insulin remains the mainstay in the management of diabetes mellitus. Insulin serves as a major antihyperglycemic agent in managing both type 1 diabetes mellitus (T1DM) and type 2 diabetes mellitus (T2DM). Hence, an insulin regimen is a widely preferred maintenance option for long-term optimal glucose control (Frias & Frias, 2017). Insulin hormones have been widely used for the past 75 years as exogenous sources for the treatment of insulin-dependent type-1 diabetes and non-insulin-dependent type 2 diabetes mellitus (Hoffman & Ziv, 1997).

Insulin is secreted inside the body from the pancreatic gland. The pancreas plays a dual role: it acts as an exocrine gland and an endocrine gland. The exocrine action includes the secretion of pancreatic juice and other vital digestive enzymes aiding in the absorption and digestion of food. The γ-cells of the pancreas are involved in the secretion of pancreatic juices. The endocrine action of the pancreas includes the secretion of hormones, namely, glucagon, insulin, and somatostatin, via α-cells, β-cells, and δ-cells, respectively. Insulin is a polypeptide hormone made of two amino acid chains (A-chain: 21 amino acids; B-chain: 30 amino acids) linked together by disulfide linkages made of 51 amino acids (Qaid & Abdelrahman, 2016).

The first peptide hormone discovered was insulin (Weiss et al., 2014). The biosynthesis of insulin is a highly regulated process that depends on the stimulation of glucose present and is augmented by the production of cyclic adenosine monophosphate (cAMP) (Thevis et al., 2009). The biosynthesis of insulin involves a cascade of processes, starting from the synthesis of preproinsulin to the generation of active insulin hormones. Insulin messenger RNA (mRNA) is encoded to produce preproinsulin, which is made up of proinsulin and a 24 amino acid signal peptide. During the translocation of preproinsulin into the endoplasmic reticulum, the signal peptide is nicked in the presence of a signal peptidase to liberate proinsulin. The released hormone, a single-chain polypeptide hormone with a molecular weight of 9388 Da, has the native structure vital for insulin synthesis (Thevis et al., 2009). Proinsulin is made up of an amino-terminal B-chain and a carboxyl-terminal A-chain connected with C-peptide. In the next step, the C-peptide is cleaved from proinsulin, and the A-chain and B-chain are interconnected by disulfide linkages to generate active insulin hormones. Upon the completion of the biosynthesis process, the produced insulin is stored in the granules of the β-cells of the pancreas, which are then, upon stimulation, released via exocytosis into the pancreatic capillaries, thereby entering the circulation. Figures 1 and 2 illustrate the biosynthesis of the insulin peptide and its release into circulation.

The metabolism of insulin is widely studied, and it is found that the liver and kidneys are the major sites involved in the active degradation of insulin. The half-life of insulin is very short, approximately 4 min, but it is effective enough to regulate the hyperglycemic condition of the biological system (Hoffman & Ziv, 1997). Insulin release into the circulation is found to occur in a pulsatile fashion, with peek formations for every 1.5–2 h (Hoffman & Ziv, 1997). The beta-cell secretion of insulin is highly regulated by a number of feedback mechanisms, of which the concentration of glucose in the blood influencing insulin release is the supreme factor. Insulin is involved in regulating the plasma glucose level by either enhancing the movement of glucose into the cells or enhancing the storage of excess glucose in a polymeric complex unit called glycogen. When insulin is released, it stimulates both Na^+-dependent and Na^+-independent glucose transporters—sodium-glucose cotransporter (SGLT) and glucose transporter (GLUT) isoforms, respectively—

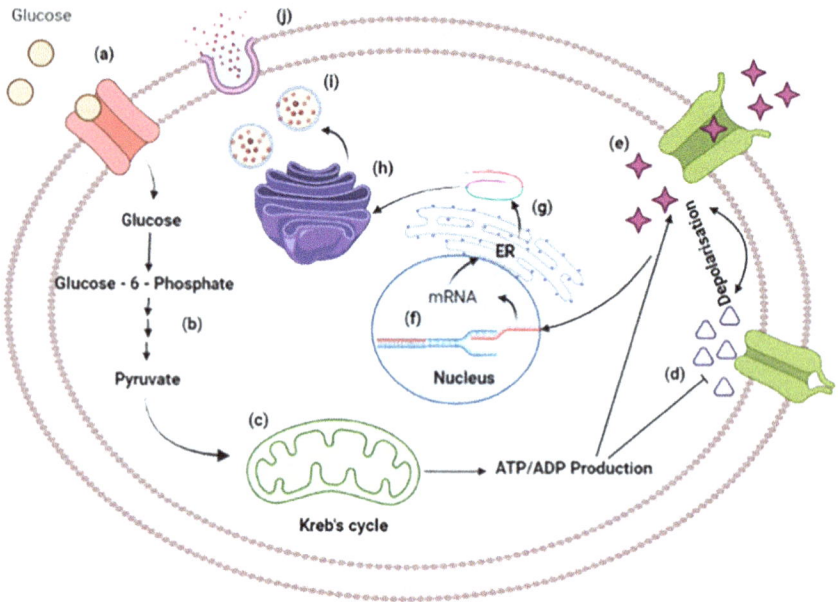

Fig. 1 Illustration of the biosynthesis of insulin in a beta cell. (**a**) Uptake of glucose into the cell via glucose transporters (GLUTs). (**b**) Glucose undergoes its metabolic pathway to yield its end product pyruvate. (**c**) The pyruvate enters the mitochondria for the Krebs cycle and undergoes oxidative phosphorylation to release adenosine triphosphate (ATP) molecules. (**d**) Increased ATP production can cause ATP-sensitive potassium channels to close. (**e**) The closure of potassium channels causes depolarization and opens the calcium channel. (**f**) Calcium influx caused the activation of gene transcription to produce mRNA (preproinsulin). (**g**) The mRNA is translated in the endoplasmic reticulum to give proinsulin. (**h**) Proinsulin migrates to the Golgi apparatus to form active insulin and is stored in granules. (**i**) Increased glucose level causes the exocytosis of the insulin granules, which enters the circulation for action

present in the cells to express on the plasma membrane, which facilitate the uptake of plasma glucose (Weiss et al., 2014).

In a normal conditions, after the intake of a meal, the insulin is rapidly released into the circulation and attains maximum concentration within 30–45 min and then gradually declines to basal insulin level after 2–3 h. In a diabetic condition, insulin analogs are available for the maintenance of the blood glucose level. Insulin analogs are compounds with similar physical and chemical properties to the insulin hormone present endogenously. These compounds are designed in such a way that they mimic the mechanism of action of natural insulin. These analogs are classified into three categories based on the time of action, namely, rapid-acting, intermediate-acting, and long-acting insulin analogs. In this review, one can get a detailed description of insulin and its analogs, the history of insulin discovery, routes of administration, inconveniences in insulin medications, and advancements in the routes of delivery.

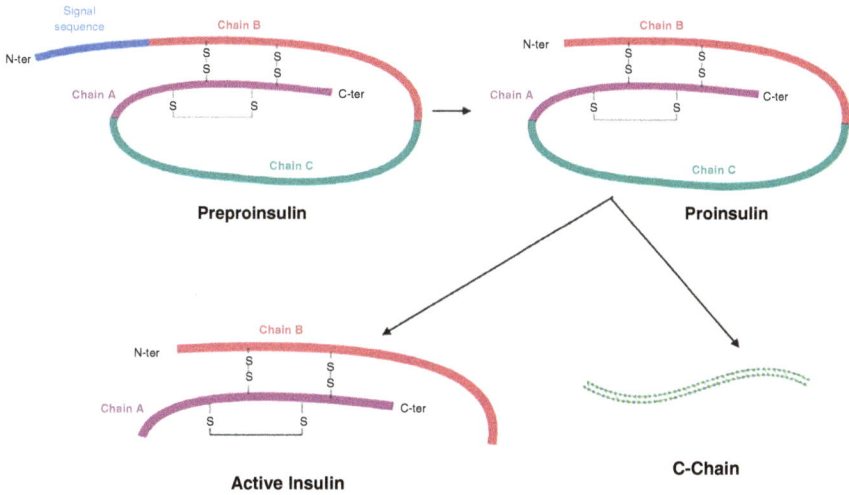

Fig. 2 Active insulin production. Preproinsulin is made up of signal peptides A-chain and B-chain connected by C-peptide. It then enters the endoplasmic reticulum, where the signal sequence is removed and releases proinsulin to the cytoplasm. The proinsulin enters the Golgi apparatus, and the C-peptide is released to produce active insulin peptides

2   History of Insulin Discovery

The foremost illustration of diabetes was made in early 552 BC in an Egyptian text book, "Ebers Papyrus" (Quianzon & Cheikh, 2022; Oubré et al., 1997). After that, in 129–199 AD, a Greek physician named Aretaeus of Cappadocia coined the term "diabetes," from the Greek word "siphon," as he noticed that diabetes causes a constant flow of urine (Marwood, 2022). Thus, prior to the advent of insulin, American physicians Frederick M. Allen and Elliott P. Joslin prescribed fasting and low-calorie diets to diabetic patients (Mazur, 2011). For a certain period of time, from 1915 until the discovery of insulin in 1922, they suggested "starvation diets"— diets based on repeated fasting and prolonged undernourishment—as a therapeutic option for managing diabetes mellitus, not for complete recovery (Mazur, 2011), thereby prolonging the life of patients and reducing the complications associated with the disease, such as reduced glucosuria, acidosis, coma, and many more.

Pancreatic islets were first mentioned by a German student, Paul Langerhans, in his thesis in 1869. However, at that time, the role of the pancreas was not known. Later, when von Mering and Minkowski, in 1889, removed the pancreas from dogs, the animals suffered from severe diabetes and were found to be dead in the following days, giving suggestive proof that the absence of a pancreas can lead to diabetic conditions (Karamitsos, 2011). This was the spark that made the researchers focus on the pancreas and its functional units, to treat and create a breakthrough in diabetes management. Further, in 1893, E. Hedon, a French physician, was experimenting with a complete pancreatectomy on a dog, and then he grafted a small piece of pancreas under the skin. The results of the experiment showed that the dog acquired diabetes only when the graft under the skin was removed, suggesting that the pancreas caused internal secretions (Karamitsos, 2011). Based on the previous attempts and findings, they came to the conclusion that the pancreas is a secretagogue, and hence many researchers (approximately 400) made attempts to treat diabetic patients with pancreatic extracts. However, there were not many publications made on this due to the failure of the experiments and owing to unsatisfying results.

Among these experiments, certain notable discoveries in attenuating diabetic conditions led to the next step in diabetes management. In 1900, G.L. Zulzer, a young German physician, managed to prepare alcoholic extracts of the pancreas and inserted them into rabbits and dogs to check antidiabetic action. Further, he administered the extract to a critical diabetic patient in a private health care center. The patient showed marked clearance in blood glucose levels, but there was a shortage of extracts for further treatment. He named this pancreatic extract "Acomatol" (Karamitsos, 2011). Acomatol, apart from antidiabetic action, showed remarkable adverse effects, like vomiting, fever, and convulsions (Bliss, 1982). Zulzer's findings concluded that pancreatic extract administration can effectively halt glucosuria and ketonuria in diabetic patients without being subjected to restricted diets and starvation (Karamitsos, 2011).

Then in 1909, J. Forschbach obtained Zulzer's extract and administered it to patients in Minkowski's clinic; however, he observed many toxic adverse events and concluded with negative comments. Later in 1912, Zulzer got a patent for his extraction use. Then in order to overcome the toxic effects, he made further intensive studies until 1914, when he joined the army for World War I. Meanwhile, in 1911–1912, another medical student, E.L. Scott, conducted further research. He suggested that the shortcomings of the extract may be due to external protease enzyme secretion, so he ligated the pancreatic ducts to avoid contact with the external secretions. However, the results he obtained were not very satisfying (Scott, 1912). In the same period, in 1912, G.R. Murlin and B. Kramer prepared pancreatic extracts and administered them to dogs, which showed positive results (Murlin & Kramer, 1913). Then in 1915, I. Kleiner also analyzed the functions of pancreatic extracts but could not progress in his research due to a lack of funding (Kleiner, 1919).

Nicolae Paulescu, a physiology professor in a medical school in Canada, led the stepping stone for the discovery of insulin but was unrecognized due to the politics of the time. He started his experiments in 1916 using pancreatic extracts, and his results showed superior results to the others. In February 1922, Paulescu successfully administered his extract to patients under the name "pancrein." Then he applied for a patent for the manufacturing and usage of pancrein. However, the work quality in Canada was very slow compared to that of the USA, which was the main reason for the delay in the patent issue for Paulescu's work (Karamitsos, 2011). His contribution to the discovery of insulin was later recognized after his death.

Simultaneously with N. Paulescu's work comes the appearance of our Nobel Prize winners for the discovery of insulin, Frederic Banting, Charles Best, J.R. Macleod, and J.B. Collip. Banting was a physician and assistant professor at London University. While Banting was reading an article by M. Barron explaining that lithiasis of the pancreas can lead to atrophy and loss of the exocrine function of the pancreas, he came up with a sudden reason that if one could successfully atrophy the exocrine part of the pancreas, then we can get its endocrine secretions without any impurities. This idea lighted up the successful path to insulin discovery. Then under the guidance of J.R. Macleod, a professor in Ontario, after many objections, Banting started his work in 1921. J.R. Macleod appointed Charles Best to assist Banting in his work. Facing many failures initially, the duo learned from their mistakes and started to outgrow them, and finally, they succeeded in their pancreatic duct ligation surgeries. The dogs that survived the surgeries for about 53 days were chosen. The ligated pancreas was taken away and placed in an ice bath with sodium chloride, and they prepared an extract with the pancreas obtained. When the extract was injected into a diabetic dog, its blood glucose levels started to decrease. But on the following day, the dog got into a coma and died without having high glucose levels. This was a promising result, which led to the confirmation of the hypothesis, and the extract was named "isletin." They also tested the administration of isletin through different routes, like rectal and subcutaneous; however, the results were not satisfying. Then in 1921, J.B. Collip, Professor of Biochemistry, joined the team. J.B. Collip managed to create a new methodology for the preparation of the extract with less or no impurities, which was shown to stop the ketogenesis process and also restore glycogen levels. They started the administration of insulin to patients and received positive results. The American pharmaceutical company Eli Lilly signed a contract to commercially produce insulin for treatment processes, which led to the production of sufficient insulin stocks by February 1923 in the USA, followed by the Medical Research Council in London (Karamitsos, 2011). A summary of the history of insulin discovery is given as a timeline in Table 1.

3   Insulin Analogs

Insulin analogs are those compounds created to imitate the physiological role of insulin in insulin deficit living systems. Human insulin developed by recombinant deoxyribonucleic acid (DNA) technology was the first insulin analog made used in the formulation of insulin early in the 1990s. In later discoveries, bolus or prandial insulin and basal insulin were intended for the management of diabetic conditions. Rapidly-acting and short-acting insulin are also referred to as bolus insulin as they are administered at the time of meal intake in order to control the peaking blood glucose concentration in postprandial conditions (Roach, 2008). Intermediate-acting and long-acting insulin are also called basal insulin, which aims to decrease circulating insulin concentrations (Roach, 2008). Soluble human insulin is referred to as regular insulin (Roach, 2008). Figure 3 gives a schematic representation of the

Table 1 Timeline of events occurring during the insulin discovery process

S. no.	Year	Investigators	Events	References
1	552 BC	Egyptian book "Ebers Papyrus"	The first illustration of diabetes.	Quianzon and Cheikh (2022), Oubré et al. (1997).
2	129–199 AD	Aretaeus	The term "diabetes" was coined, indicating that the disease condition is characterized by a constant flow of urine.	Marwood (2022)
3	1869	Paul Langerhans	Identification of the pancreatic gland; however, its functions were not known.	
4	1889	Von Mering and Minkowski	Accidental removal of the pancreas in dogs led to the development of diabetes, resulting in death.	Karamitsos (2011).
5	1893	E. Hedon	Pancreatectomy and the grafting of the pancreas under the skin—proving internal secretions made by the pancreas.	Karamitsos (2011).
6	1900	G.L. Zulzer	Prepared an alcoholic extract of the pancreas and named it "Acomatol." He concluded that the pancreatic extract can halt glucosuria and ketonuria.	Karamitsos (2011)
7	1909	J. Forschbach	He analyzed Zulzer's extract and published his results.	
8	1912	G.L. Zulzer	Obtained a patent for the use of his extract, Acomatol, then made extensive investigations on overcoming the toxic effects of his extract.	
9	1911–1912	E.L. Scott	Hypothesized that the presence of protease could be a major reason for the shortcomings in the use of the extract. To confirm this, he ligated the pancreatic duct to avoid contact with the external secretions; however, his findings were not satisfying.	Scott (1912)
10	1912	G.R. Murlin and B. Kramer	Prepared pancreatic extracts and tested them on dogs.	Murlin and Kramer (1913).
11	1915	I. Kleiner	Analyzed the functions of the pancreatic extracts.	Kleiner (1919).
12	1916	Nicolae Paulescu	Prepared pancreatic extracts with comparatively higher purity and efficacy.	Karamitsos (2011)

(continued)

Table 1 (continued)

S. no.	Year	Investigators	Events	References
13	1921	Frederic Banting and Charles Best	Came up with a reasoning that the removal of the exocrine part of the pancreas could be a better solution for overcoming impurities in the extract. After a number of failures, they prepared an extract with low or no impurities and named it "isletin." Banting and Best conducted their experiments through the funding of J.R. Macleod. But J.R. Macleod also contributed some efforts to this investigation.	Karamitsos (2011)
		J.B. Collip	Further, J. B. Collip joined the group and developed a new methodology to prepare pancreatic extracts with no or negligible amount of impurities.	
14	1922	Nicolae Paulescu	Named his extract "pancrein" and successfully administered it to patients. However, his works and discoveries were recognized only after his death.	
15	1923	Frederic Banting, Charles Best, J.B. Collip, and J.R. Macleod	Awarded with noble price.	
		Eli Lilly and Company	Produced insulin in a large scale and commercialized it.	

process by which the subcutaneous route of insulin analog administration enhances glucose utilization by the cells.

4   Insulin Lispro

Insulin lispro was the first short-acting insulin analog discovered in 1994 (Howey et al., 1994), which got approval for human therapy by the year 1996. Lispro is the only analog that is a monomer. When compared to normal human insulin, its only difference is that the positions of proline and lysine, B28 and B29 respectively, are found to be reversed in lispro (Hoffman & Ziv, 1997). Upon subcutaneous administration of this analog, it exhibits insulin-mimetic action with similar pharmacodynamic and pharmacokinetic properties. Being a monomer, this analog is found to actively bind to zinc molecules and elicit more rapid pharmacodynamic properties than regular human insulin (Howey et al., 1994). The conjugation of insulin lispro

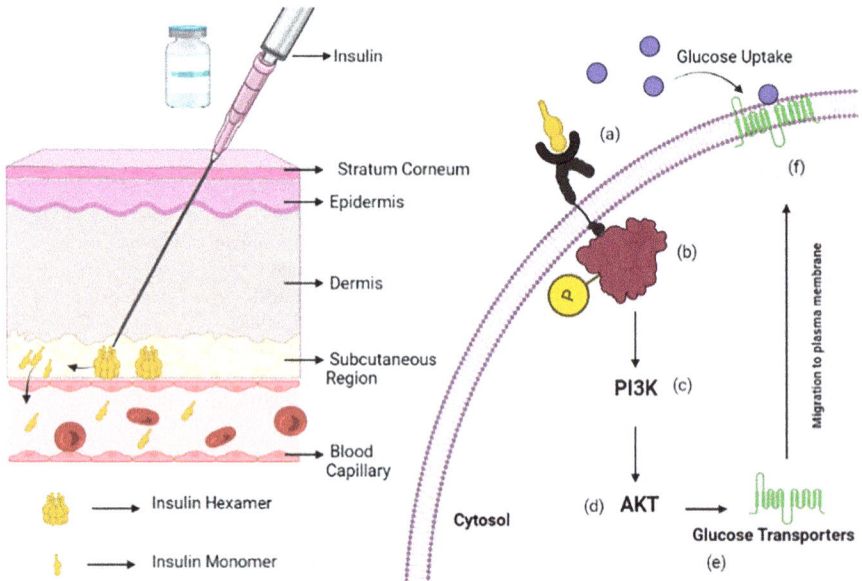

Fig. 3 Illustration of the mechanism by which insulin administration in the subcutaneous region brings reduced glucose levels in circulation. (**a**) The insulin monomer binds to the insulin receptor. (**b**) The binding of ligand to the receptor initiates the phosphorylation of tyrosine. (**c**) Phosphorylation activates phosphatidylinositol 3-kinase (PI3K). (**d**) PI3K activates AKT serine /threonine kinase, which causes (**e**) the migration of glucose transporters to the plasma membrane (**f**) to enhances the transport of glucose into the cell for metabolism

with zinc can help in the stabilization of the analog in a solution by forming hexamers. However, being a monomer, insulin is found to be less avidly bound to zinc molecules, resulting in the easy dissociation of hexamers, which in turn leads to a rise in serum insulin concentration. The major effect of insulin lispro is that it leads to a reduction of the plasma glucose level, mainly by concealing hepatic glucose production, and it also has the effect of slightly increasing the utilization of glucose. As for the pharmacokinetic properties of insulin lispro, on subcutaneous administration, the drug starts its action within 15 min, attaining a peak concentration approximately within 1 h, and the clearance rate of the analog is noted at 2–4 h (Chaluvaraju et al., 2012).

5   Insulin Aspart

Insulin aspart is a rapid-acting and short-acting insulin analog. This is synthesized using recombinant DNA technology. In insulin aspart, the proline present in position 28 is replaced by charged aspartic acid. This analog averts the formation of hexamers, thereby enhancing the absorption rate and causing a rapid onset of action. This analog is administered through the subcutaneous route. The absorption rate of insulin aspart is double the speed of human insulin absorption and reaches a maximum concentration that is much higher than human insulin, but the duration of its action is comparatively shorter (Lindholm et al., 1999). The short duration of the action of insulin is considered both an advantage and a disadvantage because, though it is effective in controlling postprandial hyperglycemic conditions, it may be insufficient to control late postprandial hyperglycemia, resulting in a glycemic condition before the next meal (Lindholm et al., 1999). Insulin aspart is advised to be administered right before the intake of meals to avoid hypoglycemic conditions (Lindholm et al., 1999).

6   Insulin Glulisine

Following the development of insulin lispro and insulin aspart, insulin glulisine was developed and got approved for clinical use (Roach, 2008). In this analog, position 3-asparagine and position 29-lysine in the B-chain of human insulin are replaced by lysine and glutamic acid, respectively. Insulin glulisine is not conjugated with zinc to improve its stability. Thus, it shows a rapid absorption rate, which exhibits a more rapid onset of action. Moreover, the chemical configuration of this analog alters the isoelectric point, resulting in a low isoelectric point (pH 5.1), which causes increased solubility, thereby enhancing its absorption rate (Chaluvaraju et al., 2012). However, when compared to other rapidly acting insulin, the onset of action of glulisine is only 10 min earlier, which makes no big appreciable difference between aspart and lispro insulin. Insulin glulisine is administered subcutaneously in the abdominal wall, thighs, or upper arms.

7   Insulin Glargine

Insulin glargine is a long-acting insulin analog, available since 2001 (Roach, 2008). During glargine synthesis, the two arginine amino acid residues are added to the C-terminal for the elongation of the B-chain, and asparagine in position 21 of the A-chain is replaced by glycine (Roach, 2008). These changes in structural configurations led to the shift of the isoelectric point to lower pH (pH 5.1), making it less soluble at physiological pH and more soluble at acidic pH. Properties like low isoelectric point and presence of glycine which stabilizes the hexamer structure, delays the absorption of insulin glargine and displays a satisfactory basal insulin level without elevation in the plasma insulin level for 24 h (Chaluvaraju et al., 2012; Bolli et al., 1999). Insulin glargine is administered subcutaneously and gets anchored in the subcutaneous tissues due to low absorptivity, resulting in a prolonged duration of action. This analog is not combined with other insulin analogs as it can form a cloudy precipitate, which can greatly affect the pharmacological and pharmacokinetic properties of insulin glargine (Chaluvaraju et al., 2012).

8   Insulin Detemir

Insulin detemir is a long-acting insulin analog administered subcutaneously. In insulin detemir synthesis, incorporation of the acyl group to amino acid lysine positioned at 29 of the B-chain of human insulin takes place (Chaluvaraju et al., 2012). In addition to this incorporation, an amino acid from the C-terminal of the B-chain is detached (Roach, 2008). These modifications in human insulin yield insulin detemir, which is chemically designated as desB30, B29 myristoyl human insulin (Roach, 2008). The addition of acyl group to insulin detemir increases its binding capacity with active free fatty acid binding sites and circulating serum albumin (Hoffman & Ziv, 1997). This action is assumed to be the basic step for prolonging its duration of action. Detemir forms hexamers at the site of injection thereby providing sufficient time for detemir to bind to serum albumin, which could be another reason for its prolonged action (Roach, 2008). These properties suggest that detemir can be given as basal insulin. The basal insulin levels are maintained up to 20 h from injection time, but this period can be affected by deviations in the dose administered (Chaluvaraju et al., 2012). A study on a porcine model, to determine the clearance time of insulin detemir showed that 50% of injected detemir is cleared from the body within 10.2 h (Kurtzhals, 2004). Table 2 gives an outline of the different types of insulin analogs used.

Table 2 Overview of commercially available insulin analog (Mane et al. 2012; Roach, 2008; Hoffman & Ziv, 1997)

S. no.	Analog	Trade name	Mode of action	Modifications
1	Insulin lispro	Humalog	Rapid and short acting	The proline and lysine in the B-chain of human insulin at positions 28 and 29 are reversed in insulin lispro.
2	Insulin aspart	Novolog	Rapid and short acting	The proline positioned as the 28th residue in the B-chain of human insulin is replaced by charged aspartic acid residue.
3	Insulin glulisine	Apidra	Rapid and short acting	Asparagine and lysine situated at positions 3 and 29 in the B-chain of human insulin are replaced by lysine and glutamic acid, respectively.
4	Insulin glargine	Lantus	Long acting	Addition of two molecules of arginine to the carboxyl terminal of the human insulin and replacement of asparagine at position 21 of A-chain with glycine.
5	Insulin detemir	Levemir	Long acting	Addition of acyl group to lysine residue present at position 29 of the B-chain.

9   Insulin Delivery Methods

Insulin administration through the skin, specifically the subcutaneous layer, is the most commonly preferred route of insulin. This route shows maximal absorption of insulin when compared to intradermal and intramuscular routes (Yaturu, 2013; Gradel et al., 2018). Currently, the dominating route of insulin administration is the subcutaneous route. The subcutaneous layer is considered to be a suitable site for insulin injection as the layers of fats provide a suitable environment for a slow, stable, and predictable amount of insulin absorption. Such properties as stable absorption and gradual increase in insulin peak to attain maximum concentrations make this route suitable for insulin administration. This route excludes enzymatic degradation and entero-hepatic circulation. Moreover, injection through the subcutaneous route is comparatively less painful than intradermal and intramuscular administration. But this route is associated with the development of allergic reactions and inflammation at the site of injection (Ahad et al., 2021). Hence, various alternatives were developed to support insulin delivery and the efficient management of the glycemic condition. Alternative routes were suggested for insulin delivery, such as oral, inhalation, transdermal, etc. The transdermal route of insulin administration is gaining interest for its noninvasive nature and its compliance with users. This route makes use of technologies such as microneedle technology, chemical enhancement of insulin absorption, insulin patches, sonophoresis, iontophoresis, electroporation, and vesicle formation.

10   Microneedle

Microneedle technology is a captivating technology that is helpful for the delivery of drugs under the skin via reversible microchannel formations, so that the skin is permeable to the administered drug (Jin et al., 2018). Microneedle technology makes use of four types of needles, namely, solid, hollow, dissolving, and degradable microneedles. These needles are long enough to reach the stratum corneum. Solid microneedles are found to be painless and assist in increasing the skin's permeability to the delivery of small-molecular-size proteins, molecules, and nanoparticles. The drug of choice is either coated over the microneedle or encapsulated in the microneedle, which is released on administration. Hollow microneedles are used to infuse drug into the skin. Usually, insulin and vaccines are administered using hollow microneedles. Owing to the needle's size, the advantages of using microneedle technology include less strain, less anxiety, and less tissue damage. Biodegradable and dissolving microneedles are arousing a lot of interest. They are polymers that, on administration, either degrade or dissolve and then gain entry into circulation. The drug of choice is coated onto the polymer matrix, thereby providing a large drug-loading capacity (Ling & Chen, 2013). In current years, dissolving microneedles made of sodium carboxymethyl cellulose and gelatin are used in anchoring insulin rapidly under the skin which then steadily dissociates, providing the necessary therapeutic bioavailability (Jin et al., 2018). Another study explains the development of completely injectable microneedles made of polyvinyl alcohol and poly-c-glutamic acid/polyvinyl pyrrolidone, which permits the effective release of insulin into tissues (Chen et al., 2015).

11   Chemical Permeation Enhancer

The name "chemical permeation enhancer" itself suggests that it is either a single compound or a mixture of compounds involved in enhancing the permeation of the desired compound into the skin. This method is useful for administering a number of hydrophobic compounds or drugs. An ideal permeation enhancer should have the following characteristics: (a) reversibly disrupts the stratum corneum to increase skin permeability and (b) acts as an external force to support transportation across the skin (c) to avoid injury to the underlying tissue (Prausnitz & Langer, 2008). These chemical enhancers break the highly organized lipid bilayer either by inserting an amphiphilic molecule, which leads to the distortion of the bilayer, or through the removal of lipid molecules from the bilayer, causing lipid layer packaging defects (Prausnitz & Langer, 2008). Specifically synthesized products like chemical enhancers are Azone (1-dodecylazacycloheptan-2-one) and SEPA (2-n-nonyl-1,3dioxolane) (Prausnitz & Langer, 2008). However, the chemical enhancement of even small molecules is reported to cause skin irritation. Even though chemical enhancers provide us with increased skin permeability for the transport of drugs into the skin, the main difficulty faced is the localization of the effects of administered drugs in the stratum corneum. Liposomes, specifically highly deformed liposomes, are used as chemical enhancers for insulin administration; however, this study is still in phase I clinical trials (Prausnitz & Langer, 2008). Yerramsetty et al. (2010) investigated possible enhancers for insulin administration and concluded their experiment with eight enhancers, namely, decanol; menthone; oleic acid; cycloundecanone; cis-4-hexen-1-ol; 2,4,6-collidine; octaldehyde; and 4-octanone, which were found to highly enhance insulin penetration and were importantly nontoxic, and among the mentioned compounds, the last five were new discoveries (Ahad et al., 2021) (Yerramsetty et al., 2010). Li et al., in their in vitro experiment, noted that pretreatment of the skin with trypsin can enhance the permeation of insulin into the skin. Pretreatment of rat skin with 0.25% trypsin showed 5.2-fold increase in permeability (Li et al., 2008).

12   Patches

Transdermal patches are available for certain drugs. This method provides us with consistent delivery of insulin at acceptable dosage forms. They are patient friendly, noninvasive, and convenient to use. Recently, insulin patches have been modified by incorporating nanoheaters, which has shown effective results similar to insulin administered through the subcutaneous route (Pagneux et al., 2020). King et al. (2002) made use of biphasic insulin in transdermal patch forms and tested them on streptozotocin-induced diabetic rats for diabetes management. These patches were placed on the abdominal region for 48 hours; the results showed that biphasic insulin is absorbed actively via the skin and recorded a decrease in blood glucose level of about 43.7% (King et al., 2002). Further studies by King et al. (2003) on the effect of transdermal insulin patches on the lymphatic system showed that the use of transdermal insulin patches increased insulin levels in the lymph system in a graded manner (King et al., 2003). In another study, Eudragit RS 100, butylphtalate, and ethyl cellulose were used to produce a transdermal insulin system (Mbaye et al., 2009). Qiu et al., in their study, describe a hydrogel patch device that showed sustained release of insulin and had a longer duration of action when compared to subcutaneous administration (Qiu et al., 2012).

13   Sonophoresis

Sonophoresis is a process where ultrasound waves are used for the transdermal application of drugs. An ultrasound frequency of 20–100 kHz is used for distortion of the outer stratum corneum thereby enhancing insulin penetration (Lavon & Kost, 2004; Boucaud et al., 2002). Sonophoresis has grasped the attention of researchers, but its actual mechanism is not yet fully understood; even though researchers have suggested several possible mechanisms, the only mechanism that was quite satisfying is cavitation (Schlicher et al., 2006; Park et al., 2014). Jabbari et al. conducted a study on developing air ultrasonic ceramic transducers for delivering insulin transdermally. Their study concluded that insulin delivery was at its therapeutic level when the sonophoresis mode of action was employed, and the blood glucose level was reduced to its normal level (Jabbari et al., 2015). Another study, by Feiszthuber et al. in 2015, also proved that insulin delivered transdermally through sonophoresis was very much active, and the mechanism behind the entry of insulin was found to be due to the cavitation of the skin (Feiszthuber et al., 2015).

14   Iontophoresis

Iontophoresis is a process wherein drugs are administered transdermally through the continuous passage of low-voltage current for a certain period of time (Prausnitz & Langer, 2008). Electricity is the driving force for the transportation of the drug across the stratum corneum. Iontophoresis uses a voltage gradient to increase the permeability of the molecules into the skin. These are effective enough to allow the transit of the desired molecules or drugs across the stratum corneum and their entry into the systemic circulation. This method is used for the administration of drugs based on charge and drugs with small molecular weight (Prausnitz & Langer, 2008). Iontophoresis is convenient as it is possible to withdraw or stop the delivery of drugs anytime. However, to achieve the maximal level of drug delivery, maximum current must be applied to the skin; this action creates a limitation in the use of this method due to the precipitation of skin irritation and pain (Prausnitz, 1996). Transdermal delivery of insulin to porcine skin was achieved through the combined use of iontophoresis and certain chemical enhancers, which synergistically enhance drug delivery (Rastogi et al., 2010; Rastogi & Singh, 2005). The uptake of ions is found to be improved upon the pretreatment of the skin with alcohol before iontophoresis.

15   Electroporation

In electroporation, short- but high-voltage electric pulses are used in order to reversibly disrupt the layers of the skin for the transfer of various agents. Electroporation is found to effectively disrupt the lipid bilayers (Denet et al., 2004). Although the electric pulse is applied only for a few milliseconds, this pulse generates an electrophoretic driving force, which can persist for hours in assisting in the transdermal transfer of various drugs, peptides, vaccines, etc. (Prausnitz & Langer, 2008). Normally, the stratum corneum shows resistance to electricity; however, upon electroporation, the resistance of the lipid bilayers and the stratum corneum drops, providing deeper penetration of the desired compounds into the skin (Prausnitz & Langer, 2008). In a study, the electroporation method is employed in the delivery of insulin in rabbits transdermally. The results of the study showed that upon electrophoretic administration of insulin, the blood glucose level was reversed back to normal condition (Mohammad et al., 2016). The limitations of the method of transdermal application of drugs include pain generation and muscle stimulation, which are due to the electric field (Pliquett & Weaver, 2007); also, the complexity of the device makes it less preferable for human use (Prausnitz & Langer, 2008). Studies also prove that electroporation and iontoporation, in combination, can act synergistically in insulin permeation, resulting in increased plasma insulin levels.

16   Vesicle Formation

Liposomes are the most commonly used vesicles for drug delivery. They gained a lot of attention due to their wide applications in the noninvasive method of drug delivery. Various studies show the combined effect of liposomes with two or more enhancers can provide a greater extent of blood glucose-lowering activity. A study by Ogiso et al. showed that liposomes synergistically worked with enhancers such as D-limonene and taurocholate, resulting in a continuous blood-glucose-lowering activity for about 10 h (Ogiso et al., 1996).

Another such compound involved in vesicle formation is "transferosomes." They are compounds with high elasticity that are made of phospholipid units and edge activators (Rai et al., 2017). These transferosomes are found to be good carriers of insulin across the skin if applied in acceptable quantities. Marwah et al. studied the preparation of transdermal transferosome insulin gel for diabetes management. The gel was prepared using the rotary evaporation sonication method and was found to have an insulin trapping efficiency of 78% and an insulin release property of

83.11%, suggesting that it can be a promising way of delivering insulin and other small protein peptides (Marwah et al., 2016).

17   Diabetes Is a Pain

The regular use of insulin in diabetes is an important aspect of the management of the disorder. However, the routine administration of insulin by various modes can cause various dermal complications. According to van Munster et al., (2014), the common skin complications associated with insulin injections are lipohypertrophy, lipoatrophy, and erythema (van Munster et al., 2014).

Lipodystrophy is a disorder that is either acquired or congenitally obtained. Lipodystrophy is a condition wherein the body fails to produce and maintain normally required levels of body fat tissues. There is marked degeneration of the adipose tissue at the site of injection. Lipodystrophy may result from various factors, such as insulin resistance, hypertriglyceridemia, nonalcoholic fatty liver disease, and metabolic syndrome. This condition can lead to the development of either lipoatrophy or lipohypertrophy. Lipoatrophy is characterized by a reduction in the size of fat droplets without loss in the count of adipocytes (van Munster et al., 2014). This condition is clinically characterized by a visible cutaneous layer indent and detectable atrophy of the subcutaneous fats (Tsadik et al., 2018). This atrophy may be resultant of immune reactions at the site of injection due to possible impurities, which further leads to the release of inflammatory markers like tumor necrosis factor-α (TNF- α), mast cells, and macrophages (van Munster et al., 2014). On the other hand, lipohypertrophy is characterized by the enlargement of the subcutaneous fat tissues, representing a tumor-like swelling of the fatty tissues at the site of injection (Fujikura et al., 2005). In insulin-injecting patients, lipohypertrophy prevails as a recognized complication affecting approximately more than 50% of the diabetic population (Blanco et al., 2013). Lipohypertrophy may occur due to repeated stimulation of the adipocytes produced by direct injury caused by the syringes or insulin itself. A study by Afewercki et al. on the effects of insulin-induced lipodystrophy on children and adolescents in 2018 gives a detailed view of the prevalence of insulin-induced lipodystrophy (Tsadik et al., 2018). The results of the study showed that upon insulin administration the probability of acquiring lipodystrophy, lipohypertrophy, and lipoatrophy account for 58.5%, 97.1%, and 2.9%, respectively. Their study showed that children, those people who need a high dose of insulin, and those who neglect the rotation of injection sites seem to be greatly affected by lipohypertrophy. This study also showed that patients with lipohypertrophic conditions showed nonoptimal glycemic control, which was three-fold higher than in nonlipohypertrophic patients.

Erythema is clinically characterized as a visible rubor at the injection site. Skin erythema results from various factors, like infection and inflammation of the subcutaneous layer and hypersensitive reactions due to the presence of particular preservatives in insulin formulations, catheters, and needles (van Munster et al., 2014). In a study by Hilde et al. (2014), surprising facts were revealed that the incidence of erythema is not influenced by the use of alcohol for disinfection, washing of hands, and reuse of needles (van Munster et al., 2014). This fact was contrary to theoretical data, thereby showing that aseptic inflammation plays an important role in the development of erythema.

Amyloid is a group of soluble proteins abnormally produced by the bone marrow. This protein aggregates to form insoluble fibrils, which are resistant to degradation and breakdown. These aggregates are found to be deposited on various organs, inducing amyloid plaque formation. These amyloid plaques are associated with number of pathological conditions varying based on the organ on which it gets deposited (Rambaran & Serpell, 2008). A study by Surabh Arora et al. (2021) explains the major skin complication, lipohypertrophy; its prevalence; and the basis of its occurrence. Further in this study, the authors also reported a rare insulin-induced amyloidosis complication, which was confirmed using a diagnostic tool "apple green birefringence" on polarized microscopy followed by anti-insulin antibody staining (Arora et al., 2021).

Fat necrosis is one rare condition noted for insulin-injection-induced complications (Hanson et al., 2014). A case report by Hanson et al. (2014) showed a middle-aged diabetic patient admitted for severe pain in the right iliac fossa; all her vitals and biochemical parameters are normal, and the cause of the pain was mysterious. Then using a magnetic resonance imaging (MRI) technique, the presence of a fat necrosis nodule was spotted, and she was advised for immediate surgery. Postsurgery observation reports indicated that there is no pain, and the glycemic control was under control. The histopathological studies on the excised fat nodule showed fat cell necrosis characterized by the presence of macrophages and other macro-foreign bodies inside the fat cells (Lee & Adair, 1920).

Insulin administration through injection itself is considered to cause pain and discomfort. Based on a survey (the DAWN survey), estimation of which was made based on the responses, 20% of the responders tend to skip insulin injection and 10% limited the number of injections (Peyrot & Rubin, 2006). The basal reason for the omission and restriction of daily insulin injections is anxiety, distress, and phobia of injections (Aronson, 2012). Pain due to injections is mainly related to three basic factors, namely, the length of the needle used, the needle diameter, and the context of the injection (Aronson, 2012). Thus, in order to mitigate the pain and discomfort in the use of insulin injections, the abovementioned factors were revised and innovated in order to meet the compliance of patients. A study by Aronson (2012) gives a detailed review of this field. In this review, the modifications made to overcome barriers in insulin injection anxiety and distress are discussed. The traditional needle length used was 12.7 mm, but later subcutaneous administration of insulin in glycemic control made it possible to reduce the size of the injections. This provided less pain, trauma, and accurate drug delivery. The diameter of the injection needles used was reduced from 28 gauge to 32 and 33 gauge to attain a thinner needle. The tips of the needles were pointed and designed with bevel cuts to lessen the force to be applied when injecting the needle.

18   Alternative Insulin Delivery Routes

It is essential to understand the pharmacokinetic properties of insulin to determine the effective route for the administration of insulin. As insulin is a protein in nature, it is easily degraded in the gastrointestinal tract; hence, intravenous, subcutaneous, and intramuscular routes are the preferred routes for insulin administration to obtain its optimal absorption (Hoffman & Ziv, 1997). After absorption, the absorbed insulin is swiftly distributed to the extracellular fluids for utilization by the cells. The bioavailability of orally administered insulin is very low; approximately less than 5% is absorbed, which makes it difficult to determine its hypoglycemic effect (Hoffman & Ziv, 1997). Keeping the highlights of the pharmacokinetic action of insulin, the suggested routes for insulin administration and innovations made in each route to overcome the disabilities are discussed below. The employment of injections is very much necessary due to the poor oral bioavailability of insulin.

19   Inhalable Insulin

The suggestion of preparing inhalable insulin was made by German researchers in 1924 (Mohanty, 2017). In our body, the nasal passage has the largest absorptive surface area, which accounts for approximately 150 cm^2 (Bahman et al., 2019). The mucosal membrane in the nasal cavity is highly vascularized and aids in the direct uptake of insulin proteins into circulation thereby bypassing first-pass metabolism (Hinchcliffe & Illum, 1999). The intranasal route of insulin administration is considered to be the most favorable alternative for invasive insulin. Having these facts in mind, many new ideas were brought up for the conversion of insulin into powdered form, with particle sizes suitable for inhalation. Inhalable insulin is available in powdered form, which is inhaled using an inhaler. Due to its ability to be rapidly absorbed and to rapidly commence its action, the intranasal route of insulin administration is suggested to be equally effective in diminishing postprandial hyperglycemic states (Gaddam et al., 2021). Hypoglycemic episodes are not noticed in patients undergoing intranasal insulin therapy as this route mimics normal physiological pulsatile endogenous insulin secretion during food intake (Gaddam et al., 2021). Having the potential to bypass the blood-brain barrier, nasal insulin can directly reach the central nervous system and produce action. There is an inconclusive statement that nasal insulin administration in patients with diabetes, along with other memory-related disorders like dementia, Alzheimer's, etc., can symbiotically improve diabetes and memory disorders (Gaddam et al., 2021). The first inhalable insulin, Exubera (rapidly acting insulin), was developed and marketed by Pfizer in 2006 after obtaining the approval of the Food and Drug Administration (FDA) (FDA Approves First Ever Inhaled Insulin Combination Product for Treatment of Diabetes – Pharmaceutical Processing World, 2022). However, due to poor marketing, it was soon withdrawn from the market. Currently, another drug, namely Afrezza, is being used, which was developed by MannKind using technosphere technology and which obtained approval from the FDA in 2014 (http://www.fda.gov/NewsEvents/Newsroom/PressAnnouncements/ucm403122. htm). When inhaled, the technosphere insulin enters the lungs and gets dispersed at its neutral pH and enters the circulation via the mucosal layer (Goldberg & Wong, 2015). However, there are a number of limitations in the use of inhalable insulin, and also minor adverse effects are noted in some patients during the trial. The common adverse effects noted are related to the nose, such as rhinitis, minimal nose bleeding, soreness, dripping, and sneezing, and patients also reported upper respiratory tract infections, headaches, dizziness, weakness, hypoglycemia (negligible), rash, and gastrointestinal symptoms (Avgerinos et al., 2018). Limitations include low permeability due to large particle size, the possibility of undergoing enzymatic degradation due to the action of aminopeptidase, and an increase in mucociliary clearance. Inhaled insulin is contraindicated for patients who smoke and patients with pulmonary diseases (Shah et al., 2016). Furthermore, patients were asked to undergo a pulmonary test before they started with the therapy, and also after the treatment initiation, the pulmonary function tests were routinely taken at 6 months interval (Shah et al., 2016).

20 Oral Insulin

The oral route of insulin administration is the easiest and the most convenient method of insulin intake and is more similar to endogenously released insulin. The oral route of insulin delivery can be of great use in improving the portal levels of insulin and preventing the occurrence of peripheral hyperinsulinemic conditions, which are associated with complications like neuropathy and retinopathy (Khafagy et al., 2007). To exhibit its action, the orally administered insulin has to maintain its structure, integrity, and conformation before reaching systemic circulation (Wong et al., 2016). However, there are a lot of challenges in preparing a successful oral insulin formulation, such as a decrease in the bioavailability of insulin due to its hydrophobic nature, the degradation of insulin by the proteolytic enzymes present in the gastrointestinal tract, and low permeation through the intestinal wall due to large particle size (Shah et al., 2016). The main proteolytic enzymes involved in digestion are pepsin, trypsin, chymotrypsin, and carboxypeptidase. The activity of these enzymes is high in the small intestines, and their activity is less in the ileum; this property can be helpful in the formulation of oral insulin analogs with protective layers so that the intact active protein will not be degraded and be available for entering the systemic circulation (Wong et al., 2016). Thus, strategies to improve the oral bioavailability of insulin target the incorporation of enzyme inhibitors, absorption enhancers, mucoadhesive polymers, and chemical modifications. A study by Yamamoto et al., based on experimentation on various enzyme inhibitors, showed increased availability of oral insulin in the large intestine for absorption. In this study, they showed a positive result by using three inhibitory enzymes; sodium glycocholate, bacitracin, and camostat mesilate, to enhance the bioavailability of oral insulin in the large intestine (Yamamoto et al., 2022). Another discovery was duck ovomucoid, a novel class of trypsin and chymotrypsin enzyme inhibitors, synthesized from ovomucoid present in egg white. This enzyme inhibitor delayed the degradation of insulin for 1 h. Thus, when used with other proteolytic inhibitors, it provides insulin available for optimal absorption by the large intestine (Agarwal et al., 2000). However, the use of enzyme inhibitors can cause adverse effects, like a deficiency in gastric enzymes, which can lead to protein malabsorption. On absorption enhancement, various enhancers help in the permeation of insulin across the lipid bilayer of the epithelial lining of the gastrointestinal tract. Some traditional absorption enhancers are bile salts, surfactants, calcium ion-chelating agents, and fatty acids (palmitic acid) (Mesiha et al., 1994). In chemical modification, the proteins and peptides are coated with a ligand molecule, which can react with the endogenous receptor molecules present in the gastrointestinal tract. Thus, protein or peptide uptake is facilitated by receptor mediation. There are two oral insulin tablets, Eligen and IN 105, that are formulated and currently undergoing the third phase in clinical trials (Wong et al., 2016). Another review, by Chun Y. Wong et al., gives us detailed information on the usage of microparticles, microcapsules, and microspheres in assisting a successful oral delivery of insulin in diabetes management. This review elaborates on the various natural and synthetic polymers and inorganic compounds that can be possibly employed in microparticle insulin formulation (Wong et al., 2018).

21   Intradermal Insulin

The intradermal route of insulin administration can be a better alternative to painful invasive routes of insulin administration. Michael Hultstrom et al. mention that for managing diabetes mellitus, this method of administration could play a promising role. They highlight the use of a micro-electro-mechanical system (MEMS) device for intradermal insulin administration. The main drawback or challenge that intradermal insulin faces is that the stratum corneum layer in the skin acts as an insuperable barrier to insulin entry. MEMS uses microneedle technology, where the microneedle is long enough to deliver the drug to blood-capillary-rich areas but also not long enough to disturb the nociceptors present. This acts as a major advantage since insulin is delivered into circulation without any hindrance and is less invasive and less painful. This route can be suggested as an alternative to the subcutaneous route if sufficient bioavailability can be achieved in a discreet, painless, and minimally invasive way when compared to the subcutaneous route (Hultström et al., 2014). Another study, by Efrat Kochba et al. (2016), shows that the intradermal route of insulin administration using the MicronJet needle device led to an improved pharmacokinetic profile of insulin than through earlier ways. They studied the pharmacokinetics and pharmacodynamics of intradermally injected insulin in human subjects and also checked the safety profile and tolerability of these injections. Their study showed that the intradermal injection of insulin by the MicronJet device gave positive results, like a good safety profile and no significant associated risks, when compared to the subcutaneous route (Kochba et al., 2016).

22   Transdermal Insulin

The transdermal route is a fascinating route for insulin delivery as this technique can be helpful for diminishing the pain and risk of acquiring skin infections on account of subcutaneous injections (Khafagy et al., 2007). This method represents an essential alternative to both oral and invasive routes of drug administration. Earlier people followed the traditional ways of applying various medicinal substances topically for medical indications; however, in the modern era, this practice has been enhanced and followed by preparing topical formulations, which were used for local indications. The first transdermal route of medicine formulated was scopolamine, indicated for motion sickness, available as transdermal patches and approved by the FDA in 1979 (Prausnitz & Langer, 2008). The transdermal route of drug administration is preferable as it avoids algia at the site of injection and surpasses first-pass metabolism in invasive route of administration. Moreover, they are convenient for self-administration and are cost-effective. Transdermal patches can provide quite long periods of continuous release of the drug transdermally. An in vivo study showed the antidiabetic activity of insulin on percutaneous administration, which is similarly effective to that of insulin administered through the subcutaneous route (Wu et al., 2010).

Transdermal smart gels are gaining a lot of attention in the drug delivery system. The use of polymer-based gel loaded with insulin acts as a closed-loop system in maintaining the blood glucose level. Phenylboronic acid (PBA) is employed in this smart gel production technology. Phenylboronic acid has the ability to form complexes with 1,2- and 1,3-*cis*-diols present in carbohydrates and forms reversible boronate-ester bonds in an aqueous solution (Matsumoto et al., 2017). The binding capacity and highly specific nature of phenylboronic acid are due to its unique stereochemistry. This property made it possible to use phenylboronic acid in the development of diagnostic and therapeutic applications. Using biomedical engineering technology, a new type of PBA "smart gel," formed by integrating PBA with an optimally amphiphilic acrylamide gel matrix, resulted in the formation of a gel surface. This gel is macroscopic and is equilibrated with glucose completely. Thus, the alteration of the hydration capacity under favorable conditions results in the permeability of loaded insulin from the surface to the systemic circulation. A study by Akira et al. showed that the release of insulin from the insulin smart gel into the skin using a silicone catheter was highly regulated in accordance with the glucose patterns observed in the system (Matsumoto et al., 2017). Their aim was to develop an artificial pancreas to replace the gap produced by the endogenously present pancreas.

23   Nanotechnology in Insulin Administration

The employment of nanotechnology has widely benefited the delivery of insulin by overcoming the oral insulin delivery system. Various natural and synthetic nanoparticles are synthesized, which are used as a vehicle for the transportation of insulin into circulation (Shah et al., 2016). Considering the improvement of insulin delivery efficiency and patient compliance as the main aim, the use of nanoparticles has been employed in both subcutaneous and oral insulin formulations (Zhang et al., 2021). Nanoparticles are used to enhance the bioavailability of insulin, prevent the degradation of insulin, and improve the capacity to regulate insulin release into circulation (Zhang et al., 2021). A study by Shilo et al. demonstrates that the use of insulin-gold nanoparticles in subcutaneous administration showed the controlled release of insulin into circulation for prolonged glucose control (Shilo et al., 2015). The authors concluded that the utilization of gold nanoparticles in insulin delivery prevented the enzymatic degradation of insulin, which was the main reason for prolonged insulin action. Oral insulin in harmony with nanotechnology helps in overcoming barriers in the oral route of insulin administration. Nanotechnology surpasses these barriers for the efficient work of insulin. The encapsulation of insulin inside the nanoparticle aids in bypassing the barriers, namely, enzymatic and chemical degradation. Sakloetsakun et al. elaborate on self-nano-emulsifying drug delivery system (SNEDDS) technology for the successful delivery of oral insulin. In this study, they formulated oral insulin with thiolated chitosan and used the SNEDDS delivery system, which showed satisfying results. The thiolated chitosan provided the necessary characteristics, such as particle size, entrapment efficiency, drug release, and stability (Sakloetsakun et al., 2013). This union of thiolated chitosan and SNEDDS can be a promising idea for improving the efficacy of oral insulin. The nanoparticles of insulin coated with polymer-based lipid hybrid were converted into microparticles and encapsulated in enteric capsules. These capsules prevent the degradation of insulin under a gastric environment and reach the intestines, where they increase the cell permeability of the epithelial cells and enhance their uptake (Yu et al., 2022). Nanoparticles that are pH sensitive are also investigated, namely, poly-N-isopropylacrylamide polymer derivatives (Karnoosh-Yamchi et al., 2022) and hydroxypropyl methylcellulose phthalates.

24   Conclusion

The present situation clearly explains that diabetes is a constantly growing epidemic condition worldwide. The above-discussed innovative strategies help in maintaining and managing the disorder "diabetes mellitus." In conclusion, consecutive progress is being made in the subject of diabetes mellitus for better, healthy, and smooth management. However, there is a need for more research to fulfill the gap in diabetes management. It is essential to spread awareness among people about the newer methods of pain-free insulin administration. Further studies are needed in this field of research to overcome the disadvantages of the present treatment strategies to have a better therapy option.

References

Agarwal, V., Reddy, I., & Khan, M. (2000). Oral delivery of proteins: Effect of chicken and duck ovomucoid on the stability of insulin in the presence ofα-chymotrypsin and trypsin. *Pharmacy and Pharmacology Communications, 6*(5), 223–227. https://doi.org/10.1211/146080800128735935

Ahad, A., Raish, M., Bin Jardan, Y., Al-Mohizea, A., & Al-Jenoobi, F. (2021). Delivery of insulin via skin route for the Management of Diabetes Mellitus: Approaches for breaching the obstacles. *Pharmaceutics, 13*(1), 100. https://doi.org/10.3390/pharmaceutics13010100

Andersen, A., Christiansen, J., Andersen, J., Kreiner, S., & Deckert, T. (1983). Diabetic nephropathy in type 1 (insulin-dependent) diabetes: An epidemiological study. *Diabetologia, 25*(6), 496–501. https://doi.org/10.1007/bf00284458

Aronson, R. (2012). The role of comfort and discomfort in insulin therapy. *Diabetes Technology & Therapeutics, 14*(8), 741–747. https://doi.org/10.1089/dia.2012.0038

Arora, S., Agrawal, N., Shanthaiah, D., Verma, A., Singh, S., Patne, S., et al. (2021). Early detection of cutaneous complications of insulin therapy in type 1 and type 2 diabetes mellitus. *Primary Care Diabetes, 15*(5), 859–864. https://doi.org/10.1016/j.pcd.2021.06.004

Avgerinos, K., Kalaitzidis, G., Malli, A., Kalaitzoglou, D., Myserlis, P., & Lioutas, V. (2018). Intranasal insulin in Alzheimer's dementia or mild cognitive impairment: A systematic review. *Journal of Neurology, 265*(7), 1497–1510. https://doi.org/10.1007/s00415-018-8768-0

Bahman, F., Greish, K., & Taurin, S. (2019). Nanotechnology in insulin delivery for Management of Diabetes. *Pharmaceutical Nanotechnology, 7*(2), 113–128. https://doi.org/10.2174/2211738507666190321110721

Blanco, M., Hernández, M., Strauss, K., & Amaya, M. (2013). Prevalence and risk factors of lipohypertrophy in insulin-injecting patients with diabetes. *Sinical & Metabolism, 39*(5), 445–453. https://doi.org/10.1016/j.diabet.2013.05.006

Bliss, M. (1982). *The discovery of insulin*. The University of Chicago Press.

Bolli, G. B., Di Marchi, R. D., Park, G. D., Pramming, S., & Koivisto, V. A. (1999). Insulin analogues and their potential in the management of diabetes mellitus. *Diabetologia, 42*(10), 1151–1167. https://doi.org/10.1007/s001250051286

Boucaud, A., Garrigue, M., Machet, L., Vaillant, L., & Patat, F. (2002). Effect of sonication parameters on transdermal delivery of insulin to hairless rats. *Journal of Controlled Release, 81*(1–2), 113–119. https://doi.org/10.1016/s0168-3659(02)00054-8

Candrilli, S., Davis, K., Kan, H., Lucero, M., & Rousculp, M. (2007). Prevalence and the associated burden of illness of symptoms of diabetic peripheral neuropathy and diabetic retinopathy. *Journal of Diabetes and its Complications, 21*(5), 306–314. https://doi.org/10.1016/j.jdiacomp.2006.08.002

Cernea, S., & Raz, I. (2020). Insulin therapy: Future perspectives. *American Journal of Therapeutics, 27*(1), e121–e132. https://doi.org/10.1097/mjt.0000000000001076

Chaluvaraju, K., Niranjan, M., Manjuthej, T., Zaranappa, T., & Mane, K. (2012). Review of insulin and its analogues in diabetes mellitus. *Journal Of Basic And Clinical Pharmacy, 3*(2), 283. https://doi.org/10.4103/0976-0105.103822

Chen, M., Ling, M., & Kusuma, S. (2015). Poly-γ-glutamic acid microneedles with a supporting structure design as a potential tool for transdermal delivery of insulin. *Acta Biomaterialia, 24*, 106–116. https://doi.org/10.1016/j.actbio.2015.06.021

Denet, A., Vanbever, R., & Préat, V. (2004). Skin electroporation for transdermal and topical delivery. *Advanced Drug Delivery Reviews, 56*(5), 659–674. https://doi.org/10.1016/j.addr.2003.10.027

Deshpande, A., Harris-Hayes, M., & Schootman, M. (2008). Epidemiology of diabetes and diabetes-related complications. *Physical Therapy, 88*(11), 1254–1264. https://doi.org/10.2522/ptj.20080020

FDA News Release: FDA approves Afrezza to treat diabetes. Available at: http://www.fda.gov/NewsEvents/Newsroom/PressAnnouncements/ucm403122.htm

Feiszthuber, H., Bhatnagar, S., Gyöngy, M., & Coussios, C. (2015). Cavitation-enhanced delivery of insulin in agar and porcine models of human skin. *Physics in Medicine and Biology, 60*(6), 2421–2434. https://doi.org/10.1088/0031-9155/60/6/2421

Fox, C. (2004). Trends in cardiovascular complications of diabetes. *JAMA, 292*(20), 2495. https://doi.org/10.1001/jama.292.20.2495

Frias, P., & Frias, J. (2017). New basal insulins: A clinical perspective of their use in the treatment of type 2 diabetes and novel treatment options beyond basal insulin. *Current Diabetes Reports, 17*(10), 91. https://doi.org/10.1007/s11892-017-0926-8

Fujikura, J., Fujimoto, M., Yasue, S., Noguchi, M., Masuzaki, H., Hosoda, K., et al. (2005). Insulin-induced Lipohypertrophy: Report of a case with histopathology. *Endocrine Journal, 52*(5), 623–628. https://doi.org/10.1507/endocrj.52.623

Gaddam, M., Singh, A., Jain, N., Avanthika, C., Jhaveri, S., De la Hoz, I., et al. (2021). A comprehensive review of intranasal insulin and its effect on the cognitive function of diabetics. *Cureus., 13*(8), e17219. https://doi.org/10.7759/cureus.17219

Garg, S., Rewers, A., & Akturk, H. (2018). Ever-increasing insulin-requiring patients globally. *Diabetes Technology & Therapeutics, 20*(S2), S2-1-S2-4. https://doi.org/10.1089/dia.2018.0101

Goldberg, T., & Wong, E. (2015). Afrezza (insulin human) inhalation powder: A new inhaled insulin for the management of Type-1 or Type-2 diabetes mellitus. *P & T: A Peer-Reviewed Journal for Formulary Management, 40*(11), 735–741.

Gradel, A., Porsgaard, T., Lykkesfeldt, J., Seested, T., Gram-Nielsen, S., Kristensen, N., & Refsgaard, H. (2018). Factors affecting the absorption of subcutaneously administered insulin: Effect on variability. *Journal of Diabetes Research, 2018,* 1–17. https://doi.org/10.1155/2018/1205121

Hanson, P., Pandit, M., Menon, V., Roberts, S., & Barber, T. (2014). Painful fat necrosis resulting from insulin injections. *Endocrinology, Diabetes & Metabolism Case Reports, 2014,* 140073. https://doi.org/10.1530/edm-14-0073

Harris, R., & Leininger, L. (1993). Preventive care in rural primary care practice. *Cancer, 72*(S3), 1113–1118. https://doi.org/10.1002/1097-0142(19930801)72:3+<1113::aid-cncr2820721328>3.0.co;2-a

Hinchcliffe, M., & Illum, L. (1999). Intranasal insulin delivery and therapy. *Advanced Drug Delivery Reviews, 35*(2–3), 199–234. https://doi.org/10.1016/s0169-409x(98)00073-8

Hoffman, A., & Ziv, E. (1997). Pharmacokinetic considerations of new insulin formulations and routes of administration. *Clinical Pharmacokinetics, 33*(4), 285–301. https://doi.org/10.2165/00003088-199733040-00004

Howey, D., Bowsher, R., Brunelle, R., & Woodworth, J. (1994). [Lys(B28), pro(B29)]-human insulin: A rapidly absorbed analogue of human insulin. *Diabetes, 43*(3), 396–402. https://doi.org/10.2337/diab.43.3.396

Hultström, M., Roxhed, N., & Nordquist, L. (2014). Intradermal insulin delivery. *Journal of Diabetes Science and Technology, 8*(3), 453–457. https://doi.org/10.1177/1932296814530060

International Diabetes Federation. (2019). *Diabetes atlas* (9th ed.). International Diabetes Federation.

Jabbari, N., Asghari, M. H., Ahmadian, H., & Mikaili, P. (2015). Developing a commercial air ultrasonic ceramic transducer to transdermal insulin delivery. *Journal of Medical Signals and Sensors, 5*(2), 117–122.

Jin, X., Zhu, D., Chen, B., Ashfaq, M., & Guo, X. (2018). Insulin delivery systems combined with microneedle technology. *Advanced Drug Delivery Reviews, 127,* 119–137. https://doi.org/10.1016/j.addr.2018.03.011

Karamitsos, D. (2011). The story of insulin discovery. *Diabetes Research and Clinical Practice, 93,* S2–S8. https://doi.org/10.1016/s0168-8227(11)70007-9

Karnoosh-Yamchi, J., Mobasseri, M., Akbarzadeh, A., Davaran, S., Ostad-Rahimi, A., Hamishehkar, H., & et al. (2022). *Preparation of pH sensitive insulin-loaded nano hydrogels and evaluation of insulin releasing in different pH conditions.* Retrieved 2 July 2022, from.

Khafagy, el-S., Morishita, M., Onuki, Y., & Takayama, K. (2007). Current challenges in non-invasive insulin delivery systems: A comparative review. *Advanced Drug Delivery Reviews, 59*(15), 1521–1546. https://doi.org/10.1016/j.addr.2007.08.019

King, M., Badea, I., Solomon, J., Kumar, P., Gaspar, K., & Foldvari, M. (2002). Transdermal delivery of insulin from a novel biphasic lipid system in diabetic rats. *Diabetes Technology & Therapeutics, 4*(4), 479–488. https://doi.org/10.1089/152091502760306562

King, M., Michel, D., & Foldvari, M. (2003). Evidence for lymphatic transport of insulin by topically applied biphasic vesicles. *Journal of Pharmacy and Pharmacology, 55*(10), 1339–1344. https://doi.org/10.1211/0022357021918

Kleiner, I. (1919). The action of intravenous injections of pancreas emulsions in experimental diabetes. *Journal of Biological Chemistry, 40*(1), 153–170. https://doi.org/10.1016/s0021-9258(18)87274-x

Kochba, E., Levin, Y., Raz, I., & Cahn, A. (2016). Improved insulin pharmacokinetics using a novel microneedle device for intradermal delivery in patients with type 2 diabetes. *Diabetes Technology & Therapeutics, 18*(9), 525–531. https://doi.org/10.1089/dia.2016.0156

Kurtzhals, P. (2004). Engineering predictability and protraction in a basal insulin analogue: The pharmacology of insulin detemir. *International Journal of Obesity, 28*(S2), S23–S28. https://doi.org/10.1038/sj.ijo.0802746

Lavon, I., & Kost, J. (2004). Ultrasound and transdermal drug delivery. *Drug Discovery Today, 9*(15), 670–676. https://doi.org/10.1016/s1359-6446(04)03170-8

Lee, B., & Adair, F. (1920). Traumatic fat necrosis of the female breast and its differentiation from carcinoma. *Annals of Surgery, 72*(2), 188–195. https://doi.org/10.1097/00000658-192008000-00011

Li, Y., Quan, Y., Zang, L., Jin, M., Kamiyama, F., Katsumi, H., et al. (2008). Transdermal delivery of insulin using trypsin as a biochemical enhancer. *Biological and Pharmaceutical Bulletin, 31*(8), 1574–1579. https://doi.org/10.1248/bpb.31.1574

Lindholm, A., McEwen, J., & Riis, A. (1999). Improved postprandial glycemic control with insulin aspart. A randomized double-blind cross-over trial in type 1 diabetes. *Diabetes Care, 22*(5), 801–805. https://doi.org/10.2337/diacare.22.5.801

Ling, M. H., & Chen, M. C. (2013). Dissolving polymer microneedle patches for rapid and efficient transdermal delivery of insulin to diabetic rats. *Acta Biomaterialia, 9*(11), 8952–8961. https://doi.org/10.1016/j.actbio.2013.06.029

Mane, K., Chaluvaraju, K., Niranjan, M., Zaranappa, T., & Manjuthej, T. (2012). Review of insulin and its analogues in diabetes mellitus. *Journal of Basic and Clinical Pharmacy, 3*(2), 283–293. https://doi.org/10.4103/0976-0105.103822

Marwah, H., Garg, T., Rath, G., & Goyal, A. K. (2016). Development of transferosomal gel for trans-dermal delivery of insulin using iodine complex. *Drug Delivery, 23*(5), 1636–1644. https://doi.org/10.3109/10717544.2016.1155243

Marwood, S. F. (2022). *Diabetes mellitus--some reflections*. PubMed. Retrieved 2 July 2022, from https://pubmed.ncbi.nlm.nih.gov/4574717/.

Matsumoto, A., Tanaka, M., Matsumoto, H., Ochi, K., Moro-oka, Y., Kuwata, H., et al. (2017). Synthetic "smart gel" provides glucose-responsive insulin delivery in diabetic mice. *Science Advances, 3*(11), eaaq0723. https://doi.org/10.1126/sciadv.aaq0723

Mayer, J., Zhang, F., & DiMarchi, R. (2007). Insulin structure and function. *Biopolymers, 88*(5), 687–713. https://doi.org/10.1002/bip.20734

Mazur, A. (2011). Why were "starvation diets" promoted for diabetes in the pre-insulin period? *Nutrition Journal, 10*(1), 23. https://doi.org/10.1186/1475-2891-10-23

Mbaye, G., Ndiaye, A., Diouf, L. A., Diallo, A. S., Diedhiou, A., Sene, M., Mbodj, M., Thioune, O., Dieye, A. M., Diop, I., Cisse, A., & Diarra, M. (2009). Developpement d'une matrice ethylcellulose/eudragit pour la liberation controlee et continue de l'insuline [development of Ethylcellulose/Eudragit matrix for controlled and continuous release of insulin]. *Le Mali Medical, 24*(3), 11–16.

Mesiha, M., Plakogiannis, F., & Vejosoth, S. (1994). Enhanced oral absorption of insulin from desolvated fatty acid-sodium glycocholate emulsions. *International Journal of Pharmaceutics, 111*(3), 213–216. https://doi.org/10.1016/0378-5173(94)90343-3

Mohammad, E., Elshemey, W., Elsayed, A., & Abd-Elghany, A. (2016). Electroporation parameters for successful transdermal delivery of insulin. *American Journal of Therapeutics, 23*(6), e1560–e1567. https://doi.org/10.1097/mjt.0000000000000198

Mohanty, R. (2017). Inhaled insulin - current direction of insulin research. *Journal of Clinical and Diagnostic Research, 11*(4), OE01–OE02. https://doi.org/10.7860/jcdr/2017/23626.9732

Murlin, J., & Kramer, B. (1913). The influence of pancreatic and duodenal extracts on the glycosuria and the respiratory metabolism of depancreatized dogs. *Experimental Biology and Medicine, 10*(5), 171–173. https://doi.org/10.3181/00379727-10-108

Ogiso, T., Nishioka, S., & Iwaki, M. (1996). Dissociation of insulin oligomers and enhancement of percutaneous absorption of insulin. *Biological & Pharmaceutical Bulletin, 19*(8), 1049–1054. https://doi.org/10.1248/bpb.19.1049

Oubré, A., Carlson, T., King, S., & Reaven, G. (1997). From plant to patient: An ethnomedical approach to the identification of new drugs for the treatment of NIDDM. *Diabetologia, 40*(5), 614–617. https://doi.org/10.1007/s001250050724

Pagneux, Q., Ye, R., Chengnan, L., Barras, A., Hennuyer, N., Staels, B., et al. (2020). Electrothermal patches driving the transdermal delivery of insulin. *Nanoscale Horizons, 5*(4), 663–670. https://doi.org/10.1039/c9nh00576e

Park, D., Park, H., Seo, J., & Lee, S. (2014). Sonophoresis in transdermal drug deliverys. *Ultrasonics, 54*(1), 56–65. https://doi.org/10.1016/j.ultras.2013.07.007

Peyrot, M., & Rubin, R. (2006). Resistance to insulin therapy among patients and providers: Results of the cross-national diabetes Attitudes, wishes, and needs (DAWN) study. *Diabetes Care, 29*(4), 953–954. https://doi.org/10.2337/diacare.29.04.06.dc05-0053

Pharmaceutical Processing World. (2022). *FDA approves first ever inhaled insulin combination product for treatment of diabetes - pharmaceutical processing world.* [online] Available at: <https://www.pharmaceuticalprocessingworld.com/fda-approves-first-ever-inhaled-insulin-combination-product-for-treatment-of-diabetes/> [Accessed 28 January 2022].

Pliquett, U., & Weaver, J. C. (2007). Feasibility of an electrode-reservoir device for transdermal drug delivery by noninvasive skin electroporation. *IEEE Transactions on Bio-Medical Engineering, 54*(3), 536–538. https://doi.org/10.1109/TBME.2006.886828

Prausnitz, M. (1996). The effects of electric current applied to skin: A review for transdermal drug delivery. *Advanced Drug Delivery Reviews, 18*(3), 395–425. https://doi.org/10.1016/0169-409x(95)00081-h

Prausnitz, M., & Langer, R. (2008). Transdermal drug delivery. *Nature Biotechnology, 26*(11), 1261–1268. https://doi.org/10.1038/nbt.1504

Qaid, M., & Abdelrahman, M. (2016). Role of insulin and other related hormones in energy metabolism—a review. *Cogent Food & Agriculture, 2*, 1. https://doi.org/10.1080/23311932.2016.1267691

Qiu, Y., Qin, G., Zhang, S., Wu, Y., Xu, B., & Gao, Y. (2012). Novel lyophilized hydrogel patches for convenient and effective administration of microneedle-mediated insulin delivery. *International Journal of Pharmaceutics, 437*(1–2), 51–56. https://doi.org/10.1016/j.ijpharm.2012.07.035

Quianzon, C., & Cheikh, I. (2022). History of insulin. Retrieved 28 January 2022, from.

Rai, S., Pandey, V., & Rai, G. (2017). Transfersomes as versatile and flexible nano-vesicular carriers in skin cancer therapy: The state of the art. *Nano Reviews & Experiments, 8*(1), 1325708. https://doi.org/10.1080/20022727.2017.1325708

Rambaran, R. N., & Serpell, L. C. (2008). Amyloid fibrils: Abnormal protein assembly. *Prion, 2*(3), 112–117. https://doi.org/10.4161/pri.2.3.7488

Rastogi, R., Anand, S., Dinda, A., & Koul, V. (2010). Investigation on the synergistic effect of a combination of chemical enhancers and modulated iontophoresis for transdermal delivery of insulin. *Drug Development and Industrial Pharmacy, 36*(8), 993–1004. https://doi.org/10.3109/03639041003682012

Rastogi, S., & Singh, J. (2005). Effect of chemical penetration enhancer and iontophoresis on the in vitro percutaneous absorption enhancement of insulin through porcine epidermis. *Pharmaceutical Development and Technology, 10*(1), 97–104. https://doi.org/10.1081/pdt-200049679

Reiber, G., Lipsky, B., & Gibbons, G. (1998). The burden of diabetic foot ulcers. *The American Journal of Surgery, 176*(2), 5S–10S. https://doi.org/10.1016/s0002-9610(98)00181-0

Roach, P. (2008). New insulin analogues and routes of delivery. *Clinical Pharmacokinetics, 47*(9), 595–610. https://doi.org/10.2165/00003088-200847090-00003

Sakloetsakun, D., Dünnhaupt, S., Barthelmes, J., Perera, G., & Bernkop-Schnürch, A. (2013). Combining two technologies: Multifunctional polymers and self-nanoemulsifying drug delivery system (SNEDDS) for oral insulin administration. *International Journal of Biological Macromolecules, 61*, 363–372. https://doi.org/10.1016/j.ijbiomac.2013.08.002

Sapra, A., & Bhandari, P. (2022). Diabetes Mellitus. Ncbi.nlm.nih.gov. Retrieved 2 July 2022, from https://www.ncbi.nlm.nih.gov/books/NBK551501/.

Schlicher, R., Radhakrishna, H., Tolentino, T., Apkarian, R., Zarnitsyn, V., & Prausnitz, M. (2006). Mechanism of intracellular delivery by acoustic cavitation. *Ultrasound In Medicine & Biology, 32*(6), 915–924. https://doi.org/10.1016/j.ultrasmedbio.2006.02.1416

Scott, E. (1912). On the influence of intravenous injections of an extract of the pancreas on experimental pancreatic diabetes. *American Journal Of Physiology-Legacy Content, 29*(3), 306–310. https://doi.org/10.1152/ajplegacy.1912.29.3.306

Shah, R., Shah, V., Patel, M., & Maahs, D. (2016). Insulin delivery methods: Past, present and future. *International Journal Of Pharmaceutical Investigation, 6*(1), 1. https://doi.org/10.4103/2230-973x.176456

Shilo, M., Berenstein, P., Dreifuss, T., Nash, Y., Goldsmith, G., Kazimirsky, G., et al. (2015). Insulin-coated gold nanoparticles as a new concept for personalized and adjustable glucose regulation. *Nanoscale, 7*(48), 20489–20496. https://doi.org/10.1039/c5nr04881h

Thevis, M., Thomas, A., & Schänzer, W. (2009). Insulin. *Handbook of Experimental Pharmacology*, (195), 209–226. https://doi.org/10.1007/978-3-540-79088-4_10

Tsadik, A. G., Atey, T. M., Nedi, T., Fantahun, B., & Feyissa, M. (2018). Effect of insulin-induced lipodystrophy on glycemic control among children and adolescents with diabetes in Tikur Anbessa Specialized Hospital, Addis Ababa, Ethiopia. *Journal of Diabetes Research, 2018*, 4910962. https://doi.org/10.1155/2018/4910962

van Munster, H., PM van de Sande, C., Voorhoeve, P., & van Alfen-van der Velden, J. (2014). Dermatological complications of insulin therapy in children with type 1 diabetes. *European Diabetes Nursing, 11*(3), 79–84. https://doi.org/10.1002/edn.255

Weiss, M., Steiner, D. F., & Philipson, L. H. Insulin Biosynthesis, Secretion, Structure, and Structure-Activity Relationships. [Updated 2014 Feb 1]. In: Feingold KR, Anawalt B, Boyce A, et al., editors. Endotext [Internet]. South Dartmouth (MA): MDText.com, Inc.; 2000. Available from: https://www.ncbi.nlm.nih.gov/books/NBK279029/

Wong, C., Al-Salami, H., & Dass, C. (2018). Microparticles, microcapsules and microspheres: A review of recent developments and prospects for oral delivery of insulin. *International Journal of Pharmaceutics, 537*(1–2), 223–244. https://doi.org/10.1016/j.ijpharm.2017.12.036

Wong, C., Martinez, J., & Dass, C. (2016). Oral delivery of insulin for treatment of diabetes: Status quo, challenges and opportunities. *Journal of Pharmacy and Pharmacology, 68*(9), 1093–1108. https://doi.org/10.1111/jphp.12607

Wu, Y., Gao, Y., Qin, G., Zhang, S., Qiu, Y., Li, F., & Xu, B. (2010). Sustained release of insulin through skin by intradermal microdelivery system. *Biomedical Microdevices, 12*(4), 665–671. https://doi.org/10.1007/s10544-010-9419-0

Yamamoto, A., Taniguchi, T., Rikyuu, K., Tsuji, T., Fujita, T., Murakami, M., & Muranishi, S. (2022). Retrieved 2 July 2022, from

Yaturu, S. (2013). Insulin therapies: Current and future trends at dawn. *World Journal of Diabetes, 4*(1), 1. https://doi.org/10.4239/wjd.v4.i1.1

Yerramsetty, K., Rachakonda, V., Neely, B., Madihally, S., & Gasem, K. (2010). Effect of different enhancers on the transdermal permeation of insulin analog. *International Journal of Pharmaceutics, 398*(1–2), 83–92. https://doi.org/10.1016/j.ijpharm.2010.07.029

Yu, F., Li, Y., Liu, C., Chen, Q., Wang, G., Guo, W., & et al. (2022). *Enteric-coated capsules filled with mono-disperse micro-particles containing PLGA-lipid-PEG nanoparticles for oral delivery of insulin*. Retrieved 2 July 2022, from.

Zhang, T., Tang, J., Fei, X., Li, Y., Song, Y., Qian, Z., & Peng, Q. (2021). Can nanoparticles and nano–protein interactions bring a bright future for insulin delivery? *Acta Pharmaceutica Sinica B, 11*(3), 651–667. https://doi.org/10.1016/j.apsb.2020.08.016

Diabetes Mellitus and iPSC-Based Therapy

Dibyashree Chhetri, Rajesh Nanda Amarnath, Sunita Samal, Kanagaraj Palaniyandi, and Dhanavathy Gnanasampanthapandian

Abstract

At present, the number of diabetes mellitus (DM) patients has exceeded 537 million worldwide, and this number continues to increase. Type 1 diabetes (T1D) is caused by the autoimmune destruction of β-cells, whereas type 2 diabetes (T2D) is caused by a hostile metabolic environment that leads to β-cell exhaustion and dysfunction. The prevalence of DM type 1 and type 2 is widespread and results in fatality without treatment. As of now, first-line medications for diabetes address hyperglycemia and insulin resistance. Owing to this, there is a need to develop advanced therapies that can either protect or replace lost β-cells with stem-cell-derived β-like cells or engineered islet-like clusters. Recently, stem cells (SCs) have been considered to be renewable cell sources in the treatment of diabetes and the development of insulin-producing cells. Clinicians used adult mesenchymal stem cells (Lavinsky et al., Proceedings of the National Academy of Sciences of the United States of America 95:2920–5, 1998) and embryonic stem cells (ESC) to improve patients' condition; however, they faced ethical challenges and the danger of developing tumors in patients. Induced pluripotent stem cell (iPSC)-based genetically modified stem cell therapy is the way to fix these problems, together with diabetes medicines available now. Patient- or disease-specific cell lines are permitted under iPSC. An improved iPSC or SC-islet differentiation

D. Chhetri · K. Palaniyandi · D. Gnanasampanthapandian (✉)
Cancer Science Laboratory, Department of Biotechnology, School of Bioengineering, SRM Institute of Science and Technology, Chengalpattu, India

R. N. Amarnath
Department of Obstetrics and Gynecology, Apollo Womens Hospital, Chennai, India

S. Samal
Department of Medical Gastroenterology, SRM Medical College Hospital and Research Centre, SRM Institute of Science and Technology, Chengalpattu, India

© The Author(s), under exclusive license to Springer Nature Singapore Pte Ltd. 2023
R. Noor (ed.), *Advances in Diabetes Research and Management*,
https://doi.org/10.1007/978-981-19-0027-3_10

225

generates cells that secrete insulin, which makes it possible for autologous diabetic cell replacement and disease modeling in vitro. As SC-islets are limitless, they can be used for individualized therapy and to overcome the drawbacks of donor islets. Therefore, this chapter focuses on the potential of islet engineering, diabetes cells, and their therapeutic potential among the iPSC-derived progenitors. These studies on stem cells and regenerative medicine may result in new treatments for diabetes.

Keywords

Diabetes mellitus · β-cell · Stem cells · iPSC · SC-islets · Mesenchymal stem cells · Embryonic stem cells (ESC) · Regenerative medicine

1   Introduction

Diabetes mellitus (DM) is a chronic condition that affects people's lifestyles and general well-being worldwide (Lavinsky et al., 1998). DM affected more than 422 million people in low- and middle-income countries (Diabetes, 2021). It was one of the top ten adult causes of death in 2017, causing four million fatalities globally (Saeedi et al., 2019). In patients with this condition, the pancreas stops producing insulin. Without insulin, glucose cannot enter the cells to fuel every cell in the body. Insulin enables glucose to enter the cells and reduces blood glucose levels (Wilcox, 2005). It is believed that elevated blood glucose is one of the primary causes of glycation end products, resulting in the production of reactive oxygen species (ROS) (Yao & Brownlee, 2010). Type 1 diabetes, which accounts for 10% of diabetes occurrences, is treated with exogenous insulin injections and cadaver's islet transplantation (Kharroubi & Darwish, 2015).

Blood glucose management is essential as chronic hyperglycemia has a number of negative effects on the body, including the mouth (Al-Maweri et al., 2013). As the sickness worsens, different pathological changes in the body are observed, such as nephropathy, retinopathy, and cardiovascular issues (Padhi et al., 2020). DM is the main factor in nontraumatic lower extremity amputations, adult-onset blindness, and end-stage renal disease (Hamano et al., 2014). In most severe cases, diabetic complications can lead to life-threatening conditions (Hamano et al., 2014). Long-term use of exogenous insulin is frequently associated with high and low blood sugar levels, which can lead to heart and renal issues. The cell treatment channel based on cadaver transplantation illustrates the significant capacity of transplantation technology to increase the longevity of diabetes patients (Hering et al., 2016). However, the number of cadaveric benefactors is constrained; therefore, immunosuppression is required (Zaharia et al., 2019). Pluripotent, self-renewing human stem cells are another interesting source of diabetic cell treatment. Type I DM (T1D) and type II DM (T2D) are the most prevalent subtypes (Chen et al., 2020).

Pancreatic cells are lost dramatically in T1D, which results in an autoimmune response. T2D (90–95%) occurs globally and affects older people typically (Saeedi et al., 2019). Resistance in tissues targeting insulin and gradationally developing β-cell dysfunction are the two pathophysiological pathways associated with T2D (Saeedi et al., 2019). Although many T2D patients require insulin therapy during the course of the disease, they often do not require it in the early stages of the illness (American Diabetes, 2021). Another form of DM, known as monogenic diabetes (MD), is caused by mutations or abnormalities in a single gene that is essential for pancreatic β-cell activity (Hattersley & Patel, 2017). The two types of MD are neonatal diabetes (ND) in infants and maturity-onset diabetes of the young (MODY), which is identified primarily in the early stages of adulthood (Hattersley & Patel, 2017).

A significant factor in the etiology of various types of DM is the genetic component. However, until recently, studying human phenotypes was challenging because there were insufficient human models that reproduced pancreatic development accurately. The creation of innovative, tailored therapeutics is made possible by recent advancements in human pluripotent stem cell (hPSC) technology, which also provides new techniques to understand the flawed phenotypes linked to each kind of diabetes (Abdelalim, 2021).

From 2006, endocrine progenitors (EPs), pancreatic progenitors (Smid et al., 2019), and definitive endoderm (DE) generated from stem cells were produced (Ma et al., 2022; Dominguez-Bendala et al., 2019). Recent progress in pluripotent stem-cell-based therapies was promising to treat DM (Zhou et al., 2022). Nonetheless, a severe benefactor deficit and the underpinning threat of tissue rejection make transplantation therapy problematic. The production of β-cells from human embryonic stem cells (hESCs) and human pluripotent stem cells (hPSCs) is one implicit means of addressing the benefactor deficit of DM (de Klerk & Hebrok, 2021; Yu et al., 2007). To develop regenerative treatments for T1D and conduct introductory pancreatic research, the guided cellular differentiation of pancreatic cells has been explored laboriously (D'Amour et al., 2006; Jiang et al., 2007; Pagliuca et al., 2014).

The use of hESC-derived pancreatic progenitors has already been launched in clinical studies (phases I and II) for T1D cases in the USA (de Klerk & Hebrok, 2021). Recently, a clinical study that involves human-induced pluripotent stem cell (hiPSC)-derived pancreatic cell transplant has been completed (Sequiera et al., 2022). Therefore, hiPSC has a potential advantage over hESC, hiPSC-derived pancreatic cells have been used for transplantation and have lower immunogenic and less ethical issues (Zheng, 2016). With their infinite self-renewal capacity and the ability to develop into all cell types that make up an adult organism, iPSCs and ESCs have similar functions (Takahashi et al., 2007), which could provide a potential alternative source of comparable patient-specific material. Multiple research projects have shown that it is feasible to produce human iPSCs from various illnesses, including metabolic diseases (such as T1D and T2D) (Takahashi et al., 2007; Kudva et al., 2012) and five types of MODY (Teo et al., 2013).

Although some of these studies documented the differentiation of hiPSCs into cells that produce insulin, perfecting this process further appears to be a requirement (Gotthardt, 2011; Murakami et al., 2021). Recent research breakthroughs on pancreas rejuvenation, complaint modeling utilizing pluripotent stem cells, and an outline of prospects for the therapeutic management of DM have also been discussed previously (Murakami et al., 2021).

2   Disease Overview

DM is a disease that affects most of the adolescent population. In the year 1755, Dobson was the first person to show that diabetic urine included sugar. In 1989, Von Mering and Minkowski discovered that a pancreatectomized dosage has the potential to cause diabetes in addition to the gastrointestinal issues of DM (Karamanou et al., 2016). The severity of DM has increased over the past two decades, making it an illness as well as a worldwide health crisis (Diabetes, 2021). It is a biochemical condition in which a person's blood sugar levels remain elevated despite their best efforts. Blood glucose, which originates from the food that a person consumes, is the primary source of energy that is produced by people.

Insulin is produced in the pancreas, and it is this hormone that enables glucose from meals to enter cells and be used for energy. Typically, humans have blood glucose levels ranging from 70 to 90 mg per 100 mL when healthy. A condition known as hyperglycemia can be described as an abnormally rapid absorption of sugar. Hypoglycemia happens when a person's blood sugar levels drop below the normal range (Kahn & Lecture, 1994).

2.1   Diabetes Insipidus

Diabetes insipidus is characterized by the progression of symptoms over time, including polyuria and polydipsia. Diabetes insipidus is an extremely rare disorder that does not affect how much glucose is in the blood but produces excessive urine, similar to how diabetes mellitus causes excessive thirst. Hyperglycemia can be caused by either a decrease in the body's ability to produce insulin or an increase in insulin resistance or both (Fenske & Allolio, 2012; Al-Maskari et al., 2011; Indurkar et al., 2016).

The term "diabetes mellitus" comes from the Greek terms "diabetes," which means "to pass through," and "mellitus," which means "sweet." Together, these words refer to the condition known as diabetes mellitus. Between 250 and 300 BC, Apollonius of Memphis coined the term "diabetes," as per historical records. When ancient Greek, Indian, and Egyptian civilizations observed the characteristics of urine in patients with this condition, the term "diabetes mellitus" was born. In the year 1889, Mering and Minkowski identified the pancreas as a key player in the development of diabetes. In 1922, at the University of Toronto, three researchers, named Banting, Best, and Collip, isolated the hormone insulin from the bovine pancreas. As a result of this discovery, an effective treatment for DM was launched in the same year (Sapra, 2022).

2.2   Diabetes Mellitus: Type I (T1D)

T1D damages pancreatic β-cells as a result of autoimmunity or infectious agents. This type of diabetes is caused by the loss of pancreatic β-cells involved in the production of insulin (Yoon & Jun, 2005).

2.3   Diabetes Mellitus: Type II (T2D)

Owing to the combination of failing pancreatic β-cells and insulin resistance, T2D is considered a disorder wherein the delivery of glucose becomes unpredictable over time progressively (Blair, 2016). T2D has now been documented in every country on the planet, with Asia accounting for 60% of all DM occurrences (Hu, 2011). When a person has this illness, their blood glucose levels can become 6–20 times higher than usual, which can cause a change in their state or a loss of function (Wareham & O'Rahilly, 1998).

Consequently, people all around the world struggle with conditions known as prediabetes and gestational diabetes. If a person's blood sugar is above normal but not high enough to be diagnosed with T2D, it is called prediabetes. Gestational diabetes causes high blood sugar levels due to the production of insulin-blocking hormones in the placenta; thereby, the likelihood of developing T2D is increased (Hu, 2011).

Obesity and DM are the two major problems that have emerged among the population of the United States. These issues affect people of various ages, ethnicities, educational levels, and degrees of smoking, and they affect both genders. Obesity is the root cause of a wide variety of other health issues (Hu, 2011). According to the findings of research carried out by Hu (Hu, 2011), obesity and DM are among the leading contributors to morbidity and mortality in the USA.

Adults with obesity and weight gain have an increased risk for DM. When a person has T2D, the function of pancreatic islets notably worsens during the course of the disease. Although age, obesity, not eating enough calories, alcohol, and tobacco consumption are all independent risk factors, the probability of developing the disease is increased for each of these risk factors (Galicia-Garcia et al., 2020).

In addition to the above factors, increased and abnormal sex hormone production, lack of exercise, and gene-related factors can all contribute to the development of DM (Galicia-Garcia et al., 2020). According to Galicia et al. (Galicia-Garcia et al., 2020), the etiology of T1D and T2D is impacted significantly by various factors, including but not limited to age, obesity, insufficient calorie consumption, smoking, and alcohol usage.

3   Disease Pathogenesis

The current consensus is that the surroundings we live in, food, chemicals, mental health, and genetic damage (including the body's response against β-cells) induce T1D in susceptible individuals. When immune tolerance fails, autoantibody-producing β-cells, self-activated CD4 and CD8 T-cells, and the innate immune system are all activated, which stimulates the destruction of β-cells (Atkinson, 2012; Reed & Herold, 2015). Rodent models, similar to nonobese diabetic (NOD) rats, have provided most of our understanding with regard to the etiology of the disease (Krogvold et al., 2016). However, updates from a diabetic virus detection (DiViD) study and the Juvenile Diabetes Research Foundation (JDRF) Network for Pancreatic Organ Donors with Diabetes (nPOD) have pointed out the major differences in mice and human disease profiles and provided important details on the pathology of human disease (Campbell-Thompson et al., 2016; Foulis et al., 1986). The hallmark of DM is the histological tissue and cell inflammation in and around islets (Willcox et al., 2009; In't Veld, 2011). Although it varies between individuals, CD8T, CD68, CD4T, and CD20B cells are inflamed mostly (In't Veld, 2011; van Belle et al., 2011) (Fig. 1).

3.1   Disease Pathogenesis

There are many assumptions about the circumstances that trigger actions that lead to T1D ultimately. Infectious events, such as a viral disease, cause a pioneering β-cell death, which releases autologous β-cell antigens (Coppieters et al., 2012). In addition, β-cells with excessive MHC class I expression have been observed in T1D tissue samples, which might make them more vulnerable to assault by cytotoxic CD8 T-cells, which are tone-reactive and cause additional antigen release (Carrero et al., 2016). Antigen-presenting cells within the islets attack and destroy the islet autoantigen; transport it to the lymph nodes, which drain the pancreas; and present it to the CD4 and CD8 T-cells (Silveira & Grey, 2006). These autoantigens are detected by autologous CD4 T lymphocytes, which results in their activation and replication after the loss of tolerance in both the central and peripheral regions. Autoantibodies against islet proteins are produced as a result of β-cell activation, which also results in the development of plasma cells (DiMeglio et al., 2018).

After their invasion of the islets, these immune cells cause insulitis and kill the β-cells eventually (Padgett et al., 2013; Coomans de Brachene et al., 2018). Inflammatory cytokines (IFNγ, IL1 and 6, TNFα, etc.) are generated when an autoimmune assault begins, thereby amplifying the immune response. Some of these cytokines may cause β-cell death directly by inhibiting the ability of β-cells to secrete insulin and activate cytotoxic T lymphocytes, in accordance with previous research (Coomans de Brachene et al., 2018). In addition, cytokines increase the human leukocyte antigen (HLA) class 1 expression of the β-cells (Padgett et al., 2013). Both high levels of nitric oxide and superoxide radical production worsen β-cell damage (Coomans de Brachene et al., 2018). Unavoidably, this damage triggers more β-cell

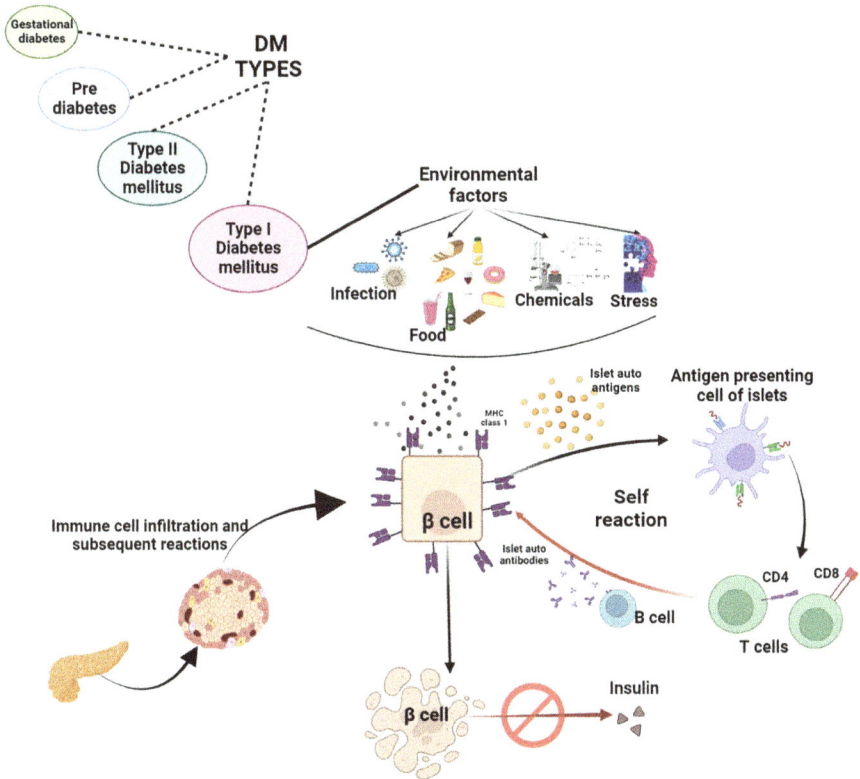

Fig. 1 Pathogenesis of diabetes mellitus among the four types of diabetes (type 1 diabetes mellitus, type 2 diabetes mellitus, prediabetes, and gestational diabetes). Type 1 diabetes mellitus pathology is autoimmune, where various factors (such as infection, food, chemicals, and stress) regulate self-reaction within the immune components and pancreatic β-cells. Consequently, autoantigens pertaining to pancreatic islet cells are produced, which target and destruct antibodies targeting β-cells of the islets, thereby leading to a decrease in the production of insulin

antigen release, which might act as a feedback loop to further the continuing β-cell apoptosis.

4 Prevalence

The International Diabetes Federation (IDF) estimated that there were 285 million diabetics worldwide (T1D and T2D combined) in 2009 (IDF Diabetes Atlas, 2009), with additional growth of 366 million in 2011 (IDF Diabetes Atlas, 2009) and 382 million in 2013 (IDF Diabetes Atlas, 2011). In 2015, 425 million people had DM compared with 415 million (IDF Diabetes Atlas, 2015) in 2017 and 463 million in 2019. By 2045, there would be 537 million individuals worldwide with DM (IDF Diabetes Atlas, 2021; Shaw et al., 2010).

According to research based on this demographic, the prevalence of diabetes is 1.5 times higher among people with intellectual disabilities (ID) than in the general population. In comparison with the general population, adolescents and females with ID have a higher prevalence of DM. Patients with ID who also have DM problems may have delays in diagnosis or in receiving treatment (Wang et al., 2021). The global projections of the prevalence of DM for the years 2010 and 2030 also made the prediction on the basis of a larger body of research, which suggests an increasing burden of DM, particularly in emerging nations (Cuypers et al., 2021).

Cell loss, inadequate insulin synthesis, and recurrent high blood sugar have been identified as the causes of T1D. Despite this fact, exogenous supplementation is a life-saving treatment, but maintaining normal blood sugar remains difficult. Tissue grafting in the form of islet cells or the entire pancreas can also treat T1D (Shapiro et al., 2006).

5 Stem Cell and Diabetes Mellitus

Of all the DM types, MD is a condition caused by a rare abnormality in the gene-specific to pancreas development (Zhang et al., 2022). As opposed to the standard insulin injection treatment, more recent remedial strategies focus on restoring endogenous insulin product. The cell treatment option that is used at present in T1D therapy is allogeneic islet transplantation. The requirement of an appropriate count of functioning cadaveric islets to achieve normal glucose levels, however, places a constraint on this method.

In addition, its use in children with T1D and the early stages of the disease has been hampered by the need for immunosuppressants over the long term. Another unidentified side effect is the isolation of the islets' ability to survive, grow, and function after transplantation (Latres et al., 2019). As a result, several β-cell symptom alleviation approaches are being considered for therapeutic use (Abdelalim et al., 2014).

For cell therapy and disease modeling, hPSCs are an invaluable resource, and they may be divided into two groups: (1) hiPSCs, which are produced by physical cell reprogramming, and (2) hESCs, obtained from the embryo's inner cell mass. Human pluripotent stem cells (hPSCs) can be differentiated and cultured into any cell type to provide a continuous supply of the required cell type (Zaharia et al., 2019; Takahashi et al., 2007). A possible option for cell treatment is hPSC-derived islet cell types, which have the benefits of being customizable and scalable.

Ten years ago, human cells were used for the first time to create iPSCs (Zaharia et al., 2019). The stem cell field has been redefined completely as a result of this technology, and it became possible to examine individual cases in vitro to investigate diseases and test potential treatments (Yamanaka, 2012). Two groups disclosed techniques in 2014 for creating iPSC-derived β-cells that respond to glucose (Pagliuca et al., 2014; Saarimaki-Vire et al., 2017). Up to this point, iPSC-derived β-cells have been employed extensively to examine the pathogenic processes underlying many MD, including instances of neonatal DM (Balboa et al., 2018), young-onset mature-onset DM, Wolfram syndrome (WS), and tRNA methyltransferase 10 homolog A (TRMT10A) deficiency (Shang et al., 2014). In addition, there have been attempts to generate stem-cell-derived β-cells from T1D (Pagliuca et al., 2014) patients or from a fulminant type of the severe insulin-dependent DM that is now prevalent in Japan (Hosokawa et al., 2017).

With regard to the β-cells produced due to the vulnerability of iPSCs to pro-inflammatory cytokines (IL-1, IFNγ, and IFN), these cells have not yet been confirmed fully as a model for studying the mediators of β-cell death in T1D. These cytokines induce endoplasmic reticulum (ER) stress (Brozzi et al., 2015), upregulate HLA class I (HLA-ABC) (Marroqui et al., 2017), produce chemokines, and cause apoptosis in β-cells, which contribute to β-cell malfunction and death in T1D (Takahashi et al., 2007).

6   Types of Stem Cell Therapy for Diabetes

T1D has been controlled predominantly through the transplantation process, which aids in retaining patients' pancreatic functions (Wisel et al., 2016). It is also considered to be a safe method, preventing any sort of trauma or risk to patients (Xiao et al., 2016).

Despite its lesser risk, the T1D transplantation method has drawbacks and a number of major obstacles, including donor islet deficiency, implicated resistance to the drug and immunosuppressive agents, and diminished graft function (Qi et al., 2010; Shapiro et al., 2000). According to the Edmonton Protocol, a conventional islet transplant necessitates at least two donors. In summary, this massive application is now impractical. Therefore, it is crucial to develop novel and efficient treatments for DM and its common consequences. The previous findings motivate attempts to create functional β-cells by differentiating stem cells or triggering endogenous rejuvenation (Zhong & Jiang, 2019).

7   β-Cells and Their Function

The endocrine region in the pancreas comprises isles of Langerhans and ovoid cells that are dispersed throughout the organ. Through the coordinated release of hormones, such as glucagon, insulin, and somatostatin, the islets contribute to maintaining blood glucose homeostasis. As commonly believed, islet cells contribute to the etiology of DM. Cells make and secrete insulin, a hormone important in controlling plasma glucose levels (Ilegems et al., 2013).

Therefore, low insulin levels cause hyperglycemia and DM due to the loss of cell mass or function (Chen et al., 2017). In T1D, the function of β-cells decreases progressively over time in prediabetes, and the β-cell number begins to decline rapidly before diabetes onset. Despite this, insulin intake can resume cellular function temporarily, prolong autoimmunity, and increase cell load, which relapses cellular fatigue and cell death inevitably. In metabolic-pathway-mediated T2D disease, increased insulin requirements in the early stages of insulin resistance led to the regulation of cellular function and cell mass. However, if the cells' work is overloaded, differentiation, dysfunction, and even cell death follow (Takahashi et al., 2007; Eizirik et al., 2020).

7.1   iPSC

The discovery of iPSCs marks a breakthrough in their role in science and regenerative medicine (Yamanaka, 2012). The formation of iPSCs relies on the alteration of four major genes (*OCT4*, *KLF4*, *SOX2*, and *c-MYC*), which led to the transformation of human or mouse fibroblasts into stem cells with properties similar to ESCs (Takahashi et al., 2007). By using these transformed cells, β-type pancreatic cells have been cultured in vitro with the ability to secrete insulin and C-peptide, thereby terminating the need for a donor with diabetes (Maehr et al., 2009). In 2009, iPSCs that are capable of differentiating into the checkpoint inhibitors (CPI) of T1D patients were using the same gene mutation approach. However, the lack of function in vivo and in vitro and the inability to express β-cell genes properly were major setbacks (Pagliuca et al., 2014).

In addition, the origin of iPSCs was reported to influence β-cell differentiation. Jeon and associates (Jeon et al., 2012) evaluated embryonic fibroblasts and pancreatic epithelial cells (PEs) from nonobese diabetic (NOD) mice, wherein β-cells from PEs were reported to be more expressive to pancreatic-specific genes, thereby normalizing the blood glucose level when transplanted.

Another group of researchers (Pagliuca et al., 2014) reported on the transformation of human pluripotent stem cells into human pancreatic β-cells. Not only these insulin-producing cells (IPCs) from nondiabetic (ND) donors were able to remain monohormonal, but they were also able to produce insulin in response to glucose in vivo. Furthermore, a recent approach to generating CPI from T1D patients was investigated, wherein both types of iPSCs displayed identical cell markers for β-cells (Millman et al., 2016). In a similar manner, this method of self-IPC collection for the treatment of diabetic patients aids in problems relating to graft rejection (Fig. 2).

Induced pluripotent stem cells are similar to pluripotent stem cells (PSCs) as both these cells originate from somatic cells. Owing to ethical standards, human ESCs are limited in research and clinical settings; therefore, it is important to identify viable alternatives. Recent research has focused on iPSCs as an alternative to hESCs due to their ability to form cell lineages of different types (Yamanaka, 2012; Lopez-Yrigoyen et al., 2019; Minagawa et al., 2018). By using transcription factors, somatic cells can be transformed into iPSCs. The implementation of this approach has led to the generation of T1D-specific iPSCs from T1D patients (Huangfu et al., 2008; Okita et al., 2008).

KlF4 and c-Myc are carcinogenic, which suggests that iPSCs have the potential to develop into malignancies. Therefore, a new approach implementing only OCT4 and SoX2 with a histone deacetylase (valproic acid) inhibitor has been designed (Okita et al., 2008). In addition, the use of retroviruses or lentiviruses to introduce

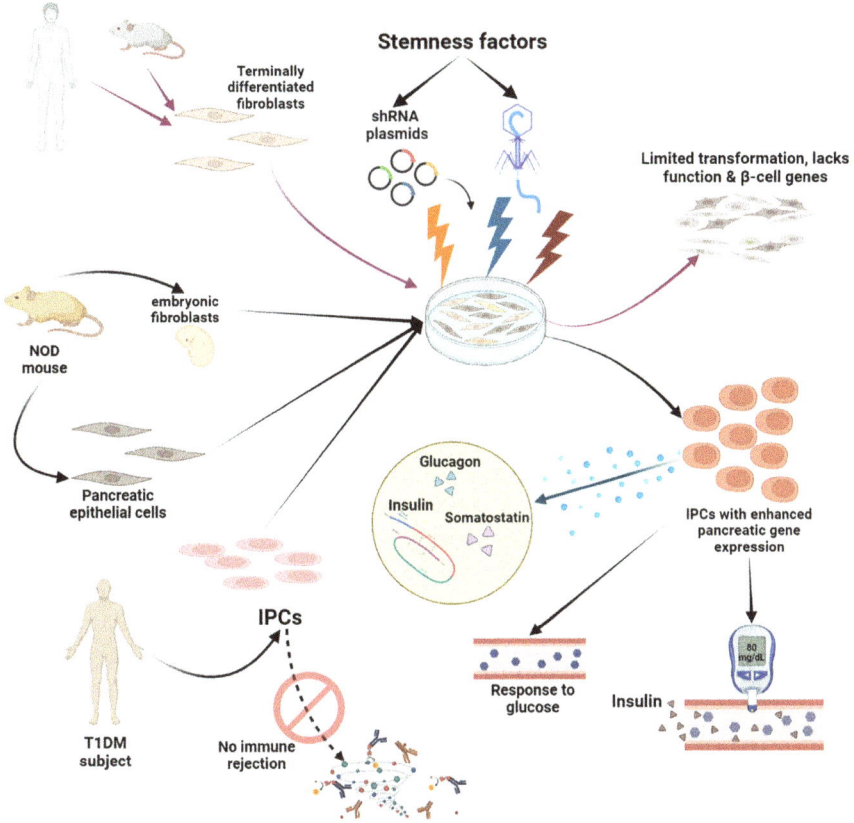

Fig. 2 Diabetes mellitus and iPSC: treatment. The treatment of type 1 diabetes mellitus by using induced pluripotent stem cells that are derived from embryonic fibroblasts (obtained through the reexpression of stemness markers) has great advantages after their transplant, such as optimum insulin production; production of all the necessary hormones, namely, insulin (especially C-peptide), glucagon, and somatostatin in a feedback manner; and proper response to blood glucose. Moreover, iPSC from the same donor has no complications with compatibility or immune rejections. On the other hand, terminally differentiated fibroblast cells do not transform into pancreatic β-cells completely

transcription factors may affect the integration of the virus into the host genome, increasing the risk of tumorigenesis. By troubleshooting this risk, an infection strategy has been adopted, which leads to plasmid expression in iPSCs that lack viral integration (Kroon et al., 2008).

Despite the availability of these important advances, researchers remain puzzled by the similarities between iPSCs and hESCs. In addition, tremendous progress in downregulating key signaling networks and regulators that determine cell destiny has been observed. These efforts have resulted in the creation of glucose-sensitive and insulin-generating pancreatic ancestral cells that were dispersed previously throughout the mouse pancreas (Giorgetti et al., 2009). In addition to keratinocytes and fibroblasts, other sources have been utilized for the generation of iPSCs, which have a reprogramming efficiency similar to keratinocytes and fibroblasts (Takahashi et al., 2007). The production of iPSCs from several organs has an increasing interest in the development of insulin-producing cells. Nevertheless, the production of insulin-producing iPSCs remains a major obstacle in DM. During the early stages of T1D development, insulin-producing cells are destroyed. A study reported the formation of islet-like cell masses from iPSCs that secrete insulin, and some cells were able to grow and produce C-peptide when stimulated with glucose (Enderami et al., 2017). These cells are more sensitive to glucose, thereby showing a high rate of apoptosis. A human-based study led to the formation of skin-derived iPSCs from the cells of T1D patients who were capable of producing insulin, C-peptide, gluca-gon, and somatostatin (Maehr et al., 2009).

The transfection of mature skin cells with OCT4, SOX2, and KlF4 was used to induce the iPSCs previously (Millman et al., 2016; Haque et al., 2019). However, the effective system will further target transplant cells produced from iPSCs derived from T1D patient somatic cells. Consequently, the utilization of both immune cells and cells produced from iPSCs for the treatment of T1D may aid in preventing autoimmunity and provide optimum insulin-generating cells ultimately (D'Amour et al., 2006).

8   Direct Differentiation: Pancreatic β-Cells

To promote the discreteness of iPSCs into pancreatic lineage cells, a technique was employed to reproduce and clone typical experimental phases of the pancreas in vitro by addressing the expression of critical transcription factors involved in pancreatic development. The fertilized egg will develop into insulin-expressing β-cells through different experimental stages, including terminal endoderm, primi-tive intestinal tract, anterior posterior portion, pancreatic endoderm, and endocrine precursors (Pagliuca et al., 2014).

By using similar approaches for hESC/iPSC isolation, essential signaling pathways were activated or inhibited (Jiang et al., 2007; Pagliuca et al., 2014; Maehr et al., 2009; Kroon et al., 2008; Toyoda et al., 2015). Moreover, researchers have been creating β-type pancreatic cells that produce and release insulin without using stimulants, such as KCl (Jiang et al., 2007).

The drawback of these cells is that unlike mature β-cells, in response to changes in blood sugar, these cells are limited to insulin secretion. In addition, pancreatic β-cells express glucagon and somatostatin simultaneously. Furthermore, gene expression studies revealed that β-like cells derived from hESC/iPSCs resemble embryonic β-cells rather than adult β-cells.

Several experiments have documented the production of embryonic pancreatic endothelial cells generated from hESCs/iPSCs that are capable of differentiating into all pancreatic cell types. After transplantation (3–4 months) into immunocompromized mice, these cells in vivo developed into mature β-cells that are capable of glucose-stimulated insulin production (Pagliuca et al., 2014; Balboa et al., 2018).

8.1   Islets Derived from Stem Cells: Insights from NGS of iPSCs

Several iPSC studies have used next-generation sequencing (NGS) (Maxwell et al., 2020; Augsornworawat et al., 2020) and microarray (Hrvatin et al., 2014) strategies to probe into differently expressed genes in islet cell types that express other genes, similar to α-cells and β-cells. By using single-cell NGS, two researchers, Balboa et al. (Balboa et al., 2018) and Maxwell (Maxwell et al., 2020), observed an increase in the β-cell gene in SC-β-like cells and SC β-cells. Similarly, Wang et al. (Wang et al., 2019) connected regulated PDX1-binding genes negatively (such as MNX1, CES1, and MEG3) to a decrease in differentiation efficiency. Millman et al. (Millman et al., 2016) found a connection between TAP1 somatic cells. The gene is most often utilized for reprogramming somatic cells by using retrovirus, Sendai virus, and batch reprogramming (Takahashi et al., 2007; Kishore et al., 2020).

Sui and colleagues used somatic cell nuclear transfer to generate T1D NT-ESCs from skin fibroblasts (Ma et al., 2018; Sui et al., 2018). Patients' iPSCs were segregated into stem cells (SCs) and gonadal cells, and cell proliferation, differentiation efficiency, and insulin secretion function were evaluated. Once variations have been identified, they often manifest after the fate of the pancreatic cells has been defined. Cells were compared collectively with nondiabetic or transgenic isogenic controls. To repair these effects and treat diabetes, mutations are created through genetic engineering. T1D and T2D are the two most prevalent forms of DM, accounting for 10% and 90% of all occurrences of DM, respectively. Owing to risk variables (such as epigenetics, the environment, and lifestyle), it is challenging to comprehend the pathophysiology of these disorders. Despite major physiological differences, especially in islets, numerous animal models of T1D and T2D are available.

Consequently, stem cell technologies have been utilized in human cases to imitate T1D and T2D and get a deeper knowledge of the origin of the diseases. Previous probes utilized pancreatic and duodenal homeobox 1 (PDX1) NK6 homeobox 1 (NKX6–1) pancreatic progenitors (Smid et al., 2019), which are accountable for all cell fates in the pancreas. Glucose-6-phosphatase 2 (G6PC2), GLRA1, and insulin-like growth-factor-binding protein-like 1 (IGFBPL1) are upregulated in cells selectively, whereas G6PC2, glycine receptor alpha 1 (GLRA1), and IGFBPL1 are downregulated specifically. As cystic fibrosis-related diabetes (CFRD) and MD are less prevalent forms of diabetes, their primary patterns are few.

Therefore, iPSCs from patients are vital for elucidating the disease's pathogenesis and for developing therapeutics. The production of pancreatic ductal epithelial cells from iPSCs for cystic fibrosis has provided a foundation for in vitro drug screening (Sui et al., 2018). Approximately, two examples of T2D induced by a single gene mutation were evaluated by Harris et al. (Harris et al., 2018). iPSCs generated from MODY, neonatal diabetes, and WS patients make it possible to mimic DM in people. Recent publications have investigated mutations in some genes, each of which is related to a unique kind of DM. However, the *SUR1* mutation caused K^+ ATP channel inactivation, resulting in congenital hypersecretion (CHI) and probably MODY, which increases the number of primary-functioning SC-like cells (Lithovius et al., 2021). Transplants are recognized for estimating cell growth in vivo and determining whether cells can regulate already existing DM or prevent the development of diabetes in mice. In conclusion, human iPSC differentiation techniques permit the in vitro eradication of DM and other diseases.

9   In Vivo Cell Therapy: Animal Models' Influences

To produce an applicable cell treatment, in vivo exploration of animal models is needed for confirmation purposes. Maxwell et al. (Maxwell et al., 2020), in a mouse model, used SC-islets and SC β-cells independently for transplantation by inoculating cells into a renal capsule before or after the development of DM in rodent models (Wang et al., 2019). Rectal cases are often performed to confirm that transplanted cells are responsible for euglycemia in renal cyst transplants. Several researchers validated the post-transplantation conformation of SC-islets and SC - β-cells by using immunostaining. In addition to the kidney capsule, transplantation sites included the portal vein, intramuscular, and subcutaneous (Ilegems et al., 2013; Stock et al., 2020; Yoshihara et al., 2020).

Yoshihara et al. (2020) maintained glucose homeostasis for 50 days to overexpress PD-L1 in immunocompetent DM mice by using engineered human islet-like organoids. With combined efforts to test in live animals, it may be possible to develop cell therapy for DM (Fig. 3).

10   Conclusion

The clinical generation of differentiated ESCs and iPSCs has reached a considerable scale, and recent developments in macroencapsulation may permit the safe use of these teratogenic cells therapeutically. The subcutaneous implants are removed and deconstructed readily, whereas macroencapsulation prevents the encapsulated cells from adopting the body. The outcomes of Viacyte™ preliminary examination of its VC-01 pancreatic endothelium transplant will determine the viability of ESC- and iPSC-based DM treatments in the near future. The majority of data points to a decreased reliance on exogenous insulin and/or antidiabetic medications as a cause of DM. In this aspect, cervical spondylotic myelopathy (CSM) may be beneficial for DM patients particularly, which, similar to those with intermediate DM, have tremendous difficulties regulating their blood sugar levels with traditional medications.

In terms of fundamental research, stem-cell-based therapies for DM have advanced significantly over the past decade. Now that it is feasible to generate viable pancreatic tissue from hiPSCs, these cells might cease the use of human donors. In

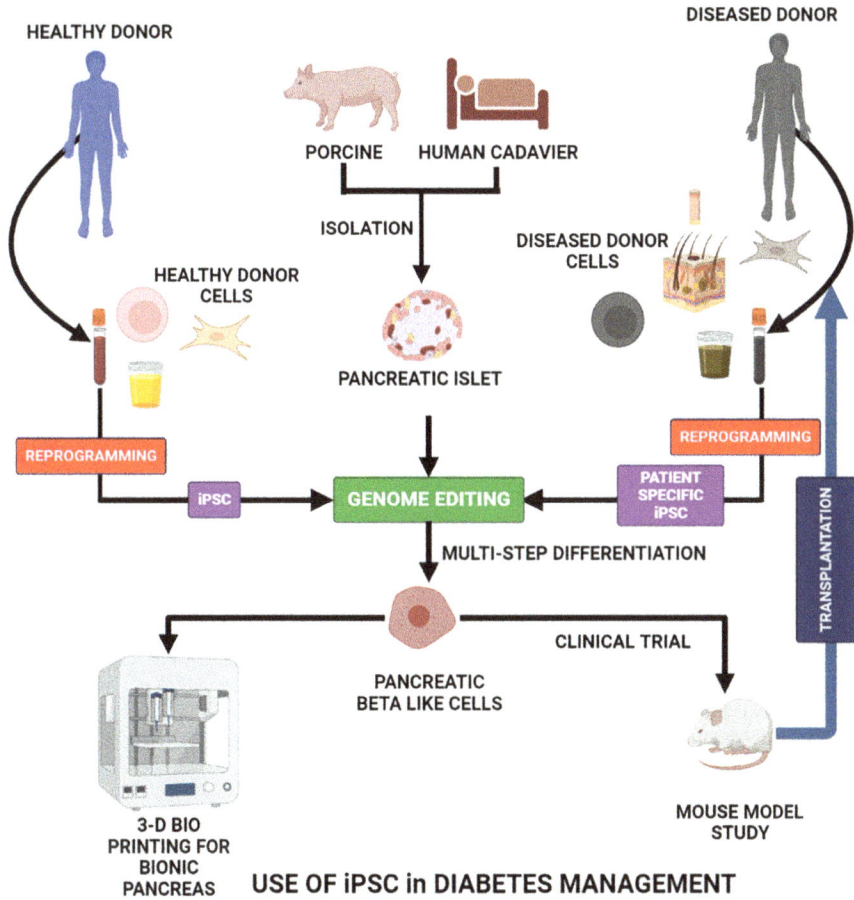

Fig. 3 Organ-derived or bioengineered cells for the treatment of diabetes mellitus

the ensuing decade, it is anticipated that the efficacy and safety of several experimental DM therapies will be anticipated. Induced pluripotent stem cell technology, which enables the in vitro production of various human cell types, has permitted the development of a fully humanized T1D model that replicates the pathophysiology of the human disease. Researchers are now able to develop the cellular target of autoimmunity in the setting of certain HLA haplotypes as a result of recent breakthroughs in procedures for producing iPSC-derived β-cells with mature functions. The successful generation of a T1D model based on human iPSCs will facilitate an explicit understanding of the pathophysiology of the illness and aid in the development of further β-cell preservation approaches. In addition to simulating various genuine disease processes, these models will provide an environment for testing case-specific therapies. Importantly, T1D stem cell models will pave the way for reducing our reliance on rodent disease models and facilitate the translation of preclinical therapy techniques into clinical trials.

Author Contribution

DC wrote the manuscript; RNA provided the clinical importance of the study. KP and DG edited the manuscript.

Acknowledgment We would like to thank Mr. M. Vijayaraman for their English language editing.

References

Abdelalim, E. M. (2021). Modeling different types of diabetes using human pluripotent stem cells. *Cellular and Molecular Life Sciences, 78*(6), 2459–2483. https://doi.org/10.1007/s00018-020-03710-9

Abdelalim, E. M., Bonnefond, A., Bennaceur-Griscelli, A., & Froguel, P. (2014). Pluripotent stem cells as a potential tool for disease modelling and cell therapy in diabetes. *Stem Cell Reviews and Reports, 10*(3), 327–337. https://doi.org/10.1007/s12015-014-9503-6

Al-Maskari, A. Y., Al-Maskari, M. Y., & Al-Sudairy, S. (2011). Oral manifestations and complications of Diabetes mellitus: A review. *Sultan Qaboos University Medical Journal, 11*(2), 179–186.

Al-Maweri, S. A., Ismail, N. M., Ismail, A. R., & Al-Ghashm, A. (2013). Prevalence of oral mucosal lesions in patients with type 2 diabetes attending hospital universiti sains Malaysia. *Malays J Med Sci, 20*(4), 39–46.

American Diabetes, A. (2021). 2. Classification and diagnosis of Diabetes: Standards of medical Care in Diabetes-2021. *Diabetes Care, 44*(Suppl 1), S15–S33. https://doi.org/10.2337/dc21-S002

Atkinson, M. A. (2012). The pathogenesis and natural history of type 1 diabetes. *Cold Spring Harb Perspect Med, 2*(11), a007641. https://doi.org/10.1101/cshperspect.a007641

Augsornworawat, P., Maxwell, K. G., Velazco-Cruz, L., & Millman, J. R. (2020). Single-cell transcriptome profiling reveals beta cell maturation in stem cell-derived islets after transplantation. *Cell Reports, 32*(8), 108067. https://doi.org/10.1016/j.celrep.2020.108067

Balboa, D., Saarimaki-Vire, J., Borshagovski, D., Survila, M., Lindholm, P., Galli, E., et al. (2018). Insulin mutations impair beta-cell development in a patient-derived iPSC model of neonatal diabetes. *eLife, 7*, e38519. https://doi.org/10.7554/eLife.38519

Blair, M. (2016). Diabetes mellitus review. *Urologic Nursing, 36*(1), 27–36.

Brozzi, F., Nardelli, T. R., Lopes, M., Millard, I., Barthson, J., Igoillo-Esteve, M., et al. (2015). Cytokines induce endoplasmic reticulum stress in human, rat and mouse beta cells via different mechanisms. *Diabetologia, 58*(10), 2307–2316. https://doi.org/10.1007/s00125-015-3669-6

Campbell-Thompson, M., Fu, A., Kaddis, J. S., Wasserfall, C., Schatz, D. A., Pugliese, A., et al. (2016). Insulitis and beta-cell mass in the natural history of type 1 Diabetes. *Diabetes, 65*(3), 719–731. https://doi.org/10.2337/db15-0779

Carrero, J. A., Ferris, S. T., & Unanue, E. R. (2016). Macrophages and dendritic cells in islets of Langerhans in diabetic autoimmunity: A lesson on cell interactions in a mini-organ. *Current Opinion in Immunology, 43*, 54–59. https://doi.org/10.1016/j.coi.2016.09.004

Chen, C., Cohrs, C. M., Stertmann, J., Bozsak, R., & Speier, S. (2017). Human beta cell mass and function in diabetes: Recent advances in knowledge and technologies to understand disease pathogenesis. *Mol Metab, 6*(9), 943–957. https://doi.org/10.1016/j.molmet.2017.06.019

Chen, S., Du, K., & Zou, C. (2020). Current progress in stem cell therapy for type 1 diabetes mellitus. *Stem Cell Research & Therapy, 11*(1), 275. https://doi.org/10.1186/s13287-020-01793-6

Coomans de Brachene, A., Dos Santos, R. S., Marroqui, L., Colli, M. L., Marselli, L., Mirmira, R. G., et al. (2018). IFN-alpha induces a preferential long-lasting expression of MHC class I in human pancreatic beta cells. *Diabetologia, 61*(3), 636–640. https://doi.org/10.1007/s00125-017-4536-4

Coppieters, K. T., Dotta, F., Amirian, N., Campbell, P. D., Kay, T. W., Atkinson, M. A., et al. (2012). Demonstration of islet-autoreactive CD8 T cells in insulitic lesions from recent onset and long-term type 1 diabetes patients. *The Journal of Experimental Medicine, 209*(1), 51–60. https://doi.org/10.1084/jem.20111187

Cuypers, M., Leijssen, M., Bakker-van Gijssel, E. J., Pouls, K. P. M., Mastebroek, M. M., Naaldenberg, J., et al. (2021). Patterns in the prevalence of diabetes and incidence of diabetic complications in people with and without an intellectual disability in Dutch primary care: Insights from a population-based data-linkage study. *Primary Care Diabetes, 15*(2), 372–377. https://doi.org/10.1016/j.pcd.2020.11.012

D'Amour, K. A., Bang, A. G., Eliazer, S., Kelly, O. G., Agulnick, A. D., Smart, N. G., et al. (2006). Production of pancreatic hormone-expressing endocrine cells from human embryonic stem cells. *Nature Biotechnology, 24*(11), 1392–1401. https://doi.org/10.1038/nbt1259

de Klerk, E., & Hebrok, M. (2021). Stem cell-based clinical trials for Diabetes mellitus. *Front Endocrinol (Lausanne), 12*, 631463. https://doi.org/10.3389/fendo.2021.631463

Diabetes. (2021). Available from: https://www.who.int/health-topics/diabetes#tab=tab_1.

DiMeglio, L. A., Evans-Molina, C., & Oram, R. A. (2018). Type 1 diabetes. *Lancet, 391*(10138), 2449–2462. https://doi.org/10.1016/S0140-6736(18)31320-5

Dominguez-Bendala, J., Qadir, M. M. F., & Pastori, R. L. (2019). Pancreatic progenitors: There and Back again. *Trends in Endocrinology and Metabolism, 30*(1), 4–11. https://doi.org/10.1016/j.tem.2018.10.002

Eizirik, D. L., Pasquali, L., & Cnop, M. (2020). Pancreatic beta-cells in type 1 and type 2 diabetes mellitus: Different pathways to failure. *Nature Reviews. Endocrinology, 16*(7), 349–362. https://doi.org/10.1038/s41574-020-0355-7

Enderami, S. E., Mortazavi, Y., Soleimani, M., Nadri, S., Biglari, A., & Mansour, R. N. (2017). Generation of insulin-producing cells from human-induced pluripotent stem cells using a stepwise differentiation protocol optimized with platelet-rich plasma. *Journal of Cellular Physiology, 232*(10), 2878–2886. https://doi.org/10.1002/jcp.25721

Fenske, W., & Allolio, B. (2012). Clinical review: Current state and future perspectives in the diagnosis of diabetes insipidus: A clinical review. *The Journal of Clinical Endocrinology and Metabolism, 97*(10), 3426–3437. https://doi.org/10.1210/jc.2012-1981

Foulis, A. K., Liddle, C. N., Farquharson, M. A., Richmond, J. A., & Weir, R. S. (1986). The histopathology of the pancreas in type 1 (insulin-dependent) diabetes mellitus: A 25-year review of deaths in patients under 20 years of age in the United Kingdom. *Diabetologia, 29*(5), 267–274. https://doi.org/10.1007/BF00452061

Galicia-Garcia, U., Benito-Vicente, A., Jebari, S., Larrea-Sebal, A., Siddiqi, H., Uribe, K. B., et al. (2020). Pathophysiology of type 2 Diabetes mellitus. *International Journal of Molecular Sciences, 21*(17), 6275. https://doi.org/10.3390/ijms21176275

Giorgetti, A., Montserrat, N., Aasen, T., Gonzalez, F., Rodriguez-Piza, I., Vassena, R., et al. (2009). Generation of induced pluripotent stem cells from human cord blood using OCT4 and SOX2. *Cell Stem Cell, 5*(4), 353–357. https://doi.org/10.1016/j.stem.2009.09.008

Gotthardt, M. (2011). A therapeutic insight in beta-cell imaging? *Diabetes, 60*(2), 381–382. https://doi.org/10.2337/db10-1591

Hamano, K., Nakadaira, I., Suzuki, J., & Gonai, M. (2014). N-terminal fragment of probrain natriuretic peptide is associated with diabetes microvascular complications in type 2 diabetes. *Vascular Health and Risk Management, 10*, 585–589. https://doi.org/10.2147/VHRM.S67753

Haque, M., Das, J. K., Xiong, X., & Song, J. (2019). Targeting stem cell-derived tissue-associated regulatory T cells for type 1 Diabetes immunotherapy. *Current Diabetes Reports, 19*(10), 89. https://doi.org/10.1007/s11892-019-1213-7

Harris, A. G., Letourneau, L. R., & Greeley, S. A. W. (2018). Monogenic diabetes: The impact of making the right diagnosis. *Current Opinion in Pediatrics, 30*(4), 558–567. https://doi.org/10.1097/MOP.0000000000000643

Hattersley, A. T., & Patel, K. A. (2017). Precision diabetes: Learning from monogenic diabetes. *Diabetologia, 60*(5), 769–777. https://doi.org/10.1007/s00125-017-4226-2

Hering, B. J., Clarke, W. R., Bridges, N. D., Eggerman, T. L., Alejandro, R., Bellin, M. D., et al. (2016). Phase 3 trial of transplantation of human islets in type 1 Diabetes complicated by severe hypoglycemia. *Diabetes Care, 39*(7), 1230–1240. https://doi.org/10.2337/dc15-1988

Hosokawa, Y., Toyoda, T., Fukui, K., Baden, M. Y., Funato, M., Kondo, Y., et al. (2017). Insulin-producing cells derived from 'induced pluripotent stem cells' of patients with fulminant type 1 diabetes: Vulnerability to cytokine insults and increased expression of apoptosis-related genes. *J Diabetes Investig.* https://doi.org/10.1111/jdi.12727

Hrvatin, S., O'Donnell, C. W., Deng, F., Millman, J. R., Pagliuca, F. W., DiIorio, P., et al. (2014). Differentiated human stem cells resemble fetal, not adult, beta cells. *Proceedings of the National Academy of Sciences of the United States of America, 111*(8), 3038–3043. https://doi.org/10.1073/pnas.1400709111

Hu, F. B. (2011). Globalization of diabetes: The role of diet, lifestyle, and genes. *Diabetes Care, 34*(6), 1249–1257. https://doi.org/10.2337/dc11-0442

Huangfu, D., Osafune, K., Maehr, R., Guo, W., Eijkelenboom, A., Chen, S., et al. (2008). Induction of pluripotent stem cells from primary human fibroblasts with only Oct4 and Sox2. *Nature Biotechnology, 26*(11), 1269–1275. https://doi.org/10.1038/nbt.1502

IDF Diabetes Atlas. (2009). Available from: https://diabetesatlas.org/resources/?gclid=Cj0KCQjw08aYBhDlARIsAA_gb0fpvIF4U-lVyapljiZs4ergUvx9VA0F9oC0y7rlSIJrbX1UhwG66ygaAj3aEALw_wcB.

IDF Diabetes Atlas. (2011). Available from: https://diabetesatlas.org/atlas/fifth-edition/.

IDF Diabetes Atlas. (2015). Available from: https://www.diabetesatlas.org/upload/resources/previous/files/7/IDF%20Diabetes%20Atlas%207th.pdf.

IDF Diabetes Atlas. 10th Edition (2021). Available from: https://diabetesatlas.org/atlas/tenth-edition/.

Ilegems, E., Dicker, A., Speier, S., Sharma, A., Bahow, A., Edlund, P. K., et al. (2013). Reporter islets in the eye reveal the plasticity of the endocrine pancreas. *Proceedings of the National Academy of Sciences of the United States of America, 110*(51), 20581–20586. https://doi.org/10.1073/pnas.1313696110

Indurkar, M. S., Maurya, A. S., & Indurkar, S. (2016). Oral manifestations of diabetes. *Clin Diabetes, 34*(1), 54–57. https://doi.org/10.2337/diaclin.34.1.54

In't Veld, P. (2011). Insulitis in human type 1 diabetes: The quest for an elusive lesion. *Islets, 3*(4), 131–138. https://doi.org/10.4161/isl.3.4.15728

Jeon, K., Lim, H., Kim, J. H., Thuan, N. V., Park, S. H., Lim, Y. M., et al. (2012). Differentiation and transplantation of functional pancreatic beta cells generated from induced pluripotent stem cells derived from a type 1 diabetes mouse model. *Stem Cells and Development, 21*(14), 2642–2655. https://doi.org/10.1089/scd.2011.0665

Jiang, J., Au, M., Lu, K., Eshpeter, A., Korbutt, G., Fisk, G., et al. (2007). Generation of insulin-producing islet-like clusters from human embryonic stem cells. *Stem Cells, 25*(8), 1940–1953. https://doi.org/10.1634/stemcells.2006-0761

Kahn, C. R., & Lecture, B. (1994). Insulin action, diabetogenes, and the cause of type II diabetes. *Diabetes, 43*(8), 1066–1084. https://doi.org/10.2337/diab.43.8.1066

Karamanou, M., Protogerou, A., Tsoucalas, G., Androutsos, G., & Poulakou-Rebelakou, E. (2016). Milestones in the history of diabetes mellitus: The main contributors. *World Journal of Diabetes, 7*(1), 1–7. https://doi.org/10.4239/wjd.v7.i1.1

Kharroubi, A. T., & Darwish, H. M. (2015). Diabetes mellitus: The epidemic of the century. *World Journal of Diabetes, 6*(6), 850–867. https://doi.org/10.4239/wjd.v6.i6.850

Kishore, S., De Franco, E., Cardenas-Diaz, F. L., Letourneau-Freiberg, L. R., Sanyoura, M., Osorio-Quintero, C., et al. (2020). A non-coding disease modifier of pancreatic agenesis identified by genetic correction in a patient-derived iPSC line. *Cell Stem Cell, 27*(1), 137–146 e6. https://doi.org/10.1016/j.stem.2020.05.001

Krogvold, L., Wiberg, A., Edwin, B., Buanes, T., Jahnsen, F. L., Hanssen, K. F., et al. (2016). Insulitis and characterisation of infiltrating T cells in surgical pancreatic tail resections from patients at onset of type 1 diabetes. *Diabetologia, 59*(3), 492–501. https://doi.org/10.1007/s00125-015-3820-4

Kroon, E., Martinson, L. A., Kadoya, K., Bang, A. G., Kelly, O. G., Eliazer, S., et al. (2008). Pancreatic endoderm derived from human embryonic stem cells generates glucose-responsive insulin-secreting cells in vivo. *Nature Biotechnology, 26*(4), 443–452. https://doi.org/10.1038/nbt1393

Kudva, Y. C., Ohmine, S., Greder, L. V., Dutton, J. R., Armstrong, A., De Lamo, J. G., et al. (2012). Transgene-free disease-specific induced pluripotent stem cells from patients with type 1 and type 2 diabetes. *Stem Cells Translational Medicine, 1*(6), 451–461. https://doi.org/10.5966/sctm.2011-0044

Latres, E., Finan, D. A., Greenstein, J. L., Kowalski, A., & Kieffer, T. J. (2019). Navigating two roads to glucose normalization in Diabetes: Automated insulin delivery devices and cell therapy. *Cell Metabolism, 29*(3), 545–563. https://doi.org/10.1016/j.cmet.2019.02.007

Lavinsky, R. M., Jepsen, K., Heinzel, T., Torchia, J., Mullen, T. M., Schiff, R., et al. (1998). Diverse signaling pathways modulate nuclear receptor recruitment of N-CoR and SMRT complexes. *Proceedings of the National Academy of Sciences of the United States of America, 95*(6), 2920–2925. https://doi.org/10.1073/pnas.95.6.2920

Lithovius, V., Saarimaki-Vire, J., Balboa, D., Ibrahim, H., Montaser, H., Barsby, T., et al. (2021). SUR1-mutant iPS cell-derived islets recapitulate the pathophysiology of congenital hyperinsulinism. *Diabetologia, 64*(3), 630–640. https://doi.org/10.1007/s00125-020-05346-7

Lopez-Yrigoyen, M., Yang, C. T., Fidanza, A., Cassetta, L., Taylor, A. H., McCahill, A., et al. (2019). Genetic programming of macrophages generates an in vitro model for the human erythroid island niche. *Nature Communications, 10*(1), 881. https://doi.org/10.1038/s41467-019-08705-0

Ma, X., Lu, Y., Zhou, Z., Li, Q., Chen, X., Wang, W., et al. (2022). Human expandable pancreatic progenitor-derived β cells ameliorate diabetes. *Science Advances, 8*(8), eabk1826. https://doi.org/10.1126/sciadv.abk1826

Ma, S., Viola, R., Sui, L., Cherubini, V., Barbetti, F., & Egli, D. (2018). Beta cell replacement after gene editing of a neonatal Diabetes-causing mutation at the insulin locus. *Stem Cell Reports, 11*(6), 1407–1415. https://doi.org/10.1016/j.stemcr.2018.11.006

Maehr, R., Chen, S., Snitow, M., Ludwig, T., Yagasaki, L., Goland, R., et al. (2009). Generation of pluripotent stem cells from patients with type 1 diabetes. *Proceedings of the National Academy of Sciences of the United States of America, 106*(37), 15768–15773. https://doi.org/10.1073/pnas.0906894106

Marroqui, L., Dos Santos, R. S., Op de Beeck, A., Coomans de Brachene, A., Marselli, L., Marchetti, P., et al. (2017). Interferon-alpha mediates human beta cell HLA class I overexpression, endoplasmic reticulum stress and apoptosis, three hallmarks of early human type 1 diabetes. *Diabetologia, 60*(4), 656–667. https://doi.org/10.1007/s00125-016-4201-3

Maxwell, K. G., Augsornworawat, P., Velazco-Cruz, L., Kim, M. H., Asada, R., Hogrebe, N. J., et al. (2020). Gene-edited human stem cell-derived beta cells from a patient with monogenic diabetes reverse preexisting diabetes in mice. *Science Translational Medicine, 12*(540). https://doi.org/10.1126/scitranslmed.aax9106

Millman, J. R., Xie, C., Van Dervort, A., Gurtler, M., Pagliuca, F. W., & Melton, D. A. (2016). Generation of stem cell-derived beta-cells from patients with type 1 diabetes. *Nature Communications, 7*, 11463. https://doi.org/10.1038/ncomms11463

Minagawa, A., Yoshikawa, T., Yasukawa, M., Hotta, A., Kunitomo, M., Iriguchi, S., et al. (2018). Enhancing T cell receptor stability in rejuvenated iPSC-derived T cells improves their use in cancer immunotherapy. *Cell Stem Cell, 23*(6), 850–858 e4. https://doi.org/10.1016/j.stem.2018.10.005

Murakami, T., Fujimoto, H., & Inagaki, N. (2021). Non-invasive Beta-cell Imaging: Visualization, Quantification, and Beyond. *Front Endocrinol (Lausanne), 12*, 714348. https://doi.org/10.3389/fendo.2021.714348

Okita, K., Nakagawa, M., Hyenjong, H., Ichisaka, T., & Yamanaka, S. (2008). Generation of mouse induced pluripotent stem cells without viral vectors. *Science, 322*(5903), 949–953. https://doi.org/10.1126/science.1164270

Padgett, L. E., Broniowska, K. A., Hansen, P. A., Corbett, J. A., & Tse, H. M. (2013). The role of reactive oxygen species and proinflammatory cytokines in type 1 diabetes pathogenesis. *Annals of the New York Academy of Sciences, 1281*, 16–35. https://doi.org/10.1111/j.1749-6632.2012.06826.x

Padhi, S., Nayak, A. K., & Behera, A. (2020). Type II diabetes mellitus: A review on recent drug based therapeutics. *Biomedicine & Pharmacotherapy, 131*, 110708. https://doi.org/10.1016/j.biopha.2020.110708

Pagliuca, F. W., Millman, J. R., Gurtler, M., Segel, M., Van Dervort, A., Ryu, J. H., et al. (2014). Generation of functional human pancreatic beta cells in vitro. *Cell, 159*(2), 428–439. https://doi.org/10.1016/j.cell.2014.09.040

Qi, Z., Shen, Y., Yanai, G., Yang, K., Shirouzu, Y., Hiura, A., et al. (2010). The in vivo performance of polyvinyl alcohol macro-encapsulated islets. *Biomaterials, 31*(14), 4026–4031. https://doi.org/10.1016/j.biomaterials.2010.01.088

Reed, J. C., & Herold, K. C. (2015). Thinking bedside at the bench: The NOD mouse model of T1DM. *Nature Reviews. Endocrinology, 11*(5), 308–314. https://doi.org/10.1038/nrendo.2014.236

Saarimaki-Vire, J., Balboa, D., Russell, M. A., Saarikettu, J., Kinnunen, M., Keskitalo, S., et al. (2017). An activating STAT3 mutation causes neonatal Diabetes through premature induction of pancreatic differentiation. *Cell Reports, 19*(2), 281–294. https://doi.org/10.1016/j.celrep.2017.03.055

Saeedi, P., Petersohn, I., Salpea, P., Malanda, B., Karuranga, S., Unwin, N., et al. (2019). Global and regional diabetes prevalence estimates for 2019 and projections for 2030 and 2045: Results from the international Diabetes federation Diabetes Atlas, 9(th) edition. *Diabetes Research and Clinical Practice, 157*, 107843. https://doi.org/10.1016/j.diabres.2019.107843

Sapra, A. (2022). *Diabetes mellitus*. Treasure Island (FL).

Sequiera, G. L., Srivastava, A., Sareen, N., Yan, W., Alagarsamy, K. N., Verma, E., et al. (2022). Development of iPSC-based clinical trial selection platform for patients with ultrarare diseases. *Science Advances, 8*(14), eabl4370. https://doi.org/10.1126/sciadv.abl4370

Shang, L., Hua, H., Foo, K., Martinez, H., Watanabe, K., Zimmer, M., et al. (2014). Beta-cell dysfunction due to increased ER stress in a stem cell model of Wolfram syndrome. *Diabetes, 63*(3), 923–933. https://doi.org/10.2337/db13-0717

Shapiro, A. M., Lakey, J. R., Ryan, E. A., Korbutt, G. S., Toth, E., Warnock, G. L., et al. (2000). Islet transplantation in seven patients with type 1 diabetes mellitus using a glucocorticoid-free immunosuppressive regimen. *The New England Journal of Medicine, 343*(4), 230–238. https://doi.org/10.1056/NEJM200007273430401

Shapiro, A. M., Ricordi, C., Hering, B. J., Auchincloss, H., Lindblad, R., Robertson, R. P., et al. (2006). International trial of the Edmonton protocol for islet transplantation. *The New England Journal of Medicine, 355*(13), 1318–1330. https://doi.org/10.1056/NEJMoa061267

Shaw, J. E., Sicree, R. A., & Zimmet, P. Z. (2010). Global estimates of the prevalence of diabetes for 2010 and 2030. *Diabetes Research and Clinical Practice, 87*(1), 4–14. https://doi.org/10.1016/j.diabres.2009.10.007

Silveira, P. A., & Grey, S. T. (2006). B cells in the spotlight: Innocent bystanders or major players in the pathogenesis of type 1 diabetes. *Trends in Endocrinology and Metabolism, 17*(4), 128–135. https://doi.org/10.1016/j.tem.2006.03.006

Smid, M., Wilting, S. M., Uhr, K., Rodríguez-González, F. G., de Weerd, V., Prager-Van der Smissen, W. J. C., et al. (2019). The circular RNome of primary breast cancer. *Genome Research, 29*(3), 356–366. https://doi.org/10.1101/gr.238121.118

Stock, A. A., Manzoli, V., De Toni, T., Abreu, M. M., Poh, Y. C., Ye, L., et al. (2020). Conformal coating of stem cell-derived islets for beta cell replacement in type 1 Diabetes. *Stem Cell Reports, 14*(1), 91–104. https://doi.org/10.1016/j.stemcr.2019.11.004

Sui, L., Danzl, N., Campbell, S. R., Viola, R., Williams, D., Xing, Y., et al. (2018). Beta-cell replacement in mice using human type 1 Diabetes nuclear transfer embryonic stem cells. *Diabetes, 67*(1), 26–35. https://doi.org/10.2337/db17-0120

Takahashi, K., Tanabe, K., Ohnuki, M., Narita, M., Ichisaka, T., Tomoda, K., et al. (2007). Induction of pluripotent stem cells from adult human fibroblasts by defined factors. *Cell, 131*(5), 861–872. https://doi.org/10.1016/j.cell.2007.11.019

Teo, A. K., Windmueller, R., Johansson, B. B., Dirice, E., Njolstad, P. R., Tjora, E., et al. (2013). Derivation of human induced pluripotent stem cells from patients with maturity onset diabetes of the young. *The Journal of Biological Chemistry, 288*(8), 5353–5356. https://doi.org/10.1074/jbc.C112.428979

Toyoda, T., Mae, S., Tanaka, H., Kondo, Y., Funato, M., Hosokawa, Y., et al. (2015). Cell aggregation optimizes the differentiation of human ESCs and iPSCs into pancreatic bud-like progenitor cells. *Stem Cell Research, 14*(2), 185–197. https://doi.org/10.1016/j.scr.2015.01.007

van Belle, T. L., Coppieters, K. T., & von Herrath, M. G. (2011). Type 1 diabetes: Etiology, immunology, and therapeutic strategies. *Physiological Reviews, 91*(1), 79–118. https://doi.org/10.1152/physrev.00003.2010

Wang, L., Li, X., Wang, Z., Bancks, M. P., Carnethon, M. R., Greenland, P., et al. (2021). Trends in prevalence of Diabetes and control of risk factors in Diabetes among US adults, 1999-2018. *JAMA, 326*(8), 704–716. https://doi.org/10.1001/jama.2021.9883

Wang, X., Sterr, M., Ansarullah, Burtscher, I., Bottcher, A., Beckenbauer, J., et al. (2019). Point mutations in the PDX1 transactivation domain impair human beta-cell development and function. *Mol Metab, 24*, 80–97. https://doi.org/10.1016/j.molmet.2019.03.006

Wareham, N. J., & O'Rahilly, S. (1998). The changing classification and diagnosis of diabetes. New classification is based on pathogenesis, not insulin dependence. *BMJ, 317*(7155), 359–360. https://doi.org/10.1136/bmj.317.7155.359

Wilcox, G. (2005). Insulin and insulin resistance. *Clinical Biochemist Reviews, 26*(2), 19–39.

Willcox, A., Richardson, S. J., Bone, A. J., Foulis, A. K., & Morgan, N. G. (2009). Analysis of islet inflammation in human type 1 diabetes. *Clinical and Experimental Immunology, 155*(2), 173–181. https://doi.org/10.1111/j.1365-2249.2008.03860.x

Wisel, S. A., Braun, H. J., & Stock, P. G. (2016). Current outcomes in islet versus solid organ pancreas transplant for beta-cell replacement in type 1 diabetes. *Current Opinion in Organ Transplantation, 21*(4), 399–404. https://doi.org/10.1097/MOT.0000000000000332

Xiao, X., Fischbach, S., Song, Z., Gaffar, I., Zimmerman, R., Wiersch, J., et al. (2016). Transient suppression of TGFbeta receptor signaling facilitates human islet transplantation. *Endocrinology, 157*(4), 1348–1356. https://doi.org/10.1210/en.2015-1986

Yamanaka, S. (2012). Induced pluripotent stem cells: Past, present, and future. *Cell Stem Cell, 10*(6), 678–684. https://doi.org/10.1016/j.stem.2012.05.005

Yao, D., & Brownlee, M. (2010). Hyperglycemia-induced reactive oxygen species increase expression of the receptor for advanced glycation end products (RAGE) and RAGE ligands. *Diabetes, 59*(1), 249–255. https://doi.org/10.2337/db09-0801

Yoon, J. W., & Jun, H. S. (2005). Autoimmune destruction of pancreatic beta cells. *American Journal of Therapeutics, 12*(6), 580–591. https://doi.org/10.1097/01.mjt.0000178767.67857.63

Yoshihara, E., O'Connor, C., Gasser, E., Wei, Z., Oh, T. G., Tseng, T. W., et al. (2020). Immune-evasive human islet-like organoids ameliorate diabetes. *Nature, 586*(7830), 606–611. https://doi.org/10.1038/s41586-020-2631-z

Yu, J., Vodyanik, M. A., Smuga-Otto, K., Antosiewicz-Bourget, J., Frane, J. L., Tian, S., et al. (2007). Induced pluripotent stem cell lines derived from human somatic cells. *Science, 318*(5858), 1917–1920. https://doi.org/10.1126/science.1151526

Zaharia, O. P., Strassburger, K., Strom, A., Bönhof, G. J., Karusheva, Y., Antoniou, S., et al. (2019). Risk of diabetes-associated diseases in subgroups of patients with recent-onset diabetes:

A 5-year follow-up study. *The Lancet Diabetes and Endocrinology, 7*(9), 684–694. https://doi.org/10.1016/s2213-8587(19)30187-1

Zhang, Q., Gonelle-Gispert, C., Li, Y., Geng, Z., Gerber-Lemaire, S., Wang, Y., et al. (2022). Islet encapsulation: New developments for the treatment of type 1 Diabetes. *Frontiers in Immunology, 13*, 869984. https://doi.org/10.3389/fimmu.2022.869984

Zheng, Y. L. (2016). Some ethical concerns about human induced pluripotent stem cells. *Science and Engineering Ethics, 22*(5), 1277–1284. https://doi.org/10.1007/s11948-015-9693-6

Zhong, F., & Jiang, Y. (2019). Endogenous pancreatic beta cell regeneration: A potential strategy for the recovery of beta cell deficiency in Diabetes. *Front Endocrinol (Lausanne), 10*, 101. https://doi.org/10.3389/fendo.2019.00101

Zhou, Z., Zhu, X., Huang, H., Xu, Z., Jiang, J., Chen, B., et al. (2022). Recent Progress of research regarding the applications of stem cells for treating Diabetes mellitus. *Stem Cells and Development, 31*(5–6), 102–110. https://doi.org/10.1089/scd.2021.0083

Influence of Ketogenic Diet on Diabetes

Natesan Sella Raja, Varsha Singh, and Subhashree Sivakumar

Abstract

The effectiveness of limiting the intake of carbohydrates in obese individuals and people with diabetes is still subject to debate, even though ketogenic diets are widespread among clinicians and patients. Studies that have been published are contentious, probably due to the overall vagueness of these diets. In addition to the inherent complexity of dietary treatments, it is challenging to compare the findings of various studies. Despite the research showing that consuming fewer carbohydrates reduces body weight and improves glucose control in people with type 2 diabetes, little is known regarding this strategy's sustainability, safety, and long-term efficacy. Ketogenic diet results in rapid and healthy weight loss and positive biomarker improvements, such as a decrease in serum hemoglobin A1c (HbA1c) in those with type 2 diabetes. Many doctors are reluctant to recommend it because it substantially raises low lipoprotein cholesterol levels. There is a significant concern about the possible long-term effects of the widespread adoption of this diet by broad parts of the public, given the popularity of the ketogenic diet, especially among persons who do not need to lose weight. In this chapter, we summarize about the ketogenic diet, the potential impact of ketogenic diets on diabetic patients, and the physiological changes experienced by an individual, followed by observational studies in diabetic patients.

Keywords

Ketogenic diet · Diabetes · Ketosis · Physiological changes · Observational studies

N. S. Raja (✉) · V. Singh · S. Sivakumar
Membrane-Protein Interaction Lab, Department of Genetic Engineering, School of
Bio-engineering, SRM Institute of Science and Technology, Kattankulathur, Chennai, Tamil Nadu,
India

1   Introduction

Various environmental, behavioral, and genetic factors combine over many years to cause a complicated disease known as diabetes. Type 1 and 2 diabetes are strongly linked to obesity and cardiovascular disease. Diabetes is sometimes referred to as a chronic progressive ailment, wherein glycemic control and wellness are predicted to deteriorate gradually in the absence of treatment (American Diabetes Association, 2017). The main objectives of conventional treatments are weight loss in overweight or obese individuals and decreasing hyperglycemia to a specific target range, typically HbA1c below 7%, through dietary changes and glucose-lowering drugs (American Diabetes Association, 2017, 2019). Massive interest has been observed in recent years in finding ways to slow the growth of diabetes and reverse it.

Ketogenic diet has a substantial impact on diabetic patients specifically, as seen by lower body weight, increased fasting insulin, lower glucose and cholesterol levels, and the reduction or elimination of diabetic medication. These are thought to occur because eating fewer carbohydrates results in lower blood sugar levels and shifts the primary energy metabolism from glucose to ketone bodies. As blood sugar levels drop, insulin resistance also improves (Gershuni et al., 2018).

2   What Is a Ketogenic Diet?

Ketogenic diet, in simple words, can be defined as a "low-carbohydrate, high-fat diet." Here, the body is constrained to acquire calories more from fat and less from carbohydrates. The diet is broadly classified into four types, namely, classic ketogenic diet, medium-chain triglyceride (MCT) ketogenic diet, modified Atkins diet (MAD), and low glycemic index diet (Dhamija et al., 2013). All these types just vary with the fat-to-carbohydrate and protein ratio. As 80 to 90% of calories can be derived from fat, fat intake is given more importance in ketogenic diet, simultaneously reducing blood glucose levels by limiting carbohydrate intake. The classic ketogenic diet is structured in such a way that the fat-to-carbohydrate and protein ratio is 4:1. This is one of the most commonly followed types of ketogenic diet and has been advised for children with epilepsy (Barzegar et al., 2021). However, following the classic ketogenic diet is time-consuming as it requires properly weighing the ingredients based on the diet chart.

MCT ketogenic diet, as its name suggests, emphasizes including medium-chain triglycerides, such as coconut oil, in our food intake. Triglycerides are classified into long-chain, medium-chain, and short-chain triglycerides. Long-chain triglycerides are a stretch of more than 12-carbon fatty acids esterified with glycerol, whereas medium-chain triglycerides hold around 6- to 12-carbon fatty acids (Bach & Babayan, 1982). The importance of medium-chain triglycerides is that they are digested and absorbed faster than long-chain fatty acids. They provide more energy, and thus, the MCT ketogenic diet consists mainly of octanoic and decanoic fatty acids, which yield more ketones (Barzegar et al., 2021; Bach & Babayan, 1982). The modified Atkins diet is considered more palatable and has fewer calorie restrictions.

Around 65% of this diet's energy is acquired from fat (Barzegar et al., 2021). However, this is less as compared to the classical ketogenic diet (the ratio of fat-to-carbohydrate and protein in the classical ketogenic diet is 4:1, which approximately provides 90% energy from fat). The modified Atkins diet also allows the uptake of more proteins. Hence, as said before, the diet is not much stringent as the previously mentioned ones (Barzegar et al., 2021; Kossoff & Dorward, 2008).

The low glycemic index diet limits the uptake of carbohydrates to 40 to 60 g/day. The use of low glycemic index food elevates sugar levels at a slower rate due to slower digestion. A low glycemic diet can be suggested for people with prediabetes or diabetes as a means of glycemic control, accompanied with reduced blood sugar levels (Barzegar et al., 2021; Wang et al., 2015). Maintaining blood glucose levels is of paramount importance for diabetic patients. Hence, the ketogenic diet is practiced, especially the low glycemic index diet. Low carbohydrate ketogenic diets are also followed in children with epilepsy to treat overweight. The basic idea behind the ketogenic diet is to instruct the body to metabolize fat for energy rather than glucose, which is the primary source of adenosine triphosphates (ATPs) under normal conditions.

Type 2 diabetes is characterized by resistance to insulin, which results in poor glucose uptake by cells. Since the cells use glucose as the primary energy source, insulin resistance will impair the production of ATPs, further leading to carbohydrates in the diet being channeled to the liver for lipogenesis (Jornayvaz et al., 2010). This mechanism of converting carbohydrates to fats increases the chance of acquiring comorbidities, especially cardiac disorders. It has been reported that insulin resistance diminishes with a decrease in dietary carbohydrates (Boden et al., 2005).

3 Physiological Changes in the Body Due to the Ketogenic Diet

As experienced in every dietary and lifestyle change, the ketogenic diet also poses physiological changes, including changes in body weight, blood pH, etc. A ketogenic diet induces acidic pH in the blood due to the circulation of ketone bodies. However, mechanisms such as carbon dioxide elimination and ion exchange in kidneys contribute to maintaining the pH to a physiologically normal range of 7.35 to 7.45. The failure of these mechanisms leads to ketoacidosis, a common manifestation of diabetes (Kakodkar, 2020). In a meta-analysis of randomized controlled trials, the effects of a low-fat diet against a ketogenic diet were studied. The study concluded that compared to low fat, the ketogenic diet alleviated the total cholesterol, triglyceride, and low-density lipoprotein (LDL) levels and increased high-density lipoprotein (HDL) levels. However, they report no comparatively significant difference in body weight and waist circumference reduction between the two diets, which postulates that the ketogenic diet can be used as an alternative to a low-fat diet to limit the metabolic risks (Hu et al., 2012; Kakodkar, 2020).

In a study by Paoli et al., 40 healthy individuals were randomly split into two groups to follow two types of diets. One group followed a low-calorie Mediterranean diet, while the other followed a ketogenic Mediterranean diet with phytoextracts. They observed a decrease in the respiratory ratio of the individuals who strictly followed the ketogenic diet. The study further revealed that the reduction in body weight was more effective in patients who followed the ketogenic Mediterranean diet. This diet also increased fat oxidation at rest, which could be attributed to effective weight loss (Paoli et al., 2012, 2013). A ketogenic diet with phytoextracts was significantly said to reduce body weight, suppress the effects of ketosis, and improve the lipid profile (Paoli et al., 2014).

It was also reported that a high-protein ketogenic diet induces satiety. In a study on the effects of a high-protein ketogenic diet, obese men were provided with low-carbohydrate and medium-carbohydrate nonketogenic diets for 4 weeks. The results reportedly say that the subjects who followed a low-carbohydrate diet experienced less hunger and more weight reduction than those who followed a medium-carbohydrate diet (Johnstone et al., 2008). It was also estimated that the increase in satiety resulted from the high protein content in the diet (Halton & Hu, 2013).

4 Ketosis and Glycolysis

Glycolysis is a pathway that converts glucose into pyruvic acid and energy, where pyruvic acid is further converted into acetyl coenzyme A (CoA) and enters the tricarboxylic acid (TCA) cycle (Chandel, 2021). In the metabolism involved in the ketogenic diet, acetyl CoA is subjected to the formation of ketones due to ketogenesis. Fatty acids are usually converted into acetyl CoA in the hepatocytes and enter the TCA cycle. Figure 1a illustrates the general pathway in which fatty acids form acetyl CoA and enter the TCA cycle to produce ATP molecules further. When the body is subjected to the ketogenic diet, where the fatty acid levels soar high, a shunted pathway to ketogenesis occurs, in which two molecules of acetyl CoA combine to form acetoacetyl CoA with the help of the enzyme thiolase. Acetoacetyl CoA is a precursor to the formation of acetoacetate (ACA) and β-hydroxybutyrate (BHB). ACA undergoes spontaneous decarboxylation to form acetone, considered a significant ketone body. Figure 1b shows the mechanism involved in elevated levels of fatty acids. Since acetone is volatile, it is eliminated through the lungs and kidneys (Masino & Rho, 2012). The increase in levels of ketones in the blood is called ketosis. Now, the body prefers ketones for energy instead of glucose, which is the main idea behind the diet. Ketones can cross the blood-brain barrier and provide energy to brain tissues. This is essential considering the absence of glucose in the body due to low amounts of carbohydrates in the diet (Jensen et al., 2020).

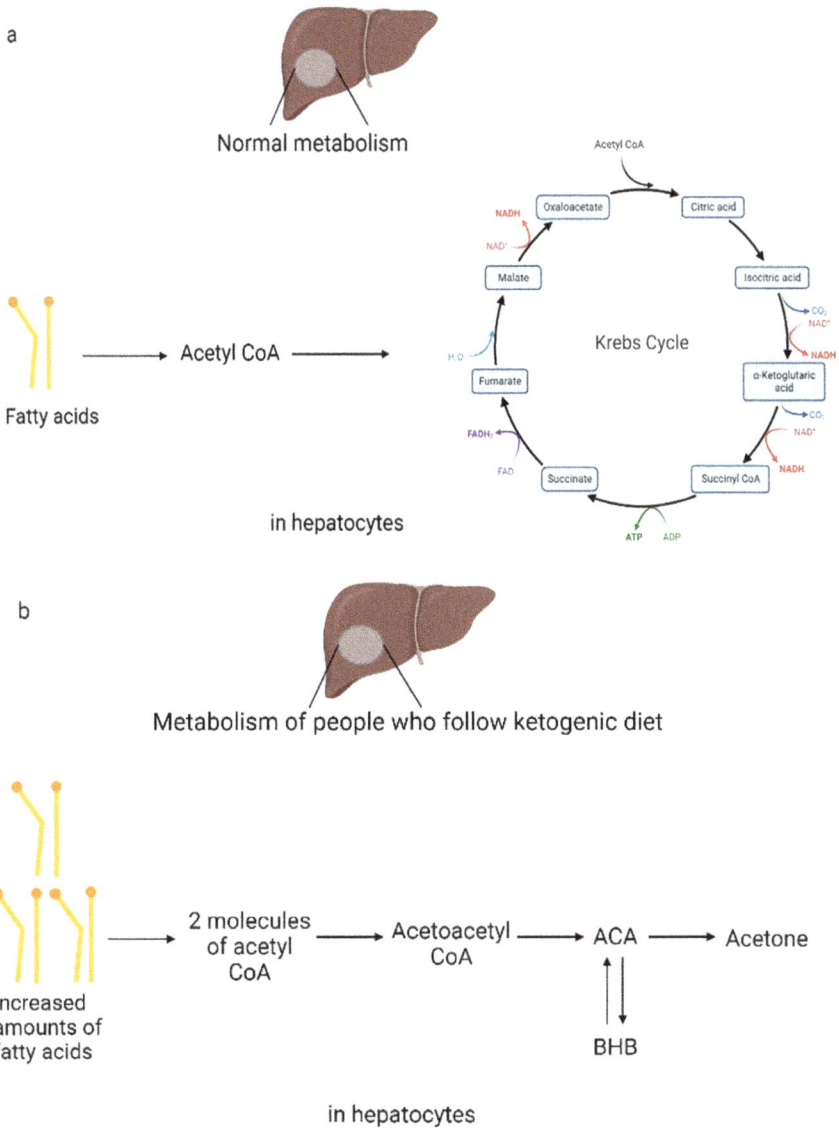

Fig. 1 Metabolism in the liver of individuals who follow (**a**) a regular carbohydrate diet and (**b**) a ketogenic diet. (**a**) Conversion of fatty acids that are present in fewer amounts to acetyl CoA, which then moves to the Krebs cycle; (**b**) higher levels of fatty acids cause the metabolic capacity of the Krebs cycle to exceed, thus diverting the pathway to ketogenesis. Acetone is a central ketone body, and it is a volatile compound. ACA and BHB are transported to the neurons and glia. Image created using BioRender

5   Ketogenic Diet and Its Observation in Diabetic Patients

As we already know, low-carbohydrate and ketogenic diets have gained widespread recognition and popularity in recent years as effective weight-loss methods, both within the scientific community and among the general population. These dietary strategies work well for weight loss. Still, there is an increasing indication that caution is required, particularly whenever these diets are practiced for an extended period by people who are very young or have certain diseases or by those who follow them for these groups of people (Ioannidis, 2018). Low-carbohydrate diets were historically recommended as a remedy for type I diabetes when no insulin was accessible. Still, such diets were very different from low-carb, high-fat ketogenic diets (Hussain et al., 2012).

Fats rather than carbohydrates are the primary energy source in the ketogenic diet. Numerous individuals with type 2 diabetes are overweight or obese, and the goal of a study by Li et al. is to periodically monitor the effects of the ketogenic diet on overweight or obese individuals who have just received a type 2 diabetes mellitus diagnosis (Li et al., 2022). This study showed that the regular ketogenic diet controls glucose and lipid levels in the blood.

Ketogenic diets result in metabolic changes that may enhance brain metabolism, restore mitochondrial ATP synthesis, reduce reactive oxygen species formation, decrease inflammation, and improve the activity of neurotrophic factors (Maalouf et al., 2007). Ketogenic diets have been demonstrated to imitate the effects of fasting and the absence of glucose/insulin signaling, which induces a metabolic shift toward consuming fatty acids (Hammami et al., 1997). Only a moderate increase in blood ketone levels can be achieved with the ketogenic diet, and a substantial dietary carbohydrate limitation is necessary to maintain persistent (therapeutic) levels of ketosis (George, 2006). The only broad recommendations for therapy indicated by dietary changes existed before the introduction of exogenous insulin for managing diabetes mellitus (type II) in the 1920s. The present low-carbohydrate dietary advice for diabetic patients was drastically different back then and focused on controlling blood glucose, typically only glycosuria (Arora & McFarlane, 2005; Klein et al., 2004).

In children experiencing type I diabetes, diabetic ketoacidosis, a potentially fatal disease, is a crucial contributor to diabetes. Insulin deficiency causes metabolic decompensation, which results in hyperglycemia and ketosis, which can be treated with insulin and water. The case of a 2-year-old girl who went to the emergency room with a 1-week past of decreased activity, polyuria, and reduced oral intake was described by Castaneda and colleagues. She had a notable medical history of epilepsy, over which she began a ketogenic diet and improved significantly. Her laboratory results were consistent with diabetic ketoacidosis; fluids and insulin were administered until the condition was corrected. To reduce the risk of her seizures, the ketogenic diet and concurrent administration of insulin glargine and insulin aspart were resumed. To maintain the influence of ketosis on seizure management, urine ketones were held at a moderate level. The patient continued to receive this combined therapy, and there were no subsequent instances of diabetic ketoacidosis or seizures (Aguirre Castaneda et al., 2012).

The study by Partsalaki et al. demonstrated that a ketogenic diet could lower body weight, waist circumference, and insulin intolerance. The waist measurement is a vital sign of central obesity and is connected to insulin resistance. This study showed that decreasing waist circumference decreased body mass, elevated related lipid metabolism indicators of the participants, and helped manage blood sugar and insulin resistance. Body mass loss was closely correlated with pattern adoption and a negative nitrogen balance brought on by caloric restriction (Partsalaki et al., 2012). In individuals with overweight or obese type 2 diabetes mellitus, the regular ketogenic diet can control weight as well as blood glucose and lipid levels. Long-term tenacity, however, is challenging. It might function as a diet treatment model. Weight loss may benefit some newly diagnosed type 2 diabetics who are overweight or obese, and some individuals may be able to manage their blood sugar levels well in the short term without taking any medication (Li et al., 2022).

A different study in Canada was conducted to examine the experiences of people with diabetes following a ketogenic diet. The ketogenic diet was continued for 6 to 19 months in this study. The primary reasons for starting the diets were to reduce or stop taking diabetes prescriptions, lose weight, and reverse diabetes. Participants have reported advantages such as better loss of weight, glycemic control, and satiety (Wong et al., 2021).

Webster and colleagues conducted another observational trial in South Africa, whereby they noted an increase in type 2 diabetes patients who are eating low-carb, high-fat diets. This study looked into the eating patterns and diabetes state of people with type 2 diabetes who have said they adhered to a low-carbohydrate, high-fat diet. This study's primary goal was to outline the food, nutrients, and eating habits that made up a low-carbohydrate, high-fat diet feasible for people with type 2 diabetes to follow daily. Most trial participants reported decreased appetite and cravings, lower HbA1c ($p < 0.001$), diabetes medications, and weight loss (Webster et al., 2019). The aforementioned observational studies show a definite trend in support of the ketogenic diet compared to controls. The combined result of these trials also illustrates how well the diet works to lower blood sugar and cholesterol levels quickly. Although ketogenic and low-carbohydrate diets are widely popular among patients, they are not listed in the medical, and nutritional therapy guidelines for type 1 diabetes in the most recent Standards of Medical Care by the American Diabetes Association (American Diabetes Association, 2018).

6   Conclusions

Ketogenic diet has been extensively implemented to treat various disorders, namely, neurological disorders, obesity, metabolic disorders, etc. Initially suggesting this to children with epilepsy, where anaerobic metabolization of ketones presented a better seizure threshold, the ketogenic diet has been widely followed in recent years to reduce the weight and control the blood sugar levels of diabetic patients. A long-standing misconception is that the ketogenic diet can affect the brain's normal functioning. This is untrue because ketones provide almost 70% of the energy needed for brain tissues. In addition, gluconeogenesis also occurs in the liver to produce glucose from noncarbohydrate moieties. Concerning its advantages, ketoacidosis, a condition where ketones rise to hazardous levels, should also be monitored. It is also to be noted that nutritional ketosis (ketones lesser than 5 mmol/L) is completely safe.

In conclusion, a low-carbohydrate diet might be possible for some type 1 diabetes patients to enhance their glycemic variability in the brief term. For type 2 diabetes, we have some data that provide evidence for the beneficial aspects of ketogenic diet. However, we acknowledge the lack of substantial proof of knowledge in this area, which calls for well-designed trials to determine a low-carbohydrate diet's long-term safety and efficacy.

References

Aguirre Castaneda, R. L., Mack, K. J., & Lteif, A. (2012). Successful treatment of type 1 diabetes and seizures with combined ketogenic diet and insulin. *Pediatrics, 129*(2), e511. https://doi.org/10.1542/PEDS.2011-0741

American Diabetes Association. (2017). 5. Lifestyle management: Standards of medical care in diabetes-2019. *Diabetes Care, 40*, S33–S43. https://doi.org/10.2337/dc17-S007

American Diabetes Association. (2018). 5. Lifestyle management: Standards of medical care in diabetes-2019. *Diabetes Care, 42*, S46–S60. https://doi.org/10.2337/dc19-S005

American Diabetes Association. (2019). 6. Glycemic targets: Standards of medical care in diabetes-2019. *Diabetes Care, 42*(Suppl 1), S61–S70. https://doi.org/10.2337/DC19-S006

Arora, S. K., & McFarlane, S. I. (2005). The case for low carbohydrate diets in diabetes management. *Nutrition & Metabolism, 2*, 16. https://doi.org/10.1186/1743-7075-2-16

Bach, A. C., & Babayan, V. K. (1982). *Medium-chain triglycerides: An update.* Retrieved from https://academic.oup.com/ajcn/article-abstract/36/5/950/4693611

Barzegar, M., Afghan, M., Tarmahi, V., Behtari, M., Rahimi Khamaneh, S., & Raeisi, S. (2021). Ketogenic diet: Overview, types, and possible anti-seizure mechanisms. *Nutritional Neuroscience, 24*(4), 307–316. https://doi.org/10.1080/1028415X.2019.1627769

Boden, G., Sargrad, K., Homko, C., Mozzoli, M., & Stein, T. P. (2005). Effect of a low-carbohydrate diet on appetite, blood glucose levels, and insulin resistance in obese patients with type 2 diabetes. *Annals of Internal Medicine, 142*(6), 403. https://doi.org/10.7326/0003-4819-142-6-200503150-00006

Chandel, N. S. (2021). Glycolysis. *Cold Spring Harbor Perspectives in Biology, 13*(5), a040535. https://doi.org/10.1101/CSHPERSPECT.A040535

Dhamija, R., Eckert, S., & Wirrell, E. (2013). Ketogenic diet. *The Canadian Journal of Neurological Sciences. Le Journal Canadien Des Sciences Neurologiques, 40*(2), 158–167. https://doi.org/10.1017/S0317167100013676

George, F. C. (2006). Fuel metabolism in starvation. *Annual Review of Nutrition, 26*, 1–22. https://doi.org/10.1146/ANNUREV.NUTR.26.061505.111258

Gershuni, V. M., Yan, S. L., & Medici, V. (2018). Nutritional ketosis for weight management and reversal of metabolic syndrome. *Current Nutrition Reports, 7*(3), 97–106. https://doi.org/10.1007/S13668-018-0235-0

Halton, T. L., & Hu, F. B. (2013). The effects of high protein diets on thermogenesis, satiety and weight loss: A critical review. *Journal of the American Nutrition Association., 23*(5), 373–385. https://doi.org/10.1080/07315724.2004.10719381

Hammami, M. M., Bouchama, A., Al-Sedairy, S., Shail, E., AlOhaly, Y., & Mohamed, G. E. D. (1997). Concentrations of soluble tumor necrosis factor and interleukin-6 receptors in heatstroke and heatstress. *Critical Care Medicine, 25*(8), 1314–1319. https://doi.org/10.1097/00003246-199708000-00017

Hu, T., Mills, K. T., Yao, L., Demanelis, K., Eloustaz, M., Yancy, W. S., Kelly, T. N., He, J., & Bazzano, L. A. (2012). Systematic reviews and meta-and pooled analyses effects of low-carbohydrate diets versus low-fat diets on metabolic risk factors: A meta-analysis of randomized controlled clinical trials. *American Journal of Epidemiology, 176*(7), 44–54. https://doi.org/10.1093/aje/kws264

Hussain, T. A., Mathew, T. C., Dashti, A. A., Asfar, S. M., Al-Zaid, N., & Dashti, H. M. (2012). Effect of low-calorie versus low-carbohydrate ketogenic diet in type 2 diabetes. *Nutrition, 28*(10), 1016–1021. https://doi.org/10.1016/j.nut.2012.01.016

Ioannidis, J. P. A. (2018). The challenge of reforming nutritional epidemiologic research. *JAMA - Journal of the American Medical Association, 320*(10), 969–970. https://doi.org/10.1001/JAMA.2018.11025

Jensen, N. J., Wodschow, H. Z., Nilsson, M., & Rungby, J. (2020). Effects of ketone bodies on brain metabolism and function in neurodegenerative diseases. *International Journal of Molecular Sciences, 21*(22), 8767. https://doi.org/10.3390/IJMS21228767

Johnstone, A. M., Horgan, G. W., Murison, S. D., Bremner, D. M., & Lobley, G. E. (2008). Effects of a high-protein ketogenic diet on hunger, appetite, and weight loss in obese men feeding ad libitum. *The American Journal of Clinical Nutrition, 87*(1), 44–55. https://doi.org/10.1093/AJCN/87.1.44

Jornayvaz, F. R., Samuel, V. T., & Shulman, G. I. (2010). The role of muscle insulin resistance in the pathogenesis of Atherogenic dyslipidemia and nonalcoholic fatty liver disease associated with the metabolic syndrome. *Annual Review of Nutrition, 30*, 273–290. https://doi.org/10.1146/ANNUREV.NUTR.012809.104726

Kakodkar, P. (2020). Ketogenic diet: Biochemistry, weight loss and clinical applications. *Nutrition and Food Science International Journal, 10*(2), 10.19080/nfsij.2020.10.555782.

Klein, S., Sheard, N. F., Pi-Sunyer, X., Daly, A., Wylie-Rosett, J., Kulkarni, K., & Clark, N. G. (2004). Weight management through lifestyle modification for the prevention and management of type 2 diabetes: Rationale and strategies. A statement from the American Diabetes Association, the north American Association for the Study of obesity, and the American Society for Clinical Nutrition. *The American Journal of Clinical Nutrition, 80*(2), 257–263. https://doi.org/10.1093/AJCN/80.2.257

Kossoff, E. H., & Dorward, J. L. (2008). The modified Atkins diet. *Epilepsia, 49*(Suppl. 8), 37–41. https://doi.org/10.1111/j.1528-1167.2008.01831.x

Li, S., Lin, G., Chen, J., Chen, Z., Xu, F., Zhu, F., Zhang, J., & Yuan, S. (2022). The effect of the regular ketogenic diet on newly diagnosed overweight or obese patients with type 2 diabetes. *BMC Endocrine Disorders, 22*(1), 1–6. https://doi.org/10.1186/S12902-022-00947-2/TABLES/3

Maalouf, M., Sullivan, P. G., Davis, L., Kim, D. Y., & Rho, J. M. (2007). Ketones inhibit mitochondrial production of reactive oxygen species production following glutamate excitotoxicity by increasing NADH oxidation. *Neuroscience, 145*(1), 256. https://doi.org/10.1016/J.NEUROSCIENCE.2006.11.065

Masino, S. A., & Rho, J. M. (2012). *Mechanisms of Ketogenic diet action - Jasper's basic mechanisms of the epilepsies - NCBI bookshelf*. Retrieved August 13, 2022, from https://www.ncbi.nlm.nih.gov/books/NBK98219/#!po=3.52113

Paoli, A., Grimaldi, K., Bianco, A., Lodi, A., Cenci, L., & Parmagnani, A. (2012). Medium-term effects of a ketogenic and Mediterranean diet on resting energy expenditure and respiratory ratio. *BMC Proceedings, 6*(Suppl 3), P37. https://doi.org/10.1186/1753-6561-6-S3-P37

Paoli, A., Rubini, A., Volek, J. S., & Grimaldi, K. A. (2013). Beyond weight loss: A review of the therapeutic uses of very-low-carbohydrate (ketogenic) diets. *European Journal of Clinical Nutrition, 67*(8), 789–796. https://doi.org/10.1038/ejcn.2013.116

Paoli, A., Parmagnani, A., & Bianco, A. (2014). Ketogenic diet and phytoextracts. Comparison of the efficacy of Mediterranean, Zone, and Tisanoreica diet on some health risk factors. *Agro Food Industry Hi Tech, 21*(4), 24–29. https://www.researchgate.net/publication/230577564

Partsalaki, I., Karvela, A., & Spiliotis, B. E. (2012). Metabolic impact of a ketogenic diet compared to a hypocaloric diet in obese children and adolescents. *Journal of Pediatric Endocrinology and Metabolism, 25*(7–8), 697–704. https://doi.org/10.1515/JPEM-2012-0131/MACHINEREADABLECITATION/RIS

Wang, M. L., Gellar, L., Nathanson, B. H., Pbert, L., Ma, Y., Ockene, I., & Rosal, M. C. (2015). Decrease in glycemic index associated with improved glycemic control among Latinos with type 2 diabetes. *Journal of the Academy of Nutrition and Dietetics, 115*(6), 898–906. https://doi.org/10.1016/J.JAND.2014.10.012

Webster, C., Murphy, T. E., Larmuth, K. M., Noakes, T. D., & Smith, J. A. (2019). Diet, diabetes status, and personal experiences of individuals with type 2 diabetes who self-selected and followed a low carbohydrate, high-fat diet. *Diabetes, Metabolic Syndrome and Obesity*. Retrieved August 12, 2022, from https://www.ncbi.nlm.nih.gov/pmc/articles/PMC6901382/

Wong, K., Raffray, M., Roy-Fleming, A., Blunden, R. D., & Brazeau, A.-S. (2021). The ketogenic diet as a standard way of eating in adults with type 1 and type 2 diabetes: A qualitative study. *Canadian Journal of Diabetes*. Retrieved August 12, 2022, from https://www.sciencedirect.com/science/article/pii/S1499267120302008

Milton Keynes UK
Ingram Content Group UK Ltd.
UKHW020752160224
437943UK00001B/9